2018 IEEE International Conference on Microelectronic Test Structures (ICMTS 2018)

Austin, Texas, USA
19 – 22 March 2018

IEEE Catalog Number: CFP18MTS-POD
ISBN: 978-1-5386-5072-1

**Copyright © 2018 by the Institute of Electrical and Electronics Engineers, Inc.
All Rights Reserved**

Copyright and Reprint Permissions: Abstracting is permitted with credit to the source. Libraries are permitted to photocopy beyond the limit of U.S. copyright law for private use of patrons those articles in this volume that carry a code at the bottom of the first page, provided the per-copy fee indicated in the code is paid through Copyright Clearance Center, 222 Rosewood Drive, Danvers, MA 01923.

For other copying, reprint or republication permission, write to IEEE Copyrights Manager, IEEE Service Center, 445 Hoes Lane, Piscataway, NJ 08854. All rights reserved.

*** *This is a print representation of what appears in the IEEE Digital Library. Some format issues inherent in the e-media version may also appear in this print version.*

IEEE Catalog Number: CFP18MTS-POD
ISBN (Print-On-Demand):   978-1-5386-5072-1
ISBN (Online):   978-1-5386-5071-4
ISSN: 1071-9032

Additional Copies of This Publication Are Available From:

Curran Associates, Inc
57 Morehouse Lane
Red Hook, NY 12571 USA
Phone: (845) 758-0400
Fax: (845) 758-2633
E-mail: curran@proceedings.com
Web: www.proceedings.com

ICMTS 2018

Proceedings of the

2018 IEEE INTERNATIONAL CONFERENCE ON MICROELECTRONIC TEST STRUCTURES

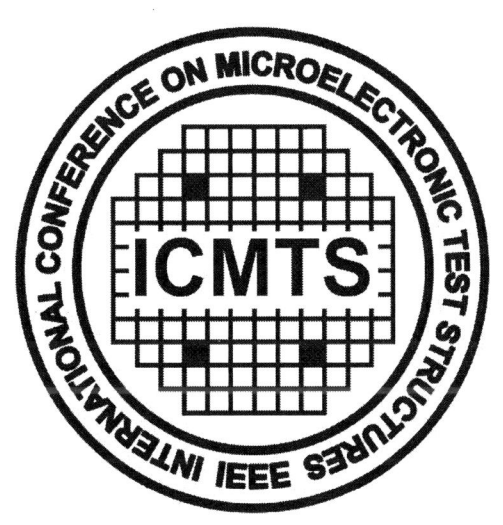

31st ICMTS

March 19-22, 2018
Courtyard by Marriott Austin Downtown/Convention Center
Austin, Texas, USA

Sponsored by the IEEE Electron Devices Society

ii

Program Schedule

Registration will open at 08:00 each day

Tuesday, March 20, 2018

09:00 - 09:10	Welcome
09:10 - 10:30	Session 1: Process Characterization
10:30 - 11:00	Break
11:00 - 12:20	Session 2: Modeling and Extraction
12:20 - 13:40	Lunch
13:40 - 14:10	Invited Talk: Jon Cheek (NXP)
14:10 - 14:50	Session 3: Reliability
14:50 - 15:20	Exhibitor Presentations
15:20 - 15:40	Break
15:40 - 17:20	Session 4: Materials Characterization
17:20 - 17:30	Close

Wednesday, March 21, 2018

08:15 - 08:45	Writing Good Papers / Getting Published
09:00 - 10:30	Session 5: Mismatch and Variability
10:20 - 10:50	Break
10:50 - 12:10	Session 6: On-Chip Characterization
12:10 - 13:30	Lunch
13:30 - 14:00	Invited Talk: Greg Yeric (ARM)
14:00 - 14:40	Session 7: Test Parallelism
14:40 - 14:50	ICMTS 2019 Presentation
14:50 - 15:20	Break
15:20 - 17:00	Session 8: Device Characterization
17:00 - 17:10	Close
18:30	Banquet at "18th Over Austin"

Thursday, March 22, 2018

08:15 - 08:45	Making Good Presentations
09:00 - 10:20	Session 9: MEMS
10:20 - 10:50	Break
10:50 - 12:10	Session 10: Noise and RF
12:10 - 12:20	Best Paper Announcement, Conference Close

Table of Contents

Session 1: Process Characterization

1.1   Test Structures for Debugging Variation of Critical Devices Caused by Layout-Dependent Effects in FinFETs
Qi Lin, Hans Pan, and Jonathan Chang   3

1.2   Passive Permutation Multiplexer to Detect Hard and Soft Open Fails on Short Flow Characterization Vehicle Test Chips
Christopher Hess and Shia Yu   7

1.3   Novel Test Structures for Extracting Interface State Density of Advanced CMOSFETs Using Optical Charge Pumping
Hyeong-Sub Song, Dong-Jun Oh, So-Yeong Kim, Sung-Kyu Kwon, Sungju Choi, Dae Hwan Kim, Dong-Hwan Lim, Chang-Hwan Choi, Dong Myong Kim, and Hi-Deok Lee   13

1.4   Test Structure to Evaluate the Impact of Parasitic Edge FET on Circuits Operating in Weak Inversion
Dale McQuirk, Chris Baker, and Brad Smith   16

Session 2: Modeling and Extraction

2.1   Comprehensive Investigation on Parameter Extraction Methodology for Short Channel amorphous InGaZnO Thin-Film Transistors
Chika Tanaka and Keiji Ikeda   23

2.2   Modeling Split-Gate Flash Memory Cell for Advanced Neuromorphic Computing
Mandana Tadayoni, Santosh Hariharan, Steven Lemke, Thibaut Pate-Cazal, Bernard Bertello, Vipin Tiwari, and Nhan Do   27

2.3   Validation of the BSIM4 Irregular LOD SPICE Model by Characterization of Various Irregular LOD Test Structures
Bob Peddenpohl, Max Otrokov, and Jeremy Wells   31

2.4   Efficient Parameter-Extraction of SPICE Compact Model Through Automatic Differentiation
Michihiro Shintani, Masayuki Hiromoto, and Takashi Sato   35

Session 3: Reliability

3.1   Test Structure Design for Model-Based Electromigration   43
Ertugrul Demircan, Mehul D. Shroff, and Hsun-Cheng Lee

3.2   Electrostatic Test Structures for Transmission Line Pulse and Human Body   48
Model Testing at Wafer Level
Robert Ashton, Stephen Fairbanks, Adam Bergen, and Evan Grund

Session 4: Materials Characterization

4.1   Reliability Analysis of the Metal-Graphene Contact Resistance Extracted   57
Through the Transfer Length Method
*Stefano Venica, Francesco Driussi, Amit Gahoiy, Satender Kataria, Pierpaolo
Palestri, Max C. Lemmeyz, and Luca Selmi*

4.2   Test Structures for Seed Layer Optimisation of Electroplated Ferromagnetic   63
Films
*C. M. Mackenzie Dover, A. W. S. Ross, S. Smith, J. G. Terry, A. R. Mount, and
A. J. Walton*

4.3   Test Structures Without Metal Contacts for DC Measurement of 2D-Materials   69
Deposited on Silicon
L. K. Nanver, X. Liu, and T. Knezevic

4.4   Test Structures for Evaluating Al_2O_3 Dielectrics for Graphene Field Effect   75
Transistors on Flexible Substrates
*Xinxin Yang, Marlene Bonmann, Andrei Vorobiev, Kjell Jeppson, and
Jan Stake*

4.5   Design of Ultraflexible Organic Differential Amplifier Circuits for Wearable   79
Sensor Technologies
*Masaya Kondo, Takafumi Uemura, Mihoko Akiyama, Naoko Namba, Masahiro
Sugiyama, Yuki Noda, Teppei Araki, Shusuke Yoshimoto, and Tsuyoshi Sekitani*

Session 5: Mismatch and Variability

5.1   A Test Structure to Reveal Short-Range Correlation Effects of Mismatch
Fluctuations in Backend Metal Fringe Capacitors   87
Hans Tuinhout, Adrie Zegers-van Duijnhoven, and Ihor Brunets

5.2   Monte Carlo Analysis by Direct Measurement Using Vth-Shiftable SRAM
Cell TEG   93
Shogo Yamaguchi, Daisuke Nishikata, Hitoshi Imi, and Kazuyuki Nakamura

5.3   Process Variation Estimation using A Combination of Ring Oscillator Delay
and FlipFlop Retention Characteristics   97
Takuma Konno, Shinichi Nishizawa, and Kazuhito Ito

5.4   NPN Mismatch Dependence on Layout   102
Cory Compton

Session 6: On-Chip Characterization

6.1   On–Chip Reconfigurable Monitor Circuit for Process Variation and
Temperature Estimation   111
Tadashi Kishimoto, Tohru Ishihara, and Hidetoshi Onodera

6.2   DFT-Enabled Within-Die AC Uniformity and Performance Monitor Structure
for Advanced Process   117
*Nui Chong, I-Ru Chen, Da Cheng, Amitava Majumdar, Ping-Chin Yeh, and
Jonathan Chang*

6.3   Versatile Chip-Level Integrated Test Vehicle for Dynamic Thermal
Evaluation   122
Suresh Parameswaran, Saravanan Balakrishnan, and Boon Ang

6.4   All-Digital On-Chip Heterogeneous Sensors for Tracking the Minimum
Energy Point of Processors   128
Shu Hokimoto, Jun Shiomi, Tohru Ishihara, and Hidetoshi Onodera

Session 7: Test Parallelism

7.1 Addressable Test Structure Design Enabling Parallel Testing of Reliability Devices   137
Lee DeBruler, Dennis Pretti, Mike Violette, Dave Peterson, Salil Mujumdar, Xia Li, and Ken Marr

7.2 Algorithm Based Adaptive Parametric Testing for Outlier Detection and Test Time Reduction   142
Veenadhar Katragadda, Martin Muthee, Arthur Gasasira, Frank Seelmann, and Jiun-Hsin Liao

Session 8: Device Characterization

8.1 Evaluation of Qss on SOI Back Si/SiO$_2$ Interface by Newly Designed Charge Pumping Method-TEG   149
Kazuma Takeda, Jiro Ida, Takayuki Mori, and Yasuo Arai

8.2 Quantitative Model of CMOS Inverter Chain Ring Oscillator's Effective Capacitance and Its Improvements in 14 nm FinFET Technology   153
Seong Yeol Mun, J. Cho, B. Zhu, P. Agnihotri, C. Y. Wong, T. J. Lee, V. Mahajan, B. W. Liu, Y. J. Shi, W. Hong, J. Ciavatti, J. G. Lee, S. B. Samavedam, and D. K. Sohn

8.3 Measurement of IGBT Trench MOS-Gated Region Characteristics Using Short Turn-Around-Time MOSFET Test Structures   157
Kiyoshi Takeuchi, Munetoshi Fukui, Takuya Saraya, Kazuo Itou, Shinichi Suzuki, Toshihiko Takakura, and Toshiro Hiramoto

8.4 Sensitivity of High-k Encapsulated MoS$_2$ Transistors to I-V Measurement Execution Time   161
Pavel Bolshakov, Ava Khosravi, Peng Zhao, Paul K. Hurley, Robert M. Wallace, and Chadwin D. Young

8.5 Total Ionizing Dose Effects on Analog Performance of 65 nm Bulk CMOS with Enclosed-Gate and Standard Layout   166
Matthias Bucher, Aristeidis Nikolaou, Alexia Papadopoulou, Nikolaos Makris, Loukas Chevas, Giulio Borghello, Henri D. Koch, and Federico Faccio

Session 9: MEMS

9.1   An On-Chip Test Structure for Studying the Frictional Behavior of Deep-RIE   173
MEMS Sidewall Surfaces
R. Ranga Reddy, Yuki Okamoto, and Yoshio Mita

9.2   Wafer Level Characterisation of Microelectrodes for Electrochemical Sensing   179
Applications
*E. O. Blair, L. Parga Basanta, I. Schmueser, J. R. K. Marland, A. Bochoux,
A. Tsiamis, C. Dunare, M. Normand, A. A. Stokes, A. J. Walton, and S. Smith*

9.3   Test Structure for Electrical Assessment of UV Laser Direct Fine Patterned   185
Material
Naoto Usami, Akio Higo, Ayako Mizushima, Yuki Okamoto, and Yoshio Mita

9.4   Open Model for External Mechanical Stress of Semiconductors and MEMS   189
R. T. Buhler and R. C. Giacomini

Session 10: Noise and RF

10.1   Importance of Complete Characterization Setup on On-Wafer TRL   197
Calibration in Sub-THz Range
*Chandan Yadav, Marina Deng, Magali De Matos, Sebastien Fregonese, and
Thomas Zimmer*

10.2   Measurement Time Reduction Technique for Input Referred Noise of   202
Dynamic Comparator
Yuki Ishijima, Shuya Nakagawa, and Hiroki Ishikuro

10.3   System Aware DUT Design for Optimum On-Wafer Noise Measurement   206
*Chih-Hung Chen, Benson Yang, Pei-Hsien Chua, Graham Brown, and Saswati
Das*

10.4   Measurement of Temperature Effect on Random Telegraph Noise Induced   210
Delay Fluctuation
A.K.M. Mahfuzul Islam, Masashi Okay, and Hidetoshi Onodera

Author Index   216

Technical Program Committee

Richard Allen	NIST	Jerome Mitard	IMEC
Yuzo Fukuzaki	Sony Corp.	Tatsuya Ohguro	Toshiba Corp.
Satoshi Habu	Keysight Technologies	Mark Poulter	Texas Instruments
Christopher Hess	PDF Solutions	Tsuyoshi Sekitani	U. Tokyo
Kjell Jeppson	Chalmers U. Technology	Luca Selmi	U. Modena and Reggio Emilia
Chang Yong Kang	Nvidia	Brad Smith	NXP Semiconductors
Mark Ketchen	Octevue	Stewart Smith	U. Edinburgh
Johan Klootwijk	Philips Research	Kiyoshi Takeuchi	U. Tokyo
Alexey Kovalgin	U. Twente	Bing-Yu Tsui	National Chiao Tung U.
Hi-Deok Lee	Chungnam National U.	Hans Tuinhout	NXP Semiconductors
Emilio Lora-Tamayo	Menéndez Pelayo Int'l U.	Bill Verzi	Keysight Technologies
Alan Mathewson	Tyndall National Institute	Anthony Walton	U. Edinburgh
Hiroaki Matsui	U. Tokyo	Larg Weiland	PDF Solutions
Colin McAndrew	NXP Semiconductors	Greg Yeric	ARM
Kevin McCarthy	U. College Cork	Chadwin Young	UT Dallas
Yoshio Mita	U. Tokyo		

Steering Committee

Bill Verzi	Keysight Technologies
Larg Weiland	PDF Solutions
Satoshi Habu	Keysight Technologies
Yoshi Mita	U. Tokyo
Anthony Walton	U. Edinburgh
Luca Selmi	U. Modena and Reggio Emilia

2018 Executive Committee

General Chair	Colin McAndrew (NXP Semiconductors)
Technical Program Chair	Brad Smith (NXP Semiconductors)
Tutorial Chair	Chadwin Young (UT Dallas)
Exhibition Chair	Scott McDade (Keysight Technologies)
Conference Manager	Catherine Shaw (C. Shaw LLC)
US Steering Committee	Bill Verzi (Keysight Technologies)
US Steering Committee	Larg Weiland (PDF Solutions)
Website	Yoshio Mita (U. Tokyo)

Session 1

Process Characterization

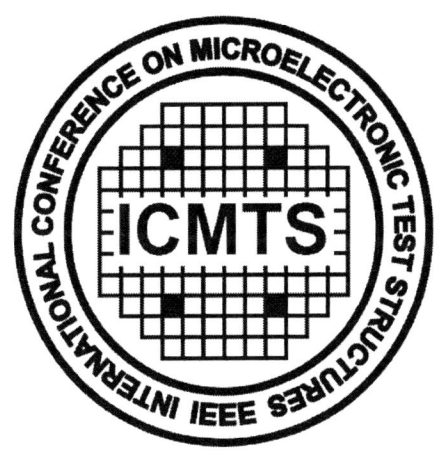

March 20, 2018

09:10-10:30

Co-Chairs:

Tsuyoshi SEKITANI, University of Osaka, Japan

Larg WEILAND, PDF Solutions, USA

978-1-5386-5072-1/18 $31.00 © 2018 IEEE

Test Structures for Debugging Variation of Critical Devices Caused by Layout-Dependent Effects in FinFETs

Qi Lin, Hans Pan, Jonathan Chang

Xilinx Inc., 2100 Logic Drive, San Jose, CA 95124
Telephone: (408) 879-4764, Fax: (408) 371–5041
Email: qilin@xilinx.com

Abstract—The increasing stress engineering in FinFETs raises concerns about performance variation caused by the strong layout-dependent effect (LDE). The challenge is that it is difficult to decouple the combination of LDEs in a layout. As a result, it is challenging for Fab to reduce the variation induced by LDE. In this paper, we present a set of test structures for monitoring and debugging the variation of critical devices caused by LDEs. These test structures were verified in 16nm FinFET technology. We also present two case studies of debugging FinFET device variation by using these test structures.

Keywords—variation, layout-dependent effect (LDE), FinFET, CMOS

I. INTRODUCTION

As CMOS scaling reaches the nanoscale regime, device behavior depends not only on traditional geometric parameters, such as channel length and width, but also on the layout implementation details of the device and its surrounding neighborhood. The increased use of stress engineering to improve device performance also adds new geometric dependencies. While layout-dependent effects (LDE) have been extensively studied on planar devices, they are still challenging to understand and model on FinFETs [1]-[3]. One difficulty is that the sources of engineered stress such as the oxide-to-oxide spacing effect (OSE), the length of oxide diffusion (LOD), and the stress memorization technique (SMT) are not necessarily linearly superimposable [4]. To resolve these issues, a dedicated model is constructed according to a specific layout, such as an SRAM transistor or analog device. Normally, the creation of a dedicated compact model combines layout extractions and direct measurements of the device. However, a common occurrence on silicon is that the base device with the minimum LDE is on target, but the devices with the specific layout are off target due to LDE variation. In this work, we developed a set of new test structures to monitor and debug the critical device variation caused by LDE. This paper is organized as follows. Section 2 reviews the design of traditional LDE test structures and presents our new LDE test structures. In Section 3, Si data from 16nm FinFET CMOS processes are discussed. Finally, the conclusions are summarized in Section 4.

II. DESIGN OF TEST SREUCTURES

To conduct the experiment, the two groups of test structures were designed and put on the test chip. The first group of test structures followed the traditional approach used in the compact model to extract layout-dependent parameters. This approach partitions the modelling into two parts: a base model and a set of layout-dependent models that modulate the base model. The layout for base model parameter extraction is normally simplified to minimize LDEs. The layout for LDE parameter extraction deviates from the layout of the base model in one LDE parameter. For example, for LOD parameter extraction, the difference between the layout-dependent model layout and base model layout is the length of oxide diffusion. The second group of test structures we designed also has two parts: a critical device with a specific layout such as an SRAM transistor or analog device used in the product, and a set of layout-dependent structures. Instead of decoupling the LDEs of the critical device layout, we superposed LDEs on a base model layout one at a time, as shown in Fig. 1. Fig. 1(A) depicts the layout of a base device. Fig. 1(B) shows the layout of the modified base device with LOD change. Fig. 1(C) shows the layout of adding the well proximity effect (WPE) on the layout in Fig. 1(B). Fig. 1(D) and Fig. 1(E) show how OSE-Y and OSE-X are added on the previous layouts, respectively. The final layout, shown in Fig. 1(F), adds the N+/P+ implant boundary on the layout in Fig. 1(E). The final layout is the same as the layout of the critical device monitored.

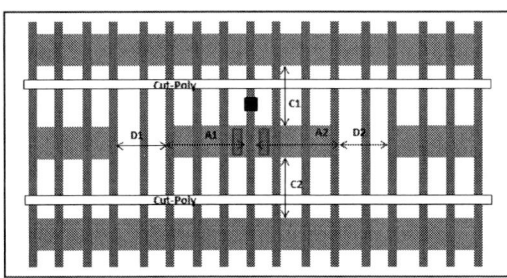

Fig. 1(A): Base device layout

978-1-5386-5072-1/18 $31.00 © 2018 IEEE

2018 INTERNATIONAL CONFERENCE ON MICROELECTRONIC TEST STRUCTURES, MARCH 19-22, 2018, AUSTIN, TEXAS, USA

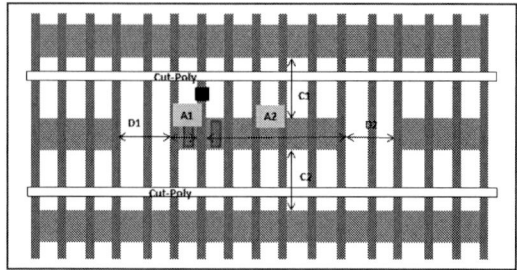

Fig. 1(B): Modify LOD of base device layout

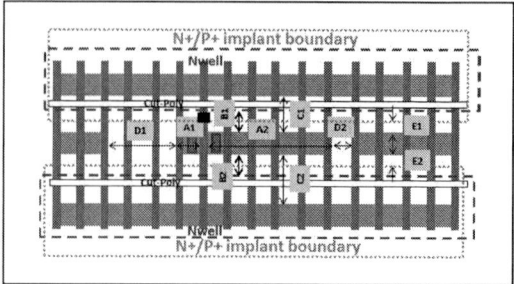

Fig. 1(F): Add N+/P+ boundary in Fig. 1(E)

III. SILICON RESULTS AND DISCUSSION

Table I shows the Si data from the 16nm FinFET based on the first group of test structures discussed in the previous section. Table I shows the Idsat of a base device nMOS and an inverter device nMOS. Both the base device and the inverter device have the same channel length L and channel width W. The Idsat of the inverter nMOS is 9% weaker than that of the base nMOS. Table II lists the Idsat change caused by each LDE related to the layout differences (LOD, WPE, OSE, and N+/P+). The sum of the values of ΔIdast is +3.5%, which is inconsistent with the Idsat difference between the base nMOS and the inverter nMOS (-9%). In other words, each decoupled LDE test structure cannot provide enough information to explain the change of the inverter nMOS Idsat.

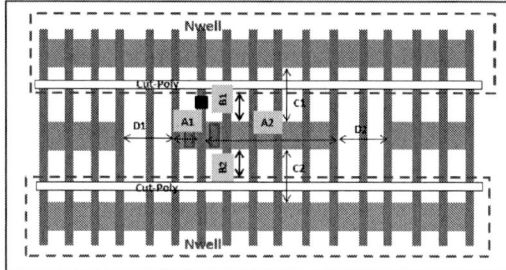

Fig. 1(C): Add WPE on layout in Fig. 1(B)

TABLE I.   SI DATA FROM BASE DEVICE AND INVERTER NMOS

	Base device	Inverter nMOS	ΔIdsat
Idsat	1.0x	0.91x	-9%

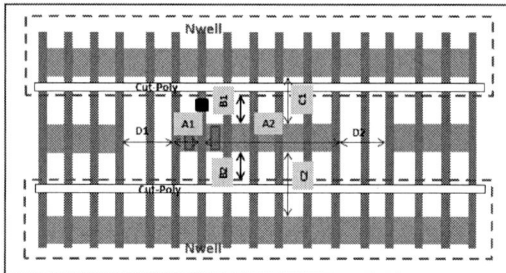

Fig. 1(D): Modify OSE-Y in Fig. 1(C)

TABLE II.   SI DATA OF ΔIDSAT CAUSED BY EACH LDE

LDE Item	LDE parameter	LDE dimensions in layout	ΔIdsat
LOD	A1	Base device: A1=1.0x	1.50%
		Inverter nMOS: A1=2.9x	
	A2	Base device: A2=1.0x	1.50%
		Inverter nMOS: A2=2.9x	
WPE-Y	B1	Base device: No WPE	2.00%
		Inverter nMOS: B1=1.0x	
	B2	Base device: No WPE	1.00%
		Inverter nMOS: B2=20.5x	
OSE-Y	C1	Base device: C1=1.0x	-3.00%
		Inverter nMOS: C1=0.27x	
	C2	Base device: C2=1.0x	0.00%
		Inverter nMOS: C2=0.9x	
OSE-X	D1,2	Base device: D1,2=1.0x	3.50%
		Inverter nMOS: D1,2=2.8x	
N+/P+	E1,2	Base device: No	-3.00%
		Inverter nMOS: E1,2=1.0x	
		Sum:	3.50%

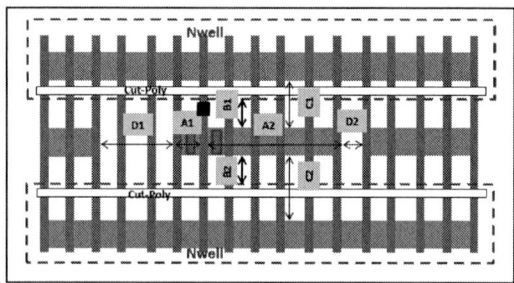

Fig. 1(E): Modify OSE-X in Fig. 1(D)

978-1-5386-5072-1/18 $31.00 © 2018 IEEE   4

Fig. 2 shows the silicon data from the second group of test structures discussed in Section 2. A set of LDE test structures starts from the base device layout and migrates to the SRAM nMOS layout by superposing LDE layouts. In Fig. 2(A), the green line is the spec at the typical corner. The blue line indicates the Si results. If Si LDEs follow the model, the gap between the green line and the red line should be constant. However, the base device is on target, but the SRAM nMOS is off target by 11%. A large increase in the gap between the lines occurs at the introduction of OSE-Y. After improving the OSE-Y process and tweaking the model, the gap between Si and spec is similar at different LDE layouts, as shown in Fig. 2(B).

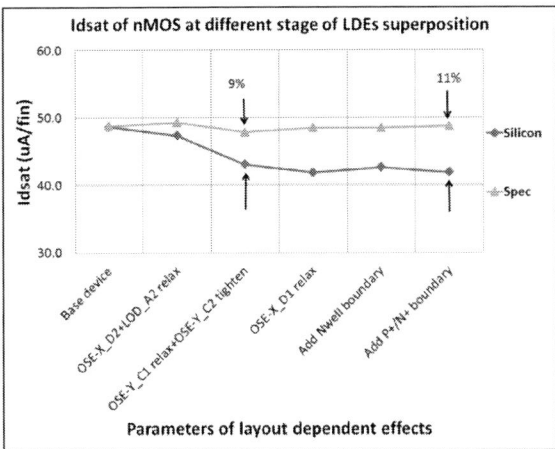

Fig. 2(A): nMOS Idsat vs device with LDE superposition at different stage. The blue line represents the Si data and the green line represents the spec. The gap between Si and spec increases after modifying OSE-Y

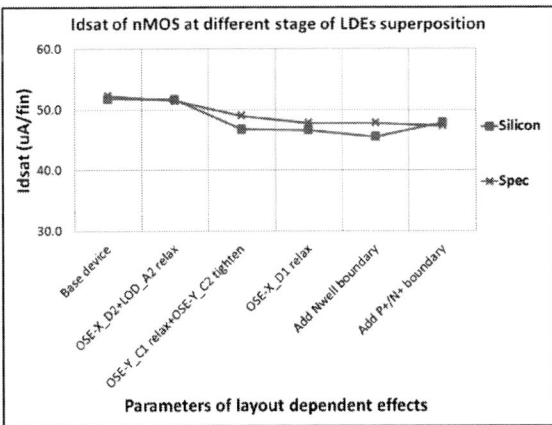

Fig. 2(B): After fixing the model and Si, the gap between Si and spec for Idsat is smaller and similar at different stage of LDE superposition. The red line represents the Si data and the purple line represents the new spec

Fig. 3 shows another case of debugging the high leakage for a critical device in speed path. In Fig. 3, the X-axis represents the normalized leakage of the base device and the Y-axis represents the normalized leakage of the LDE device. If

the Si data are above the red line, the LDE structure has higher leakage than that of the base device. Otherwise, the LDE structure has lower leakage. Fig. 3(A) shows the measured leakage of the modified base device with LOD layout change compared to the measured leakage of a base device on the same silicon. The leakages of both devices are similar, since the blue dots follow the red line. Fig. 3(B) illustrates the measured leakage of the modified base device with LOD and OSE-Y layout changes compared to the measured leakage of a base device. In Fig. 3(B), most of the blue dots are below the red line, indicating that the modified base device with LOD and OSE-Y changes has less leakage than that of the base device. Fig. 3(C) plots the measured leakage of the modified base device with LOD, OSE-Y and OSE-X layout changes compared to the measured leakage of a base device. Similar to the situation in Fig. 3(B), most of the blue dots are below the red line, so the modified device has less leakage than the base device. Fig. 3(D) presents the measured leakage of the modified base device with LOD, OSE-Y, OSE-X and WPE layout changes compared to the measured leakage of a based device. Again, this modified base device is healthy since most of the blue dots are below the red line. Lastly, in Fig. 3(E), the modified base device adds an N+/P+ implant boundary near DUT on the modified base device shown in Fig. 3(D). As a result, the measured leakage of the modified base device is larger than that of a base device since the blue dots move above the red line. After investigating the device leakage at different stages of LDE superposition, we could narrow down the root cause of high leakage in a critical device to the N+/P+ step. The process improvement related to the N+/P+ step significantly reduced the critical device leakage.

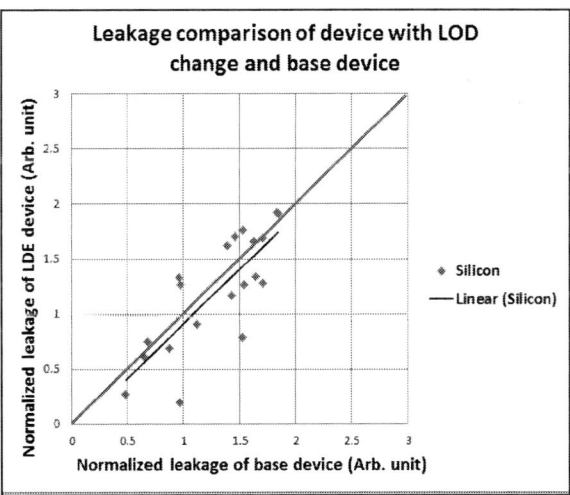

Fig. 3(A): Leakage comparison between device with LOD change and base device

978-1-5386-5072-1/18 $31.00 © 2018 IEEE

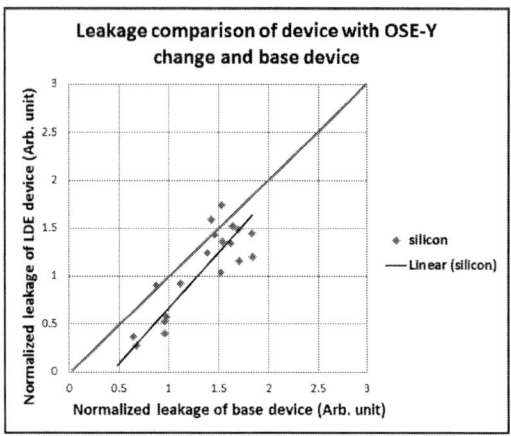

Fig. 3(B): Leakage comparison between device with LOD and OSE-Y changes and base device

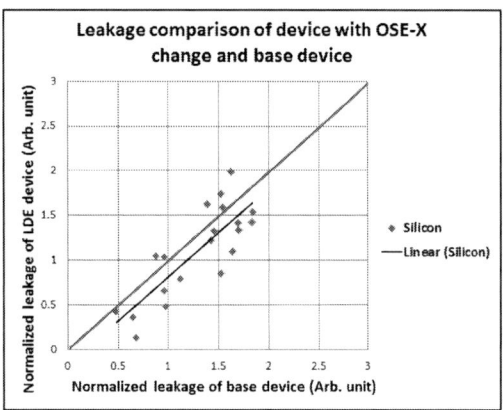

Fig. 3(C): Leakage comparison between device with LOD, OSE-Y, OSE-X changes and base device

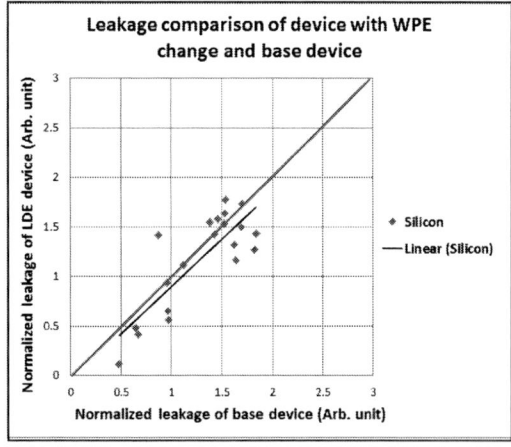

Fig. 3(D): Leakage comparison between device with LOD, OSE-Y, OSE-X, WPE changes and base device

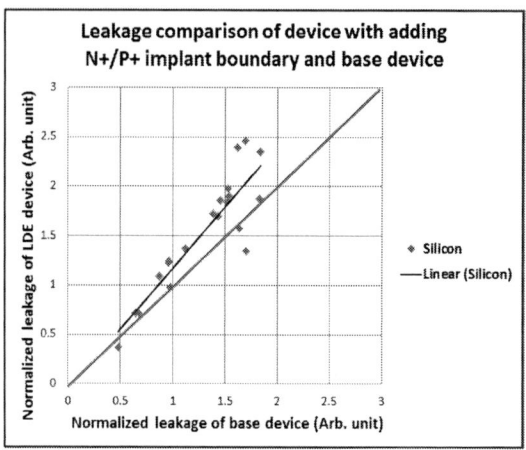

Fig. 3(E): Leakage comparison between device with LOD, OSE-Y, OSE-X, WPE and N+/P+ implant boundary changes and base device

IV. CONCLUSIONS

Using traditional LDE test structures to monitor or debug the variation of critical devices in nanoscale FinFET technology is not efficient. We developed a new set of LDE test structures which take the nonlinear combination of LDEs into account. The new set of test structures can not only monitor the variations of LDEs, but also pinpoint the process step that causes the monitored critical device to be off target on silicon.

REFERENCES

[1] C. Ndiaye, V. Huard, R. Bertholon, M. Rafik, X. Federspiel, A. Bravaix, "Layout Dependent Effect: Impact on device performance and reliability in recent CMOS nodes", *IEEE International Integrated Reliability Workshop*. pp. 24-28, 2016.

[2] D. Chen, G. Lin, T. Lee, R. Lee, Y. Liu, M. Wang, Y. Cheng, D. Wu, "Compact Modeling Solution of Layout Dependent Effect for FinFET Technology", *IEEE International Conference on Microelectronic Test Structures*, pp. 110-115, 2015.

[3] M. Bardon, V. Moroz, G. Eneman, P. Schuddinck, M. Dehan, D. Yakimets, D. Jang, "Layout-induced stress effects in 14nm & 10nm FinFETs and their impact on performance", *Symposium on VLSI Technology*, pp. 114-115, 2013.

[4] J. Faricelli, "Layout-Dependent Proximity Effects in Deep Nanoscale CMOS", *IEEE Custom Integrated Circuit Conference*, pp. 1-8, 2010

Passive Permutation Multiplexer To Detect Hard and Soft Open Fails on Short Flow Characterization Vehicle Test Chips

Christopher Hess, Shia Yu

PDF Solutions Inc., San Jose, CA 95110, USA

Abstract— Short flow characterization vehicle test chips are a major contributor to fast learning cycles especially for BEOL process steps. While hard open fails can be easily detected even in large via chains, it is very difficult to detect soft open fails like a 100 times larger via resistance of just one via within a large chain of vias. A Passive Permutation Multiplexer (PPM) is presented to optimize the signal to noise ratio for detecting soft open fails. The PPM implements a balanced routing access to a local population of resistive Devices Under Test (DUT) such as via or contact chains. Thus, soft open fail are easily recognizable as outliers of all measured resistance values within such a local population of DUTs. Compared to traditional passive multiplexers, the PPM contains up to twice as many DUTs. Furthermore, significantly larger Design of Experiments (DOE) can be implemented, since the PPM can hold more than just one DOE level within the same array.

Keywords—Test Structure; Passive Permutation Multiplexer; Short Flow Characterization Vehicle Test Chip

I. INTRODUCTION

Short flow characterization vehicles are commonly used to enable fast learning cycles of selected process steps. They are especially beneficiary to evaluate BEOL layers, since all the time consuming process steps to make devices can be skipped. However, the lack of selecting transistor results in significantly larger test structures given the number of available pads to access them. While hard open fails can be easily detected even in large via chains, it is very difficult to detect soft open fails like a 100 times larger via resistance, since such small overall resistance change of one via cannot be measured in large via chain of a million vias or more.

One option to detect soft open fails is the implementation of many small via chains as suggested in [1], which is not feasible on short flow characterization vehicle test chips due to the lack of selection transistors. Another option is to implement a small population of identical via chains on a chip and then look for a statistical outlier within that local population. However, pad count will be the main limiting factor of such an approach. Passive Multiplexers (ref. [2] & [3]) as well as Checkerboard Test Structures (ref. [4] & [5]) are used to significantly increase the count of Devices under Test (DUT) while decreasing the number of required pads. However, the routing within Passive Multiplexers is different for each DUT within the array, which causes a routing dependent gradient of the measured resistance values throughout the array, even if all the DUTs are identical. Thus, the signal to noise ratio is insufficient to detect small statistical variation

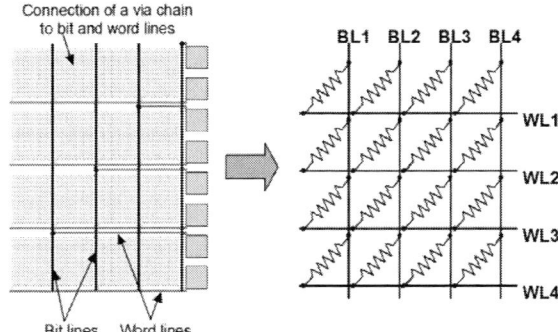

Figure 1: Layout (left) and schematic (right) of a Passive Multiplexer configuration (from [3]).

among the measurement values. Checkerboard Test Structures only target short circuit fails and have not been evaluated for detecting open fails.

The paper proposes a Passive Permutation Multiplexer (PPM), which will combine those two techniques. The PPM enables balanced routing for a small population of DUTs, thus opening the door to detect soft open fails. It describes the design concept followed by guidelines to create DUTs and Design of Experiments (DOE). Then, testing and analysis procedures are discussed before presenting some experimental results.

II. DESIGN CONCEPT

The proposed new DUT array is based upon two known arrays. First, there is a Passive Multiplexer as proposed by [2] and [3] and illustrated in Figure 1. Resistive test structures like via chains are placed as DUTs inside the array. The Passive Multiplexer splits all nets into two groups each containing half of the nets. Those two sets are arranged as rows and columns, where resistive test structures like via chains will be connected to one line of each group. Given n pads, the DUT count will be $(n/2)^2$. Obviously, routing is different for each DUT, as mentioned before.

The second known array to place DUTs is based upon a Permutation Matrix introduced by [4] and [5] and illustrated on the left of Figure 2. Compared to the Passive Multiplexer, the Permutation Matrix actually contains all possible $n/2 \cdot (n-1)$ pairs of nets. A Permutation Matrix can be created for any even number n of nets using the following equation to determine the elements within the matrix.

978-1-5386-5072-1/18 $31.00 © 2018 IEEE

2018 INTERNATIONAL CONFERENCE ON MICROELECTRONIC TEST STRUCTURES, MARCH 19-22, 2018, AUSTIN, TEXAS, USA

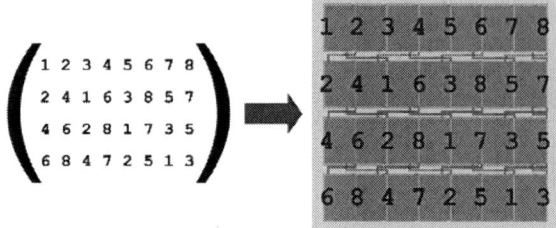

Figure 2: Permutation Matrix (left) and its implementation as Checkerboard Test Structure (right) (from [5]).

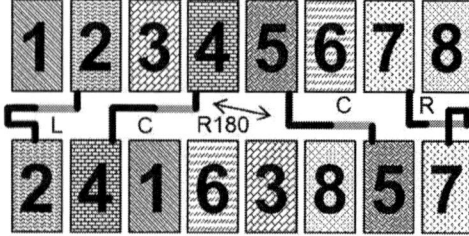

Figure 4: Basic routing elements L (left), C (center) and R (right) to connect nets between two rows of the PPM.

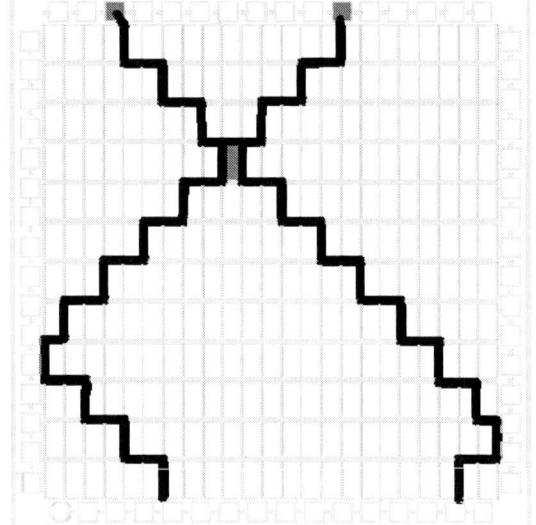

Figure 3: Routing of two nets within a Checkerboard Test Structure with 276 comb type DUTs (from [5]).

$$a[i,j] := \begin{cases} j + 2 \cdot i - 2 & where \ \frac{j}{2} \in \mathbb{N} \wedge i \leq \frac{n-j+2}{2} \\ 2 \cdot n - j - 2 \cdot i + 3 & where \ \frac{j}{2} \in \mathbb{N} \wedge i > \frac{n-j+2}{2} \\ 2 \cdot i - j - 1 & where \ \frac{j+1}{2} \in \mathbb{N} \wedge i > \frac{j+1}{2} \\ j - 2 \cdot i + 2 & where \ \frac{j+1}{2} \in \mathbb{N} \wedge i \leq \frac{j+1}{2} \end{cases} \quad (1)$$

The Permutation Matrix can be implemented as Checkerboard Test Structure containing up to twice as many DUTs for a given number n of pads. In the past, only comb type test structures have been used to catch short circuit fails. DUT and routing resistance does not really matter in that application as long as only a few short circuits occur within a single array.

Most importantly, implementing the Permutation Matrix enables a balanced routing of each net through the Checkerboard Test Structure as can be seen in Figure 3. The routing distance from a pad to a DUT is almost the same for each DUT within a row except for the left most and right most DUT. If those exceptions on the left and right can be redesigned, it should be possible to completely eliminate a routing related gradient between the DUTs within the same row.

Figure 5: Identical routing channels between rows of the PPM.

To detect soft open fails we combine those two known techniques by implementing the Permutation Matrix as a Passive Permutation Multiplexer (PPM). First, the routing between the rows has to be redesigned. Connecting the nets between two rows requires three base routing elements L, C and R as shown in Figure 4. It is important that the resistance of those base elements is identical. Thus, the elements L & R are redesigned to match the number of vias as well as the path length per layer of element C. The, element L is placed once on the left and element R is placed once on the right of a routing channel. Element C is used multiple times in the middle of the routing channel to connect all remaining nets by rotating element C by 180^0 when needed. Once the routing is completed between the first two rows, the entire routing channel can be copied between all remaining rows as can be seen in Figure 5.

The measurement points will be moved into the middle of the Permutation Matrix to further minimize the routing impact, which requires a 1 by N or 2 by N pad frame instead of a boundary pad frame. Figure 6 shows an example of a rotated PPM with 30 nets implemented to the left and right of a 2by16 pad frame. The PPM has 15 columns with 29 DUT per column totaling 435 DUTs. 7 columns of DUTs are placed to the left of the pad frame and 8 columns of DUTs are placed to the right of the pad frame. The remaining two pads are used to monitor the routing resistance for a selected net on the left and the right side of the PPM. Finally, resistive DUTs such as via or contact chains will be placed instead of the combs inside the PPM.

978-1-5386-5072-1/18 $31.00 © 2018 IEEE

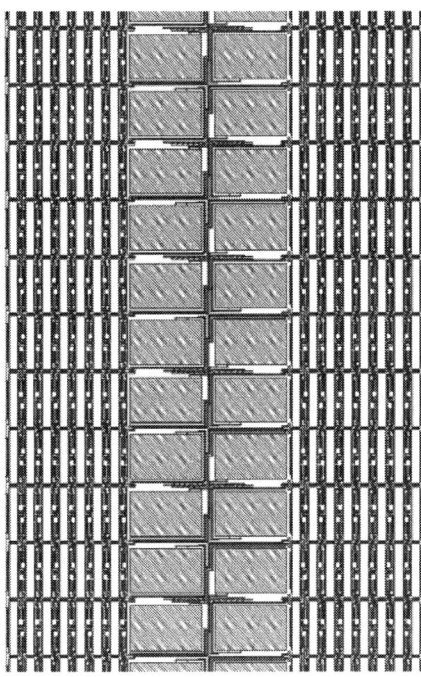

Figure 6: Passive Permutation Multiplexer (PPM) with pads in the middle of the array. It is an implementation of a Permutation Matrix rotated by 90 degrees.

III. DUT & DOE GUIDELINES

Some basic statistical guidelines have to be considered to design contact and via chains in a way that soft open fails can be detected. The resistance of a via chain as well as its measured current is given by the following equations:

$$R_{chain} = k \cdot R_{via} \tag{2}$$

$$I = \frac{VDD}{k \cdot R_{via}} \tag{3}$$

To detect outliers from a small local population of via chains it is safe to assume a Gaussian distribution of the via resistance $\Delta R_{via}/R_{via}$. The following equations define the 3σ window ΔR_{chain} of a via chain as well as its smallest corresponding measurement current ΔI:

$$\Delta R_{Chain} = 3 \cdot \sqrt{k} \cdot \Delta R_{via} \tag{4}$$

$$\Delta I = \frac{VDD}{k \cdot R_{via}} - \frac{VDD}{k \cdot R_{via} + \Delta R_{Chain}} \tag{5}$$

Resolving $\Delta I/I$ will provide a guideline for a maximum number of vias per chain. Any measured DUT resistance outside $R_{chain} \pm \Delta R_{chain}$ has a 0.3% probability of a single via soft fail with a resistance $R_{softFail} > \Delta R_{chain}$.

$$k < \left(3 \frac{\Delta R_{via}}{R_{via}} \cdot \left(\frac{1}{\frac{\Delta I}{I}} - 1 \right) \right)^2 \tag{6}$$

$$R_{softFail} > \Delta R_{Chain} = 3 \cdot \sqrt{k} \cdot \Delta R_{via} \tag{7}$$

For instance, according to equation (6) the maximum number of vias in a chain would be 14371 given a natural local via resistance variation of $\Delta R_{via}/R_{via}$=4% and also given the

ability to reliably measure a delta current of $\Delta I/I$=0.1%. According to equation (7), the 3σ resistance of the entire via chain would be ΔR_{chain}=360Ω. In this case it is easily possible to detect soft fails such as single vias with a resistance of 1kΩ or more. It is important to always check the nominal current according to equation (3) to ensure that the test equipment can reliably measure ΔI within available range settings. For instance, the expected measured average current would be around I=2.8uA for a chain of 14371 vias with an average via resistance R_{via}=25Ω (including local metal links) and VDD=1V. The tester has to be able to reliably measure ΔI=2.8nA (0.1% of I) in a 10uA range setting to detect ΔR_{chain}=360Ω within the overall resistance R_{chain}=360kΩ of the entire via chain.

Lowering VDD may enable some degree of freedom to choose a current range that best fits given test equipment. As alternative, the following approximation can be used to determine the maximum number k of vias in a chain dependent on ΔI rather than $\Delta I/I$:

$$k < \sqrt[3]{\left(3 \frac{\Delta R_{via}}{R_{via}} \cdot \frac{VDD}{R_{via} \cdot \Delta I} \right)^2} \tag{8}$$

For instance, lowering ΔI=1nA would allow 28455 vias in a chain given all the other limits as before (VDD=1V, R_{via}=25Ω, $\Delta R_{via}/R_{via}$=4%). The overall resistance of the via chain increases to R_{chain}=711kΩ (I=1.4uA), so does the 3σ resistance of ΔR_{chain}=506Ω, which is still a reasonable number to detect soft fails such as single vias with a resistance of 1kΩ or more. However, the signal to noise ratio is lower at $\Delta I/I$=0.07%, which may be acceptable for measuring a single via chain, but questionable for hundreds of via chains within an array. Thus, the following guidelines are highly recommended for a Design of Experiments (DOE) with DUTs placed inside the PPM:

- The maximum number k of vias in a chain should be chosen, so that ΔI will be about an order of magnitude above the relative channel to channel measurement error of the targeted range setting of the tester.

- At least two chains with different counts should be implemented for some DOE levels to maintain visibility on excursion wafers.

Furthermore, simulations have been run to determine some general guidelines to optimize the signal to noise ratio during the design of PPMs:

- All DUTs within the same column of the PPM have to be identical. Thus, the maximum number of DOE levels is limited to n/2 for a PPM with n nets.

- DUTs with lower DUT resistance should be placed in columns closest to the pads.

- Contacts into N type and P type active areas should be placed in separate PPMs.

- The average target DUT resistance should be two orders of magnitude larger than the overall resistance of the routing per net for PPMs with n≤30 nets.

- The ratio between the highest and lowest target DUT resistance should be less than 2 within the same PPM.

Obviously, some back and forth between tester range settings as well as statistical and general design guideline is required to pick chain sizes to create an effective DOE.

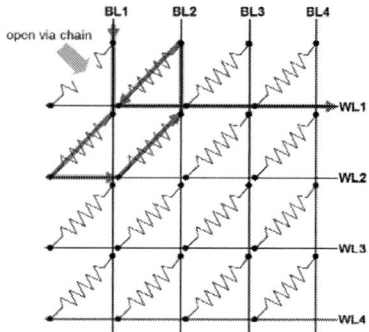

Figure 7: Potential sneak path to cover a defective DUT inside a passive multiplexer (from [3]).

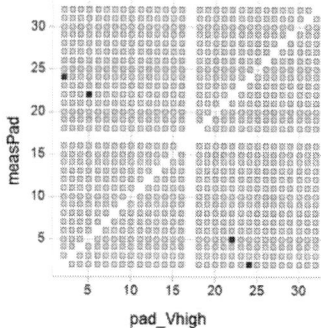

Figure 9: Bit map of measure PPM with two detected hard open fails.

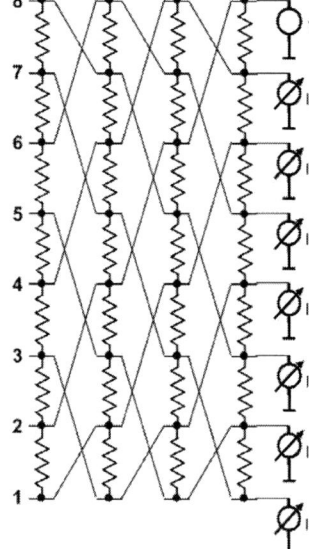

Figure 8: Test Procedure for PPM.

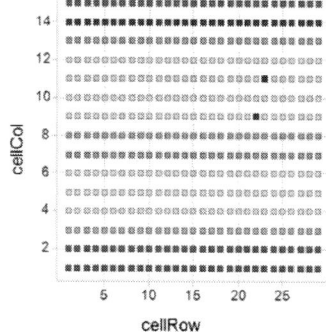

Figure 10: Location of failed DUTs inside Permutation Matrix, which is rotated inside the layout of the PPM.

IV. TESTING AND ANALYSIS PROCEDURES

Taking measurements within a passive multiplexer is not simply a matter of measuring the current flowing between two nets. As reported by [2] and [3], any passive multiplexor suffers from the potential risk, that an open DUT will be covered by a sneak path through surrounding DUTs as shown in Figure 7. If for instance the DUT in the upper left corner is open, a current can still sneak across the word line and bit lines attached to this particular DUT. To mitigate sneak paths, testing traditional passive multiplexers usually selects one bit line while executing current measurements on all word lines in parallel. The signal to noise ratio is dependent on what happens on the unselected bit lines. Keeping them floating is the recommended best known method, but not a guarantee for success.

While the PPM still carries the same potential risk of sneak paths, it mitigates them much more effectively by simply avoiding unselected nets during test. One net is set to a fixed voltage serving as driver while current is measured on all remaining nets in parallel, thus completely removing the issue of ambiguity of unselected nets (ref. Figure 8). The test

program simply consists of a walking one over all nets in a way that each net is once in driver mode. Such a test procedure actually measures each DUT twice, which furthermore enhances data quality, since those redundant measurements can detect measurement related errors during data collection. Yet, overall test time per DUT is similar for the PPM and traditional passive multiplexers.

Once current measurements have been collected for all DUTs inside a PPM, they have to be sorted by their location inside the Permutation Matrix using the algorithm and equations described in the appendix of the paper. Once sorted, cumulative distribution function (CDF) plots can be created to identify local outliers per column to identify hard and soft fails.

V. EXPERIMENTAL RESULTS

A Passive Permutation Multiplexer has been implemented with contact and via chains as resistive DUTs. Different Design of Experiment (DOE) levels have been placed in different columns of the PPM. However, all DUTs within a column are identical to extract local outliers based upon a local population of DUTs. Figure 9 shows the bitmap of measured data where hard fails can already be marked, since they are usually orders of magnitude off expected target vales. Since each DUT will be measured twice, such marked values have to be symmetric along the main diagonal of the bit map.

Next, the measured data above or below the diagonal will be sorted by the location of their corresponding DUT inside the Permutation Matrix as can be seen in Figure 10. The two hard fails are located in different rows, but within the same DOE level.

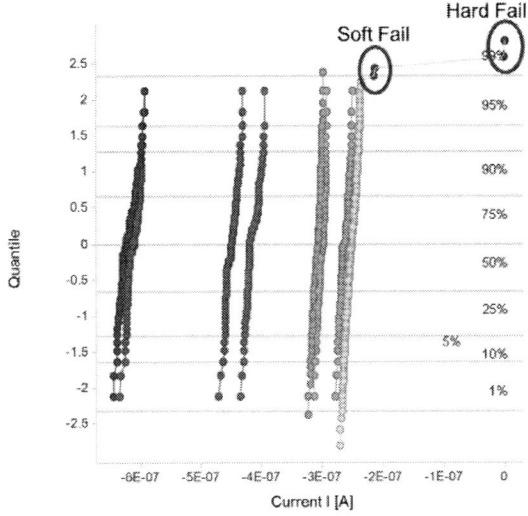

Figure 11: CDF plot of measured data indicating hard and soft open fails inside the PPM. The different colores represent different DOE levels.

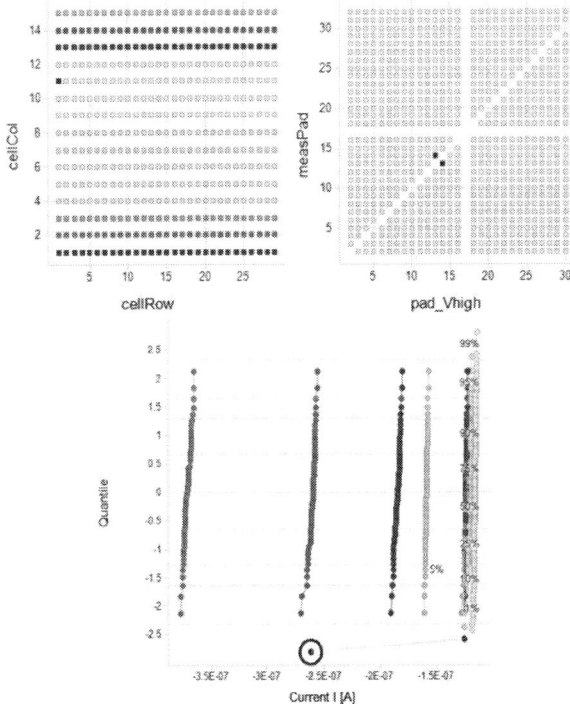

Figure 12: Location of failed DUTs (top left), bitmap (top right), and CDF plot of measured data (bottom) indicating a hard short fail inside a PPM. The different colores represent different DOE levels.

Screening for outliers has to be executed by row, since the layout of the PPM is a 90 degree rotation of the underlying Permutation Matrix. Figure 11 shows the distribution of measure resistance values of those via chains from just a few PPMs placed in close proximity inside the same die. The different colors indicate the different DOE levels. The two hard open fails are clearly visible on the top right of the figure. Also, two soft open fails can be distinguished from the otherwise very tight distribution of measured resistance values.

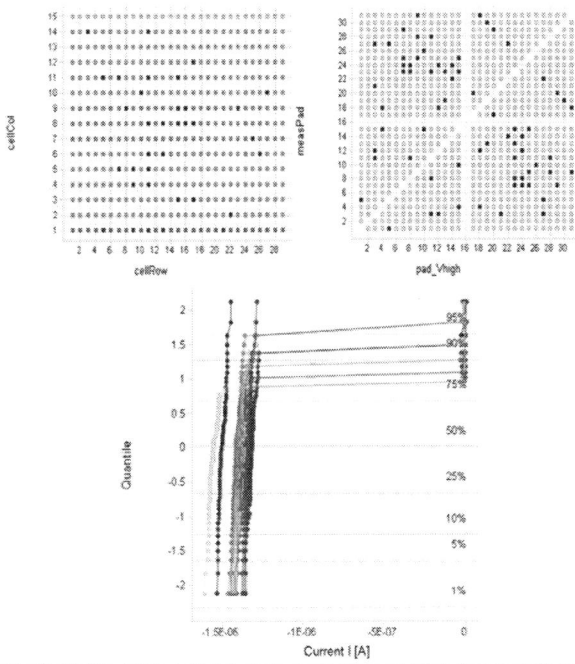

Figure 13: CDF plot of measured data indicating multiple hard open fails inside a PPM. Each color represents the distribution of a single row within the corresponding Permutation Matrix.

Also short circuit fails occur occasionally within via and contact chains, which can be recognized by higher than expected current measurement data.

Figure 12 shows such a case where several via chain links have been shorted together reducing the overall length of the chain and thus resulting in a higher measured current value.

Some ambiguity regarding multiple short circuits within Checkerboard Test Structures has been previously reported [5]. However, at the time measurements have been executed using digital testers measuring voltages on all nets due to the lack of equipment providing dozens of parallel low current measurement channels. Thus, all but the driver net are practically floating, which creates sneak paths as describe in Figure 7.

PPMs with multiple fails have been evaluated to see if changing the measurement procedure to measuring parallel currents as suggested here will in deed eliminate the issues reported in the past. Figure 13 shows on the top right a bitmap with plenty of failing DUTs. All measurements are perfectly symmetrical as it is the case for those PPMs with just a few fails. Thus, their location can be determined without ambiguity in the map on the top left. The CDF plot shows the separation of measurement values, where all fails are clearly hard fails, which are easy to localize without any ambiguity.

VI. Conclusion

Two established methods to build passive DUT arrays have been combined to provide a new Passive Permutation Multiplexer (PPM) to detect hard and soft fails within contact and via chains. Balanced routing of all DUTs within a column provides a locally identical population of DUTs with an improved signal to noise ratio by furthermore removing any floating nets during testing which in combination leads to a very tight distribution of measurement values not previously reported. Small 0.1% measurement outliers are now recognizable which have a very high probability of being just a single high resistive via or contact failure instead of a random shift of the entire chain resistance. Up to twice as many DUTs can be placed inside the PPM compared to traditional passive DUT arrays. More than just one DOE level can be implemented inside the same PPM due to the superior signal to noise ratio. Redundant measurements provide better detection of potential measurement errors while maintaining the same overall test time.

APPENDIX

LOCALIZATION OF DUTs

Resistance of a DUT is being measured between a net pair (p,q) inside Permutation Matrix. Figure 14 describes the algorithm to determine the row index i and column index j within the Permutation Matrix for a given pair (p,q) of numbers. To do so, the algorithm utilizes functions (9) and (10). Keep in mind that the implementation of the Permutation Matrix inside the PPM is rotated by 90 degrees. Thus, the meaning of row and column inside the PPM is actually swapped.

$$f(x) := \begin{cases} \dfrac{x}{2} & if \ \dfrac{x}{2} \in \mathbb{N} \wedge 1 < x \leq n \\ n - \left(\dfrac{x-1}{2}\right) & if \ \dfrac{x-1}{2} \in \mathbb{N} \wedge 1 \leq x < n \end{cases} \quad (9)$$

$$g(x) := \begin{cases} ((x-1) \bmod n) + 1 & if \ x > 0 \\ (x \bmod n) + n & if \ x \leq 0 \end{cases} \quad (10)$$

n: number elements in Permutation Matrix (equals number of nets)
p,q: pair of elements connected to a DUT with 1 <= p < q <= n
i,j: row and column indices within Permutation Matrix

```
IF q-p = 1 THEN
  i := 1
  j := p
ELSE
  j := g(f(q)-f(p))
  IF j even THEN
    x := j
  ELSE
    x := j + 1
  ENDIF
  i := g(f(q)-f(x)+1
ENDIF
```

Figure 14: Algorithm to determine row and column of a DUT within the Permutation Matrix.

REFERENCES

[1] Hess, C., Squcciarini, M., Yu, S., Burrows, J., Cheng, J., Lindley, R., Swimmer, A., Winters, S., "High Density Test Structure Array for Accurate Detection and Localization of Soft Fails", Proc. International Conference on Microelectronic Test Structures (ICMTS), pp. 131-136, Edinburgh (UK), March 2008

[2] Walton, A. J., Gammie, W., Marrow, D., Stevenson, J. T. M., Holwill, R. J., "A novel Approach for Reducing the Area Occupied by Contact Pads on Process Control Chips", International Conference on Microelectronic Test Structures, San Diego / USA, 1990

[3] Hess, C., Stine, B. E., Weiland, L. H., Mitchell, T., Karnett, M., Gardner, K., "Passive Multiplexer Test Structure For Fast and Accurate Contact and Via Fail Rate Evaluation", IEEE Transactions on Semiconductor Manufacturing, pp. 330-337, Vol. 16, No. 2, 2003

[4] Hess, C., Ströle, A., "Modeling of Real Defect Outlines and Defect Parameter Extraction Using a Checkerboard Test Structure to Localize Defects", IEEE Transactions on Semiconductor Manufacturing, pp. 284-292, Vol. 7, No. 3, 1994

[5] Hess, C., Weiland, L. H., Bornefeld, R., "Customized Checkerboard Test Structures to Localize Interconnection Point Defects", Proc. VLSI Multilevel Interconnection Conference (VMIC), pp. 163-168, Santa Clara (USA), 1997

2018 INTERNATIONAL CONFERENCE ON MICROELECTRONIC TEST STRUCTURES, MARCH 19-22, 2018, AUSTIN, TEXAS, USA

Novel test structures for extracting interface state density of advanced CMOSFETs using optical charge pumping

Hyeong-Sub Song[1], Dong-Jun Oh[1], So-Yeong Kim[1], Sung-Kyu Kwon[1], Sungju Choi[2], Dae Hwan Kim[2],
Dong-Hwan Lim[3], Chang-Hwan Choi[3], Dong Myong Kim[2], and Hi-Deok Lee[1*]

[1]Department of Electronics Engineering, Chungnam National University
[2]School of Electrical Engineering, Kookmin University
[3]Division of Materials Science and Engineering, Hanyang University
*Fax: +82-42-823-5436 E-mail: hdlee@cnu.ac.kr

Abstract— In this paper, we proposed novel test structures to evaluate the distribution of interface state density of MOSFETs by using optical charge pumping method. Unlike other measurement methods to extract interface state density (D_{it}), which have a limited range of measurable energy states and influenced by gate area and gate leakage, D_{it} can be extracted without these problems by using the proposed test structures. Test structures were fabricated using a 0.18μm CMOS process or FD-SOI technology with high-k dielectric, respectively. Optical charge pumping was performed in proposed test structures and D_{it} is extracted from 10^9 cm$^{-2}\cdot$ eV^{-1} to 10^{13} cm$^{-2}\cdot$ eV^{-1}.

Keywords—interface state density, optical charge pubping

I. INTRODUCTION

Recently, conventional measurement methods such as charge pumping or quasi-static capacitance-voltage method, which can extract the interface state density of Metal Oxide Semiconductor(MOS) devices, are no longer accurate due to the extremely reduced gate area and increased gate leakage current [1]. Accordingly, there are strong needs for accurately evaluating the interface characteristics of state-of-the-art devices. It was reported that optical charge pumping could extract the interface state density in the oxide using the relationship between the energy of the injected photon and the Fermi-level [2]. The method is based on adjusting the energy of the light source, and it is possible to extract the interface state density more precisely with the relative insensitivity to the gate area and the gate leakage. In order to measure the interface state density through optical charge pumping, injected light must reach the gate oxide region. However, unlike thin film transistors (TFT) of which interfaces are frequently measured by optical charge pumping [3], in many bulks or silicon on insulator(SOI) devices, light sources are blocked by interconnection lines and dummy metal layers, and optical charge pumping could not be evaluated properly. To solve problems caused by light blocking effect in bulk or

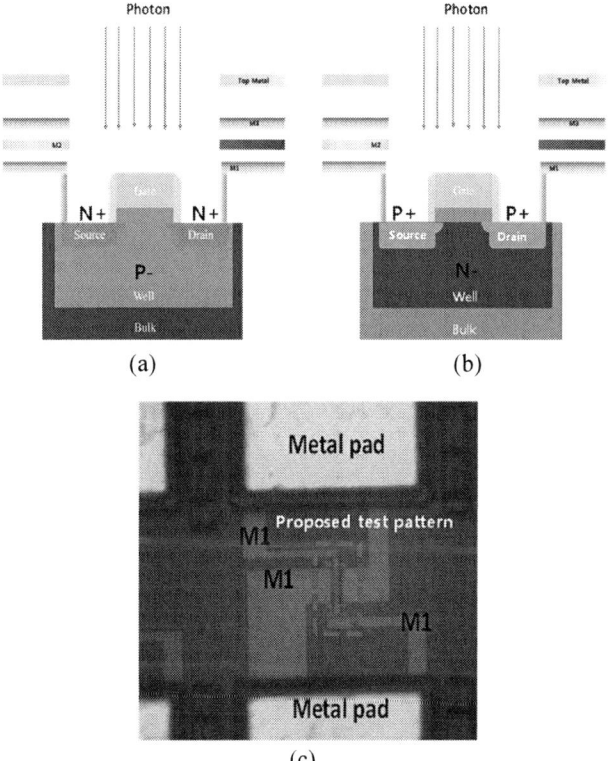

(a)   (b)

(c)

Fig. 1. Proposed test structures with symmetric source/drain doping to extract D_{it} using optical charge pumping technique (a) cross-sectional view of n-type MOS, (b) cross-sectional view of p-type MOS, and (c) actual fabricated device using a 0.18 μm process.

SOI CMOS, test structures without metal layers are proposed. That is, in order to measure the U-shaped interface state density using optical charge pumping, two test structures of different type MOSFETs, i.e., NMOSFET and PMOSFET (symmetric source/drain) are required. To eliminate this inconvenience and to measure the U-shaped D_{it} using only one test structure, a novel test structure with different source and drain doping

978-1-5386-5072-1/18 $31.00 © 2018 IEEE   13

types (asymmetric source/drain) is also proposed.

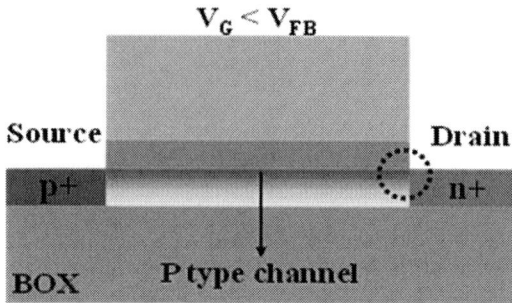

Fig. 2. Cross-sectional view of a proposed test structure with asymmetric source/drain doping to extract D_{it} using optical charge pumping technique.

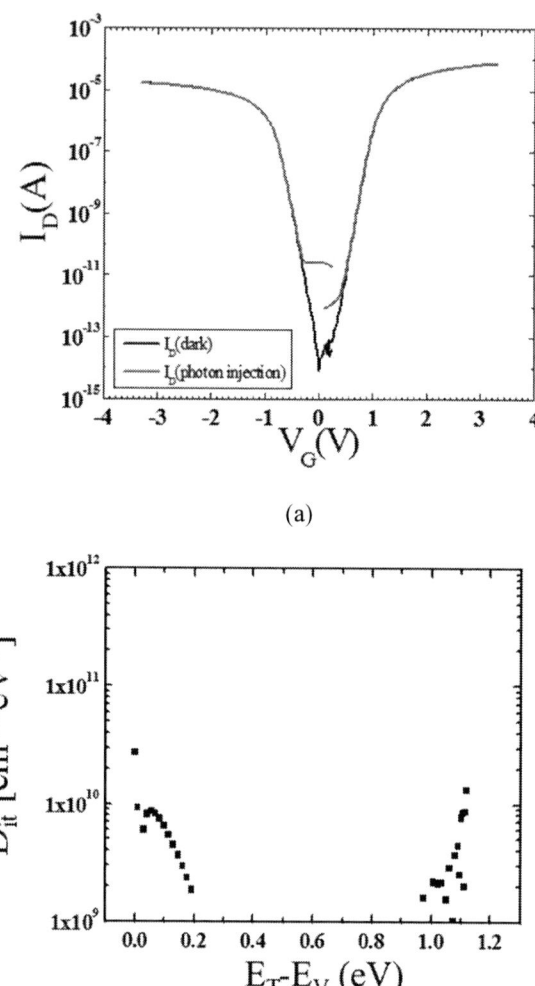

Fig. 3. Proposed test structures with symmetric source/drain doping to extract Dit using optical charge pumping technique, (a) I_D-V_G characteristics (b) D_{it} versus E_T-E_V characteristics.

II. PROPOSED TEST STRUCTURES

Figures 1 and 2 show proposed test structures for optical charge pumping. As shown in Fig. 1(a) and 1(b), the interconnection metal lines and dummy metal lines are blocked are all designed to be able to avoid the gate region to maximally expose the light near the gate oxide. In order to proceed with the actual optical charge pumping measurement, test structures were fabricated using a 0.18 µm CMOS technology as shown in Fig. 1(c). Two test structures with n-type and p-type MOS are required to observe the D_{it} of the U-shape. To improve these disadvantages, new test structures with the asymmetric source/drain doping are proposed to measure wider ranges of interface state density. The test structure is fabricated using a SOI process to minimize the body leakage component and we used a 150 nm TiN gate, a 5 nm HfAlO gate oxide, 50 nm Si p-type: 1.5×10^{15} cm^{-3}), a n$^+$ drain, and a p$^+$ source (both 3×10^{19} cm^{-3}).

III. EXPERIMENT AND RESULTS

Optical charge pumping was performed using a 1310 nm infrared laser; thus, the photon energy was approximately 0.943 eV. We applied a gate voltage ranging from 0 to -3.5 V and -0.1 V to the drain of PMOSFET, from 0 to 3.5 V and 0.1 V to the drain of NMOSFET. Then, I_D-V_G curves without (dark) and with photon injection were obtained as shown in Fig. 3. Next, ΔI_D is observed between two curves, which is caused by photon injection. Based on equation (1) - (3), D_{it} at gate oxide/channel interface is extracted as a function of energy and extracted result exhibits D_{it} distribution from 10^9 cm^{-2}· eV^{-1} to 10^{10} cm^{-2}· eV^{-1}. Since the MOSFETs used here has a thermal gate oxide, the magnitude of D_{it} is relatively very small, hence D_{it} was not extracted properly in the vicinity of the mid- bandgap.

$$\frac{1}{\eta_{dark}} \cong \left(\frac{V_{th}}{V_{GS}-V_{TN0}}\right) \ln\left(\frac{I_{D,Dark}}{I_{D0}}\right) \quad (1)$$

$$\frac{1}{\eta_{photo}} \cong \left(\frac{V_{th}}{V_{GS}-V_{TN,photo}}\right) \ln\left(\frac{I_{D,photo}}{I_{D0}}\right) \quad (2)$$

$$D_{it} = \left(\frac{\Delta C_{it,photo}}{q}\right) = \frac{C_{ox}(\Delta\eta_{dark}-\Delta\eta_{dark})}{q} \quad (3)$$

The extracted D_{it} using the optical charge pumping and the test structure with the asymmetric source/drain doping is shown in Fig. 4. As the proposed MOSFET with asymmetric source/drain has HfAlO dielectric, D_{it} was extracted about

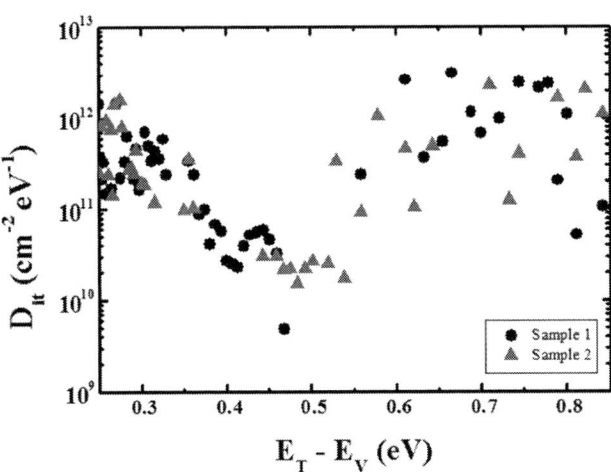

Fig. 4. D_{it} versus E_T-E_V characteristics in proposed test structure with asymmetric source/drain doping.

[3] J. H. Park et al., "Extraction of Density of States in Amorphous GaInZnO Thin-Film Transistors by Combining an Optical Charge Pumping and Capacitance-Voltage Characteristics", IEEE Electron Device Lett., vol. 30, no. 10, p.1069-1071, 2009.

two orders of magnitude higher than the test structure with the thermal oxide. Unlike test structures with the symmetric source/drain doping which required two MOSFETs to extracted the U-shaped D_{it}, test structure with the asymmetric source/drain could extract the U-shaped D_{it} with only one test structure.

IV. CONCLUSION

As the gate area and thickness in the MOS device are reduced extremely, the method of extracting D_{it} becomes not reliable. To overcome this problem, we designed novel test structures utilizing optical charge pumping. Using the proposed test structures and optical charge pumping method, it is expected that D_{it} of advanced MOSFETs can be extracted accurately and effectively

ACKNOWLEDGMENTS

This research was supported by the MOTIE (Ministry of Trade, Industry & Energy (10067808) and the KSRC (Korea Semiconductor Research Consortium) support program for the development of future semiconductor devices. The chip fabrication and EDA tool were also supported by the IC Design Education Center (IDEC), Korea.

REFERENCES

[1] K. Martens et.al , "New Interface State Density Extraction Method Applicable to Peaked and High-Density Distributions for Ge MOSFET Development", IEEE Electron Device Lett., vol. 27, no. 5, pp. 405-408, 2006.

[2] M. S. Kim et al., "Optical Subthreshold Current Method for Extracting the Interface States in MOS Systems", IEEE Electron Devices Lett., vol. 25, no. 2, pp. 101-103, 2004.

978-1-5386-5072-1/18 $31.00 © 2018 IEEE

Test Structures to Evaluate the Impact of Parasitic Edge FET on Circuits Operating in Weak Inversion

Dale McQuirk, Chris Baker, and Brad Smith

NXP Semiconductors N.V.

6501 William Cannon Dr., Austin, TX 78746

Email: chris.baker@nxp.com, dale.mcquirk@nxp.com, and brad.smith@nxp.com

Abstract—Precision analog circuit accuracy in a microcontroller product was impacted by unmodeled behavior across the temperature range. Three critical analog circuits from the microcontroller were built and tested in discrete parametric test structures. It was shown that a process with reduced parasitic edge FET leakage dramatically improved the accuracy of the analog circuits, which were operating in the subthreshold region.

Keywords— subthreshold, analog, matching, comparator, ΔVgs current reference, bandgap voltage reference, STI, models, microcontroller, hump

I. INTRODUCTION

Analog features implemented on microcontroller-based products can have aggressive electrical specifications for accuracy. Accurate models are necessary to construct circuits that will meet the electrical specification across process variation, operating voltage, and an operating temperature range from – 40 C to 125 C.

Three fundamental circuits that are used in analog designs for microcontroller products are shown in Figs. 1-3. The output variability of each circuit is critical to meet the accuracy specifications. Fig. 1 is a bandgap reference circuit and its output of interest is the voltage labeled Vout. Fig. 2 is a ΔVgs current reference circuit and its output of interest is the current labeled Iout. Fig. 3 is a high-gain comparator whose output voltage will be either Vdd or ground unless the two inputs are nearly identical. The output of interest is the difference between the input voltages (labeled Vos) at which the output flips (called Vin_trip). The bandgap voltage reference can be measured directly with standard test equipment on microcontroller products. The ΔVgs current reference and comparator circuits are not externally exposed and cannot be measured directly.

The NMOS devices for the three analog circuits operate in the subthreshold region to meet low power performance targets. In this region, the circuit outputs are described by well-known equations [1]. Equation (1) describes the output voltage of the bandgap voltage reference circuit in Fig. 1. Any mismatch of electrical characteristics between M1 and M2 would impact Vos and have a multiplicative impact on Vout. Equation (2) describes the output current of the ΔVgs current reference circuit from Fig. 2. Note that L3 and L4 cancel out since they are

equivalent. The width of M4 is six times larger than that of M3 to generate the ΔVgs. Equation (3) describes the relationship between Vout and Vos in Fig. 3. Mismatch between M5 and M6 impacts Vos

The models assume classic mismatch described in current literature [2-9]. These behaviors must be predictable and modeled correctly to achieve the desired accuracy from each of these circuits.

$$V_{out} = V_{eb2} + \left(1 + \frac{R}{R_1}\right)V_{os} + \frac{R}{R_1}\frac{kT}{q}ln\left(\frac{A_{E1}}{A_{E2}}\left(1 - \frac{V_{os}}{I_1 R_1}\right)\right) \quad (1)$$

$$I_{out} = \frac{NkT}{Rq}ln\left(\frac{W_4 L_3}{W_3 L_4}\right) = \frac{NkT}{Rq}ln\left(\frac{W_4}{W_3}\right) \quad (2)$$

$$V_{in\_trip} = V_{ref} + V_{os} \quad (3)$$

II. BACKGROUND

Fig. 4 shows voltage measurements from a bandgap voltage reference circuit implemented on a microcontroller product in low power 90 nm CMOS technology. 100 samples were measured at three temperatures. Some samples had a positive slope as a function of temperature, in agreement with models, while others had a negative slope. A larger variation in the output at cold temperature was also observed. Accuracy is limited by these unpredicted behaviors across the operating temperature. Observation of analog features that use the ΔVgs current reference and comparator circuits suggest they had similar behavior but could not be measured directly in the microcontroller.

A well-known problem [10-14] observed in MOSFET devices with STI structure is shown in Fig. 5. Colloquially known as the subthreshold "hump", this undesirable subthreshold behavior is caused by parasitic edge FETs formed at the STI interface along the channel of a MOSFET. It has the largest impact on devices operating in the subthreshold region. Since the analog circuits presented herein operate in the subthreshold region, the hump became the leading hypothesis for the cause of the high variation and unpredictable temperature slopes.

To test the hypothesis, M1 and M2 in the bandgap circuit of Fig. 1 were re-laid out with wider gate fingers while maintaining the same total device width and length. This was done to reduce the amount of edge, thereby reducing the impact of the parasitic edge FET. At the same time, the bias current was increased to reduce the sensitivity to any mismatch of M1 and M2.

Measurements from a microcontroller with the modified bandgap voltage reference are shown in Fig. 6. Note the more-consistent temperature slopes and smaller cold temperature variation. This suggests that the parasitic edge FET does, indeed, modulate the circuit's behavior. The improvements observed in the bandgap voltage reference were limited and similar changes were not feasible in the other two circuits. Therefore, process improvements were investigated.

Several mitigation methods have been proposed to eliminate the subthreshold hump [4,13]. Most involve either adding dopant or extending the gate to the channel edge to weaken the parasitic edge FET. A comparable method was used in this study and compared to the standard process. Fig. 7 shows Id-Vg results of NMOS devices comparing the standard and improved processes. Clearly, the improved process eliminates the subthreshold hump.

III. TEST STRUCTURE DESIGN

The three analog circuits from Figs. 1-3 were extracted from the microcontroller and placed into two 22-probepad frames. This enabled direct measurement of each of the circuits using high precision parametric test equipment, which was not possible in the microcontroller product. Fig. 8 shows the layout of the test structure.

The test structures were built in a low power 90 nm CMOS technology process. Wafers were manufactured through both the standard and improved processes mentioned earlier.

IV. RESULTS

Wafers were tested using a Keysight 4070-series parametric tester. The three circuits were measured at -40 C, 25 C, and 125 C. Approximately 50 sites from across the wafer were measured.

Fig. 9 shows the measurements from the ΔVgs current reference circuits for both process splits. Equation (2) predicts a positive linear temperature slope. The standard process (Fig. 9a) shows a negative temperature slope and larger than expected variation, consistent with observations made on the microcontroller. Clearly there is some mechanism not represented in the model. The improved process (Fig. 9b) shows the correct temperature slope and has less variability, more in line with the models.

For the comparator measurement, Vref was held constant at 1.0 V and Vin was swept until the output voltage toggled. Fig. 10 shows Vos, which is the difference between Vref and Vin. The standard process (Fig. 10a) showed higher than expected variation at -40 C with both positive and negative temperature slopes, again, consistent with observations from the

microcontroller. The Vos measurements from the improved process shown in Fig. 10b were not sensitive to temperature, consistent with models. In addition, the Vos variation at -40 C was reduced.

These three circuits have demonstrated similar response to the elimination of the subthreshold hump. This confirms that the cause of the unmodeled behavior is the parasitic edge FET. Even if this FET was modeled as simply having a lower Vt, that would not explain the incorrect temperature slope or higher variation at cold temperature. For digital circuits this behavior can simply be modeled as a leakage, which can be managed. For analog circuits – especially those operating in the subthreshold region – even if this behavior could be modeled, it is highly undesirable.

V. CONCLUSION

Degradation of precision analog circuit accuracy was described in this report. A parasitic FET formed at the STI MOSFET edge was shown to be the cause. Performance of three different analog circuits showed high variability and unexpected temperature slopes not predicted by models. A process improvement that weakened the edge device showed much more predictable behavior in analog circuits operating in the subthreshold region. The use of a parametric test structure containing analog circuitry taken from a microcontroller allowed high-precision measurements that were not possible on a microcontroller.

ACKNOWLEDGMENT

The authors wish to thank Heather Desjardins, Denise Rose, and Donald Hall for their invaluable help in completing the equipment setup and data collection necessary for this investigation.

REFERENCES

[1] P.E. Allen, D. R. Holberg, CMOS Analog Circuit Design", Holt, Rinehart and Winston, inc.,1987.

[2] L. Vancaillie, F. Silveira, B. Linares-Barranco, T. Gotarredona, D.Flandre, "MOSFET mismatch in weak/moderate inversion: model needsand implications for analog design," in ESSCIRC Tech. Dig., pp. 671-674, 2003.

[3] M. Chen, J. Ho, T. Huang, "Dependence of Current Match on Bach-Gate Bias in Weakly Inverted MOS Transistors and Its Modeling," IEEE J.Solid-State Circuits, Vol. 31, No. 2, pp.259-262, February 1996

[4] K. Sakakibara, T. Kumamoto and K. Arimoto, "Impact of subthreshold hump on bulk-bias dependence of offset voltage variability in weak and moderate inversion regions," Proceedings of the IEEE 2012 Custom Integrated Circuits Conference, San Jose, CA, 2012, pp. 1-4

[5] J. P. de Gyvez and H. P. Tuinhout, "Threshold voltage mismatch and intra-die leakage current in digital CMOS circuits," IEEE J. Solid-State Circuits, vol. 39, no. 1, pp. 157-168, Jan. 2004.

[6] Y. Joly et al., "Gate Voltage Matching Investigation for Low-Power Analog Applications," in IEEE Transactions on Electron Devices, vol. 60, no. 3, pp. 1263-1267, March 2013

[7] F. Forti and M. E. Wright, "Measurement of MOS current mismatch in weak inversion region," IEEE J. Solid-State Circuits, vol. 29, no. 2, pp. 138-142. Feb 1994.

[8] K. R. Lakshmikumar, R. A. Hadaway, and M. A. Copeland, "Characterization and modeling of mismatch in MOS transistors for precision analog design," IEEE J. Solid State Circuits, vol SC-21, pp. 1057-1066, Dec. 1986.

[9] M.J.M. Pelgrom, A.C.J. Duinmaijer, A.P.G. Welbers, "Matching Properties of MOS Transistors," IEEE J. Solid-State Circuits, Vol 24, No. 5, pp 1433-1440, October 1989.

[10] A. Mizumura, T. Ohishi, N. Yokoyama, M. Nonaka, S. Tanaka, H. Ammo, "A Study of 90nm MOSEFET Subthreshold Hump Characteristics Using Newly Developed MOSFET Array Test Structure," in ICMTS Tech. Dig, pp. 39-42, 2005.

[11] P. Sallagoity, "Analysis of width edge effects in advanced isolation schemes for deep submicron CMOS technologies," IEEE Transactions on Electron Devices, vol. 43, no. 11, pp. 1900, 1906, Nov. 1996

[12] D. H. Lee et al., "Failure Analysis of an Anomalous Subthreshold Current in Nano-Scale NAND Flash Memory," 2007 IEEE International Reliability Physics Symposium Proceedings. 45th Annual, Phoenix, AZ, 2007, pp. 612-613

[13] Byoung-Chul Park et al., "A Fermi Level Controlled High Voltage Transistor preventing subthreshold hump," 2008 9th International Conference on Solid-State and Integrated-Circuit Technology, Beijing, 2008, pp. 172-175

[14] A. Valletta, "Hump characteristics and edge effects in polysilicon thin film transistors,"J. Applied Physics, 104, 124511, 2008

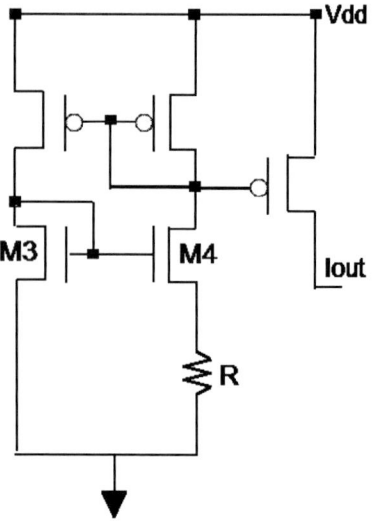

Fig. 2. Simplified schematic of a ΔVgs current reference. NMOS devices M3 and M4 operate in subthreshold region. Poor transistor modeling of temperature slope impacts the the accuracy of the output current, Iout.

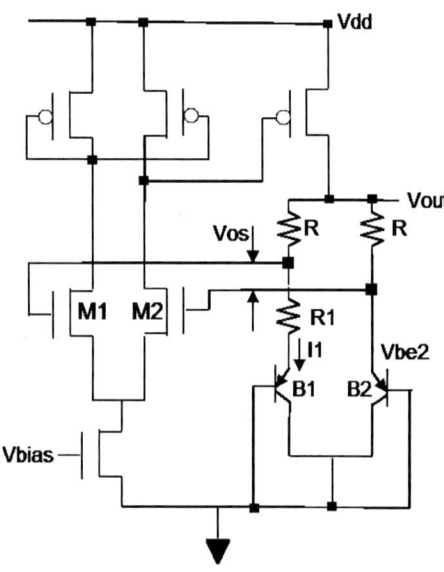

Fig. 1. Simplified schematic of a bandgap voltage reference circuit. NMOS devices M1 and M2 operate in subthreshold region. Poor transistor matching of M1 and M2 impacts the accuracy of the output voltage, Vout.

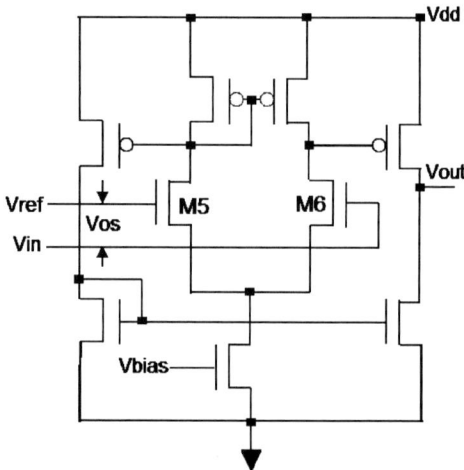

Fig. 3. Simplified schematic of a comparator circuit. NMOS devices M1 and M2 operate in subthreshold region. Poor transistor matching of M1 and M2 impacts the accuracy of the output voltage, Vout.

2018 INTERNATIONAL CONFERENCE ON MICROELECTRONIC TEST STRUCTURES, MARCH 19-22, 2018, AUSTIN, TEXAS, USA

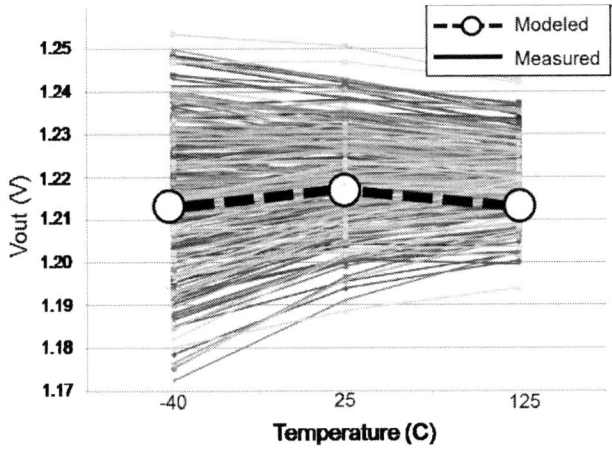

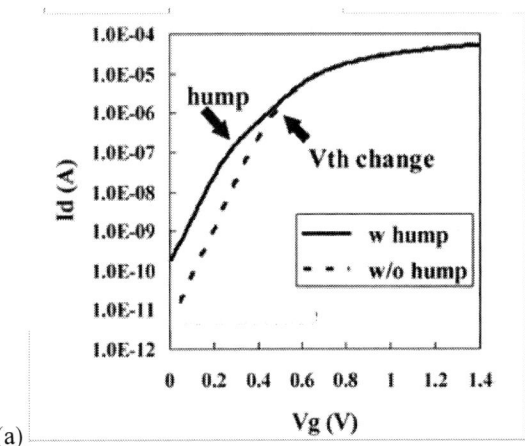

(a)

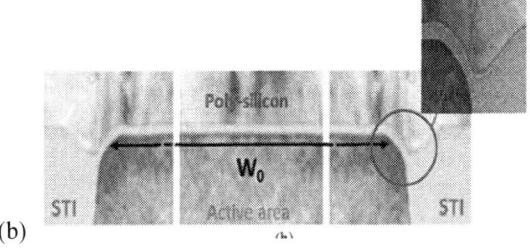

(b)

Fig. 4. Output voltage, Vout, from a bandgap voltage reference measured at -40 C, 25 C, and 125C. Model based simulation points are overlaid. Unmodeled variation at -40 C and a mixture of positive and negative temperature slopes degrades the accuracy of the circuit.

Fig. 5. (a)Transistor Id-Vg curve from [10]. The hump identified in the curve is indicative of leakage caused by a parasitic FET formed along the STI edge in a MOSFET device operating in the subthreshold region. (b) Cross-section from [6] showing the active area and STI boundary edge.

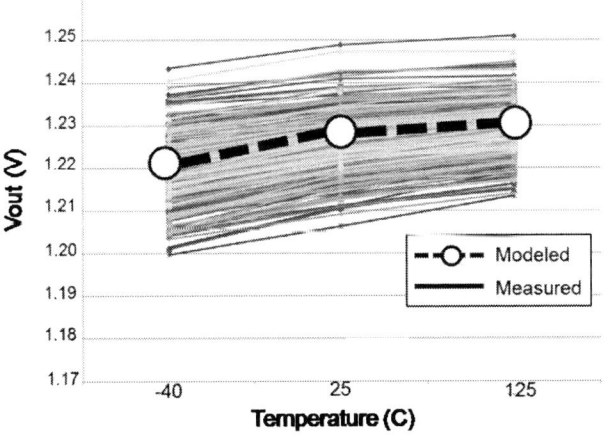

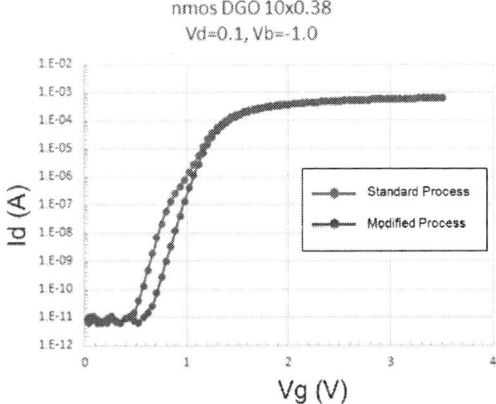

Fig. 6. Output voltage, Vout, from a modified bandgap voltage reference with wider gate fingers and higher current density make it less sensitive to the parasitic FET. Measurements show improved temperature slope consistency and less cold temperature variation.

Fig. 7. Transistor I-V curves from the controlled process experiment. Results from the new process indicate the elimination of the hump.

978-1-5386-5072-1/18 $31.00 © 2018 IEEE   19

2018 INTERNATIONAL CONFERENCE ON MICROELECTRONIC TEST STRUCTURES, MARCH 19-22, 2018, AUSTIN, TEXAS, USA

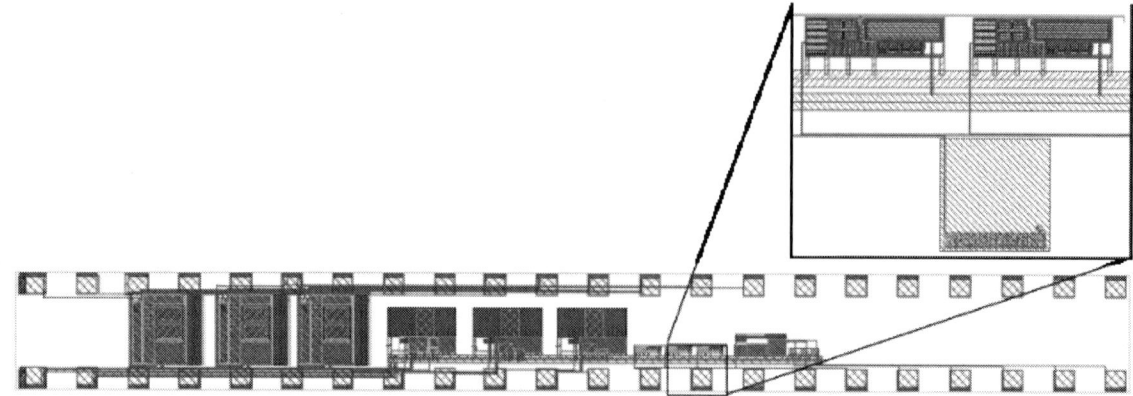

Fig. 8.   Layout the test structures containing three analog circuits extracted from a microcontroller.

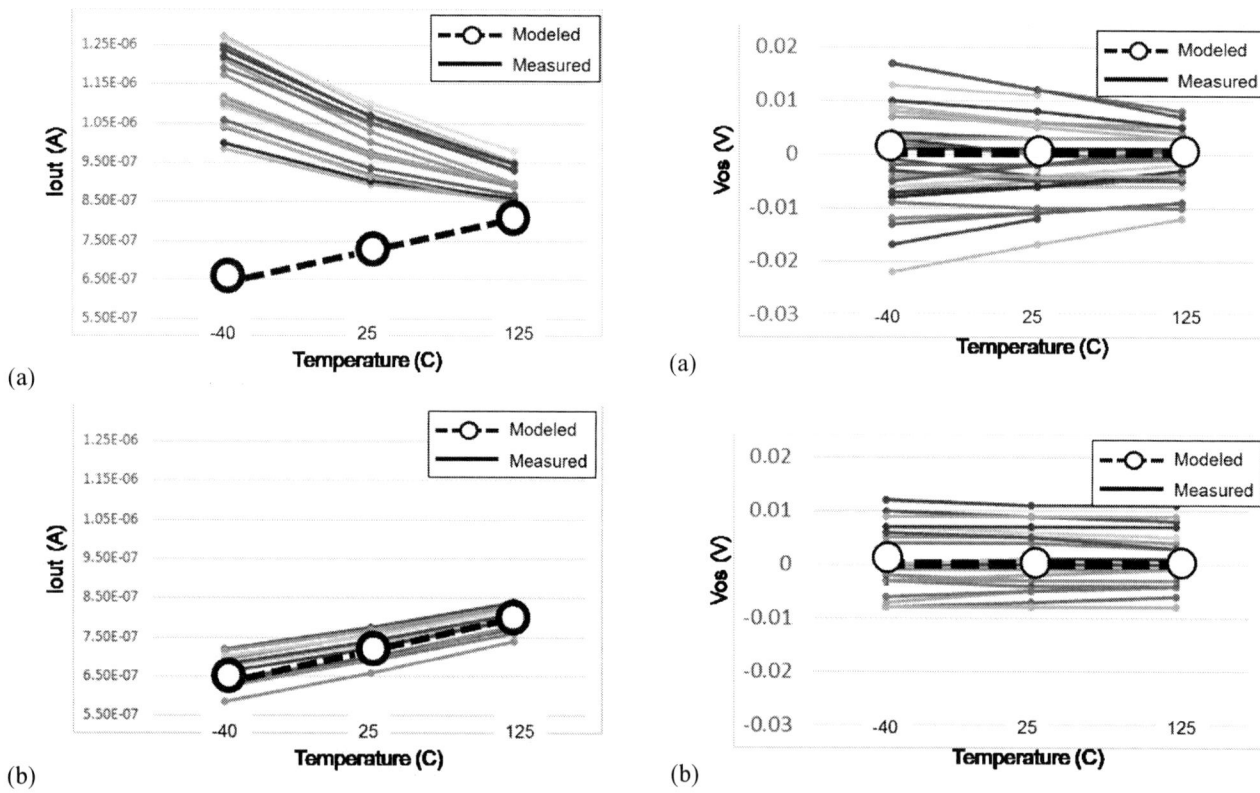

Fig. 9.   Measured results from the ΔVgs current reference test circuit. (a) Iout measured on the standard process shows an opposite temperature slope from the model. (b) Iout measured using the modified process has lower variability and has the correct temperature slope in agreement with the model.

Fig. 10. Measured results from the comparator test circuit. (a) Offset voltage, Vos, measured on the standard process shows a mixture of temperature slopes and larger variation at cold temperature. (b) Vos measured on the modified process has no temperature slope and less variation at cold temperature, in agreement with models.

Session 2

Modeling and Parameter Extraction

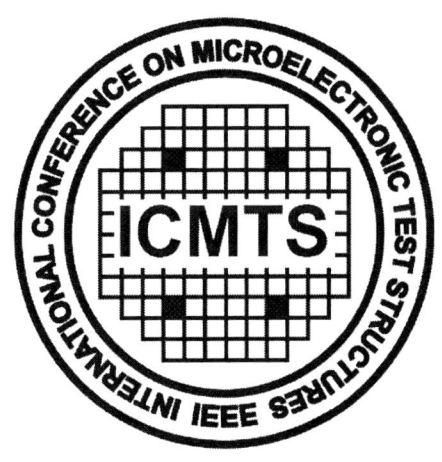

March 20, 2018

11:00-12:20

Co-Chairs:

Yoshio MITA, University of Tokyo, Japan

Colin McANDREW, NXP Semiconductors, USA

978-1-5386-5072-1/18 $31.00 © 2018 IEEE

Comprehensive Investigation on Parameter Extraction Methodology for Short Channel amorphous-InGaZnO Thin-Film Transistors

Chika Tanaka and Keiji Ikeda

Future Memory Development Department, Device Technology Research & Development Center,
Institute of Memory Technology Research & Development, Toshiba Memory Corporation
1, Komukai Toshiba-cho, Saiwai-ku, Kawasaki 212-8582, Japan
Phone: +81-44-549-2192 E-mail: chika.tanaka@toshiba.co.jp

Abstract—**We proposed a comprehensive parameter extraction method for short channel amorphous InGaZnO (α-InGaZnO) thin-film transistors (TFTs) on the basis of measurement data and TCAD simulations. Single parameter set were successfully extracted for channel length down to 500nm by using RPI α-Si TFT model with channel length modulation modeling. It makes possible to more accurate and scalable circuit performance characterization, since the extracted parameters correspond to the physical behavior of α-InGaZnO TFTs.**

Keywords— *amorphous-InGaZnO, TFT, Parameter extraction*

I. INTRODUCTION

Amorphous InGaZnO (α-InGaZnO) thin-film transistors (TFTs) are promising candidates as BEOL transistor for 3D-integrated LSIs due to its high mobility and low off-state leakage current [1, 2]. In order to implement α-InGaZnO TFTs into Si-CMOS technology, accurate and scalable compact model are indispensable. The electrical transport properties of α-InGaZnO TFTs are governed by the sub-gap density of states (DOS) over the band-gap such as the acceptor-like localized states near the conduction band minima (CBM) [3]. Thus, the adequate compact model based on these transport model is needed [4, 5].

RPI α-Si TFT model has been widely recognized as a conventional IV model for α-Si TFTs [6] with accumulation mode operation. Since RPI α-Si TFT model addresses the physical effects in α-Si TFTs to explain the DOS near the conduction band tail, it can also be adequate model for α-InGaZnO TFTs. In this study, we investigate the RPI α-Si TFT model parameter extraction procedure for α-InGaZnO TFTs with short channel regime on the basis of measurement data and TCAD simulation for the test structures.

II. DEVICE STRUCTURES AND CHARACTERISTICS

Figure 1 shows the cross-sectional TEM image of α-InGaZnO TFTs fabricated with Bottom gate (W) / top-contact (W) structure. A 40 nm SiO$_2$ gate oxide and 15 nm α-InGaZnO

channel were adopted. As listed in Table I, various channel length (L_{ch}) devices were prepared as geometries for parameter extraction. The measured I_d-V_g and I_d-V_d characteristics are shown in Fig. 2. Although S-factor was not depend on L_{ch}, DIBL was not suppressed at L_{ch} of 500 nm, as shown in Fig. 2 (c), due to the short channel effects (SCE) by channel length modulation (CLM).

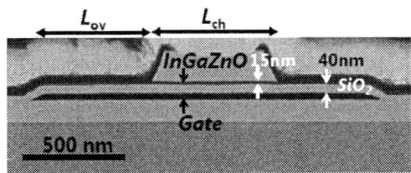

Fig. 1   Cross-sectional TEM image of α-InGaZnO TFTs fabricated with bottom gate (W) / top-contact (W) structure. A 40 nm SiO$_2$ gate oxide and 15 nm α-InGaZnO channel were adopted. L_{ov} is gate-to-S/D overlap length.

TABLE I.   LISTS OF THE TARGETED DEVICE GEOMETRIES FOR PARAMETER EXTRACTION.

Die Name	L_{ch} [nm]	W [μm]	L_{ov} [μm]
23A	700	1	0.4
23B	600	1	0.4
23D	500	1	0.4

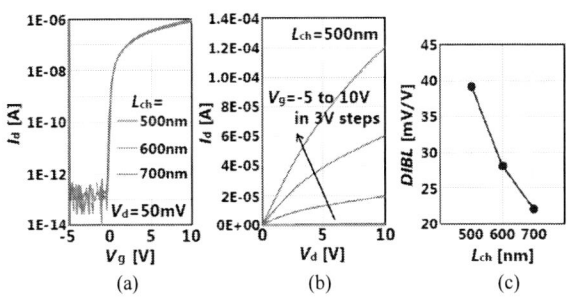

Fig. 2   (a) I_d-V_g characteristics of α-InGaZnO TFTs with L_{ch} = 500, 600 and 700 nm for V_d = 50 mV and (b) I_d-V_d characteristics with L_{ch} = 500 nm. (c) L_{ch} dependence of DIBL characteristics.

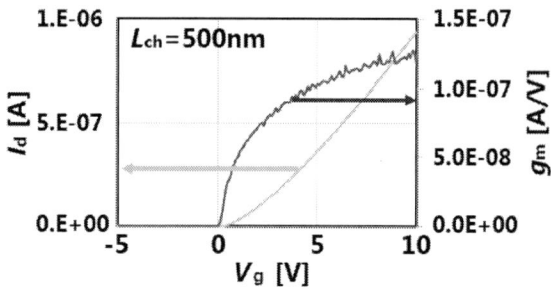

Fig. 3 g_m-V_g and I_d-V_g (linear) plots. g_m has not the maximum value for V_g.

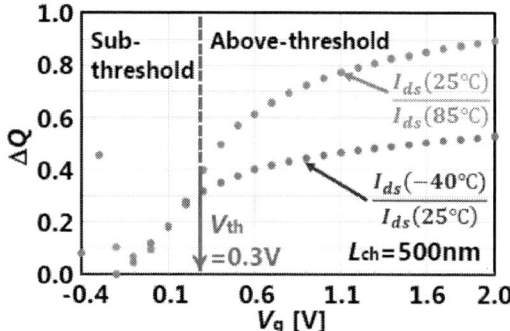

Fig. 4 ΔQ-V_g characteristics for $L_{ch} = 500$ nm. Charge increase ratio ΔQ by temperature at constant V_g was defined as $I_d(T) / I_d(T')$ of $T > T'$. We extracted $V_{th} = 0.3$ V as a border between sub-threshold and above-threshold.

III.   SPICE PARAMETER EXTRACTION FOR α-INGAZNO TFTS

A. Threshold voltage extraction

Figure 3 shows the g_m-V_g and I_d-V_g linear plots. Unlike Si-MOSFET, threshold voltage (V_{th}) extraction method from g_{mmax} is not accurate due to CLM caused by the space-charge region under the source (S) / drain (D) contacts. In this study, we defined V_{th} as the transition voltage from sub-threshold with thermal excitation to above-threshold regime by using temperature dependence of I_d-V_g characteristics. Figure 4 shows the V_g dependence of the charge increases ratio (ΔQ) by temperature, where ΔQ was defined by $I_d(T) / I_d(T')$ of $T > T'$. We extracted $V_{th} = 0.3$ V, because ΔQ for sub-threshold regime should be a constant value for any temperature regime.

B. Evaluation of parasitic resistance

In α-InGaZnO TFTs, large S/D resistance (R_{sd}) and contact resistance (R_c) are significant in the high resistivity intrinsic semiconductor layer and the contact between metal and wide bandgap semiconductor, respectively. Furthermore, S/D parasitic resistance (R_{para}) depends on gate-to-S/D overlap length (L_{ov}) by CLM, as shown in Fig. 5 (a). In addition to V_{th} shift caused by SCE (ΔV_{th}), the space-charge region formed by spreading the electric field from S/D to channel depends on L_{ov} and it adds to ΔV_{th} caused by CLM (ΔV_{th}^{CLM}) of $1-(L_{int}-\Delta) / L_{ch}$ calculated from charge-share approximation. Figure 5 (b) shows L_{ov} dependence on V_{th} change ratio for various L_{ch}.

Remarkable ΔV_{th}^{CLM} was observed only for short L_{ch} of 1 μm with long L_{ov}. Figure 6 (a) and (b) show R_{on}-L_{ch} plot for $L_{ov} = 0.8$ μm and R_{para} extracted from the measured R_{on}-L_{ch} plot with take into account the ΔV_{th}^{CLM}, respectively. It is found that the modulation of R_{para} by L_{ov} can be expressed by ΔV_{th}^{CLM}.

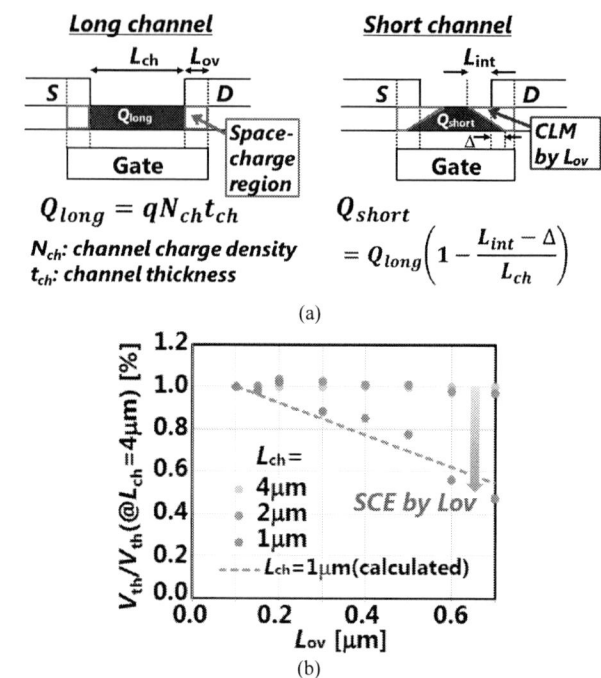

(a)

(b)

Fig. 5   (a) Schematic illustration of CLM under charge-share approximation for long and short channel α-InGaZnO TFTs. (b) L_{ov} dependence on V_{th} change ratio for $L_{ch} = 1$, 2 and 4 μm, and calculated value ($L_{ch} = 1$ μm).

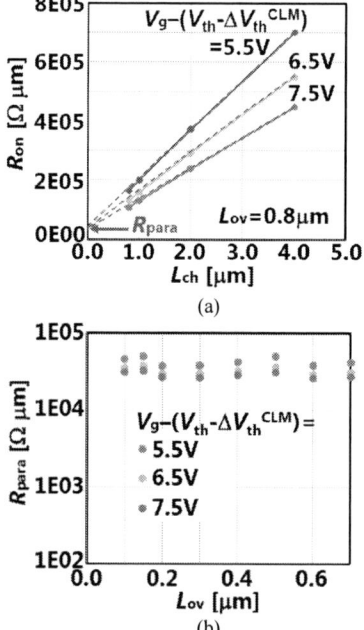

Fig. 6   (a) R_{on}-L_{ch} plot for $L_{ov} = 0.8$ μm and (b) R_{para} extracted from the measured R_{on}-L_{ch} plot with take into account the ΔV_{th}^{CLM}.

C. Mobility model parameter extraction

Here, we evaluate the impact of V_g dependence of R_{para} to the mobility model parameter by comparison of macro model with and without V_g dependence. Figure 7 shows a measured V_g dependence of R_{para}. For constant R_{para} model, R_{para} was carefully derived at saturation region ($R_{para}(V_g^{sat})$). Field effect mobility (μ_{FE}) in RPI α-Si TFT model was defined as μ_{FE}=MUBAND[(V_g-V_{th})/VAA]^{GAMMA}, where "MUBAND", "VAA" and "GAMMA" are mobility model parameter. We used the intrinsic I_d-V_g characteristics excluded R_{para} from measured IV data and "MUBAND"=0.001, V_{th} = 0.3. "VAA" and "GAMMA" extraction results are shown in Fig. 8. Single parameter set was obtained even though constant R_{para} model was used. This result suggests that CLM can expressed as the V_g dependence on μ_{FE}.

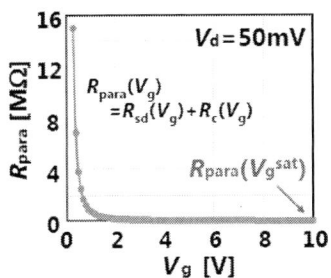

Fig. 7 V_g dependence of R_{para} (measured).

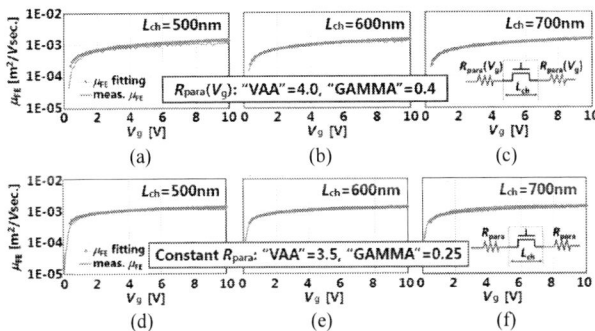

Fig. 8   Comparison of mobility model parameter extraction results. R_{para} with V_g dependence of L_{ch}=(a)500nm, (b)600nm and (c)700nm. And constant R_{para} of L_{ch}=(d)500nm, (e)600nm and (f)700nm.

D. Capacitance model parameter extraction

The parasitic capacitance (C_{para}), especially fringing capacitance component from the gate to overlapped S/D regions, is one of the key parameter determining the circuit performance of TFTs. In order to derive the fringe capacitance in the model parameters, C_{para} for short-channel TFTs was extracted from two-dimensional TCAD simulation [7] assuming that the gate to S/D overlap capacitance (C_{ov}) is sum of the parallel-plate capacitance between S/D and gate electrode (C_{ov}^p) and the fringe capacitance between S/D and the side wall of gate electrode (C_{ov}^f), where the device structure was formed in the same way as the fabricated one with different L_{ov}, gate height (H_g) and overlap length for gate

edge (L_{ovs}) as shown in Fig. 9. Figure 10 (a) also shows measured overlap capacitance (C_{ov}) for long channel devices extracted from the gate-to-channel capacitance (C_{gc}) by Split-CV measurement [8], and measured C_{ov} is almost the same irrespective of L_{ch}. Figure 10 (b) shows L_{ov}' dependence on C_{ov}^p and C_{ov}^f under the constant H_g, where $L_{ov}' = L_{ov} - L_{ovs}$. C_{ov}^p is proportional to L_{ov}', and C_{ov}^f is influenced by the fringing electric field of gate edge and only depends on H_g. Using the geometry dependence, the simulated C_{ov} for targeted device structure with L_{ov} of 400 nm extracted from TCAD data is 1.21 fF/μm. In this study, the extracted value of C_{ov} were manually put as "CGSO/CGDO" of SPICE model parameter.

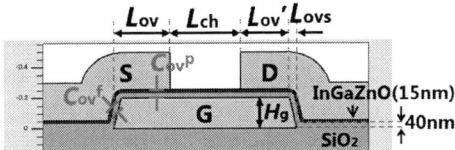

Fig. 9   TCAD structure for C_{para} extraction of targeted device geometry. H_g and L_{ovs} are gate height and gate to S/D overlap length for gate edge, respectively.

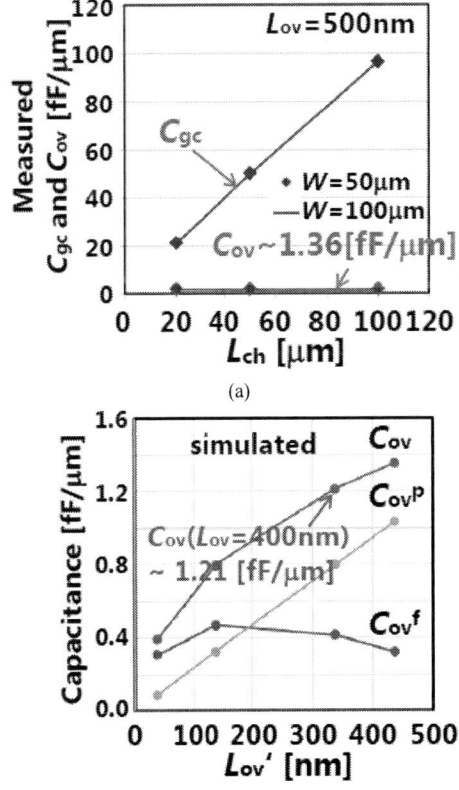

Fig. 10   (a) L_{ch} depenence of C_{gc} and C_{ov} for long channel devices with L_{ov} = 500 nm (measured) and (b) L_{ov}' dependence on C_{ov}^p and C_{ov}^f under the constant H_g, where $L_{ov}' = L_{ov} - L_{ovs}$. Extracted C_{ov} for targeted device of $L_{ov} = 400$ nm is 1.21 fF/μm (simulated).

IV. RESULTS AND DISCUSSIONS

Figure 11 (a) and (b) show the comparison of optimization results of I_d-V_g, I_d-V_d, g_m-V_g and g_d-V_d curves for L_{ch} = 500 nm using UTMOST IV [7]. When the initial values of SPICE model parameter not take into account CLM (see Fig. 11 (a)), the extracted parameter do not express the physical behaviors of α-InGaZnO TFTs. On the other hand, when parameter listed in Table 2 were used (see Fig. 11 (b)), extracted parameter were optimized to describe both sub-threshold and above-threshold region of α-InGaZnO TFTs. As a results, by using the proposed CLM modeling, good fitting with single model parameter set was achieved for L_{ch} down to 500 nm.

TABLE II. INITIAL VALUES OF SPICE MODEL PARAMETER DEFINED BY PROPOSED OPTIMIZATION PROCEDURE WITH CLM MODELING

Parameter	Value	Parameter	Value
CAPMOD	1	GAMMA	0.245 [V]
TOX	4.0E-08 [m]	EPSI	3.9
MUBAND	0.001 [m²/Vsec]	EPS	11
VT0	0.3 [V]	CGSO	1.21E-09 [F/m]
VAA	3.5 [V]	CGDO	1.21E-09 [F/m]

(Silvaco, SmartSpice LEVEL = 35 Version = 2.0)

V. CONCLUSIONS

We proposed SPICE model parameter extraction procedure for short channel α-InGaZnO TFTs on the basis of measurement data and TCAD simulations. After V_{th} was extracted by temperature dependence of I_d-V_g characteristics, V_{th} shift caused by CLM was calculated in order to extract the parasitic resistance. It is found that CLM at short channel region can properly expressed as the V_g dependence on μ_{FE} by using constant R_{para} macro model with take into account CLM. A single model parameter set extraction which can express the physical behaviors of α-InGaZnO TFTs with L_{ch} down to 500 nm was successfully demonstrated by using RPI α-Si TFT model with CLM modeling. It makes possible to more accurate and scalable circuit performance characterization, since the extracted parameters correspond to the physical behavior of α-InGaZnO TFTs.

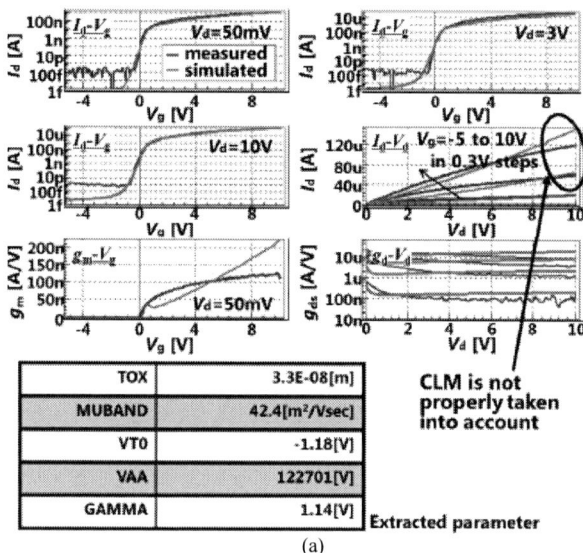

(a)

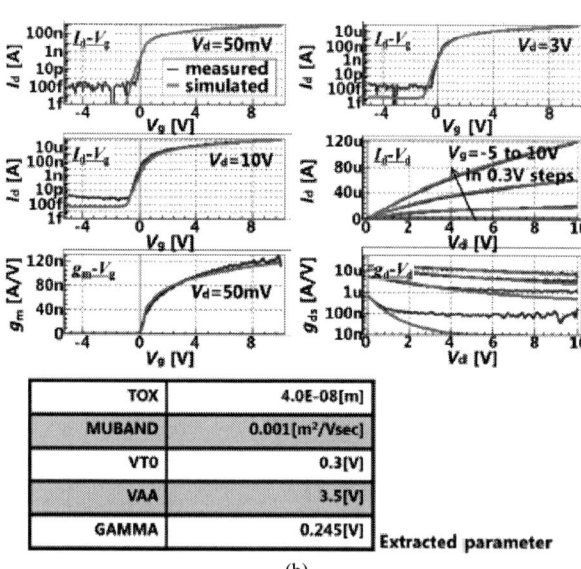

(b)

Fig. 11 Comparison of optimization results for L_{ch} = 500 nm using UTMOST IV. Initial values of SPICE model parameter were extracted by (a) conventional procedure without CLM modeling and (b) proposed procedure (listed in Table II).

REFERENCES

[1] K. Nomura, H. Ohta, A. Takagi, T. Kamiya, M. Hirano, and H. Hosono, "Room-temperature fabrication of transparent flexible thin-film transistors using amorphous oxide semiconductors.", Nature, 432, pp. 488–492, 2004.

[2] K. Kaneko, N. Inoue, S. Saito, N. Furutake, and Y. Hayashi, "A Novel BEOL Transistor (BETr) with InGaZnO Embedded in Cu-Interconnects for On-chip High Voltage I/Os in Standard CMOS LSIs", 2011 Symposium on VLSI Technology Digest of Technical Papers, 6B-3, pp.120-121.

[3] S. Lee, K. Ghaffarzadeh, A. Nathan, J. Robertson, S. Jeon, C. Kim, I-H. Song, and U-I. Chung, 98, " Trap-limited and percolation conduction mechanisms in amorphous oxide semiconductor thin film transistors", Appl. Phys. Lett., 98, 203508, 2003.

[4] C. Perumal, K. Ishida, R. Shabanpour, B. K. Boroujeni, L. Petti, N. S. Münzenrieder, G. A. Salvatore, C. Carta, G. Tröster, and F. Ellinger, "A Compact a-IGZO TFT Model Based on MOSFET SPICE Level = 3 Template for Analog/RF Circuit Designs", IEEE Electron Dvice Letters, 43, 11, pp.1391-1393 , 2013.

[5] Y. W. Jeon, I. Hur, Y. Kim, M. Bae, H. K. Jung, D. Kong, W. Kim, J. Kim, J. Jang, D. M. Kim, and D. H. Kim, "Physics-Based SPICE Model of a-InGaZnO Thin-Film Transistor Using Verilog-A", Journal of semiconductor Technology and Science, 11, 3, pp.153-161, September 2011.

[6] M. S. Shur, H. C. Slade, M. D. Jacunski, A. A. Owusu, and T. Ytterdalac, "SPICE Models for Amorphous Silicon and Polysilicon Thin Film Transistors", J. Electrochem. Soc., Vol. 144, No. 8, pp. 2833-2839, August 1997.

[7] Silvaco, http://www.silvaco.com/.

[8] C.G. Sodini, T.W. Ekstedt, and J.L. MollC, "Charge accumulation and mobility in thin dielectric MOS transistors", Solid-State Electronics, Vol. 25, Issue 9, pp. 833-841, September 1982.

Modeling Split-Gate Flash Memory Cell for Advanced Neuromorphic Computing

Mandana Tadayoni, Santosh Hariharan, Steven Lemke, Thibaut Pate-Cazal, Bernard Bertello, Vipin Tiwari, and Nhan Do

Silicon Storage Technology, a Subsidiary of Microchip Technology Inc.
450 Holger Way
San Jose, California 95134, USA
Fax: (408) 522-7353
Email: nhan.do@microchip.com

Abstract- **Split-gate flash memory technology had recently been used in neuromorphic computation where a non-volatile memory array is designed in such a way that enables high-precision tuning of individual memory elements. This work proposes for the first time a SPICE model of the two-transistor, select gate and floating gate, of the split-gate flash memory cell, implemented in a 180 nm CMOS technology, that allows the users to set the individual memory cell to any precise analog state.**

Keywords- neuromorphic, split-gate flash memory, machine learning

I. INTRODUCTION

A split-gate non-volatile memory cell is ideal for analog computing due to its flexibility in continuous tuning which can be used to perform certain useful tasks in signal processing and artificial neural network [1-2]. In recent years, the rise of Deep Neural Network (DNN) architectures has paved the way for building very effective models for performing various tasks such as object recognition, voice recognition and data mining. The training or learning aspect of machine learning is very compute intensive and is best done by GPUs and other high-end computing approaches, whereas the classification or inference aspect of machine learning can be done via low precision integer computing. The first generation of Tensor Processor Unit (TPU) is an example of a low-power, low-precision inference engine which is still a digital computing approach. The computing power needed for an inference engine could be significantly reduced by using in-memory analog computing with a Non-Volatile Memory (NVM) cell. In order to effectively design an analog inference engine based on an NVM cell, it is essential to accurately model the NVM cell in sub-threshold region.

II. SPLIT-GATE FLASH MEMORY CELL STRUCTURE AND OPERATION

Figure 1 illustrates the cross section of a pair of split-gate memory cells sharing a common source junction and its schematic representation. As shown, the split-gate memory cell consists of a floating gate, a select gate (which may also function as an erase gate), a source and a drain. This is a one-and-a-half transistors cell and its size is comparable to that of the traditional stacked-gate NVM cell for a given technology node. The channel, split between the drain and source, is a series combination of the floating gate and select gate transistors. The select gate portion of the channel isolates the floating gate portion from the drain which prevents over-erase, a common issue found in stacked-gate NVM cells. The split-gate memory is programmed using Source-Side Channel Hot Electron (SS CHE) injection and erased using poly-poly Fowler-Nordheim (FN) tunneling between floating and select gates. Such separation of erase and program enhances the immunity of disturbing the unselected bits during program and erase. It also enables tuning the very fine states of the memory cell by weakly erasing or programming. Table 1 shows how the cell is operated in digital applications. In analog computing, the program and erase conditions can be modified to tune the memory cell to precise levels.

III. MODELING THE SPLIT-GATE FLASH MEMORY CELL

In this work, a macro-model is created to accurately describe the characteristics of the memory cell, including the sub-threshold region of the operation which is needed for neuromorphic computing [1]. The equivalent circuit for the macro model is shown in Figure 2. Both select gate and the storage transistors are modeled using BSIM3v3 SPICE model [3]. The voltage, V_{fg}, and charge, Q_{fg}, of the storage transistor is modeled with the electrostatic coupling as shown in Figure 3. As indicated in the equation, the bias conditions used in erase and program are considered in the calculation of Q_{fg}. Figure 4 describes Q_{fg} at different stages of erasing and programming. Assuming that the bit cell is initially in the programmed state, $Q_{fg} = Q_{fger}$ at the end of erasing (see Case 1 in Figure 4). Similarly, Case 2 in Figure 4 indicates the cell at the end of the programming mode. As shown, unlike in digital applications where only ones and zeros are considered, the memory cell can be modeled in weakly programmed states between the end of Case 3 and Case 2 in Figure 4. In the same way, the memory cell can be modeled in weakly erased states between the initial phase (Qfginit) and Case 2 where Qfg = Q_{fger}. The final floating gate charge in the steady state is defined by $Q_{fg} =$(State)$+ Q_{fgpr}(1-$State) in Figure 3.

IV. RESULTS

Figure 5 illustrates a reasonable fitting between the simulated current-voltage characteristics of the bit cell being programmed into several fine states, and the measurement, including subthreshold region. The I-V curve on the left of Figure 5 represents the cell in the fully erased state. The cell is then programmed, starting with a very soft condition of source line voltage (Vsl) of only 5.4V. As the cell is programmed, electrons are injected into the storage element therefore lowering the floating gate potential which reduces the cell read current. Each curve from left to right represents the I-V characteristics of the cell post programming with Vsl increment of 400 mV. As described in Reference 1, the separation of the programmed

states can be tuned with Vsl increment as low as 50 mV.

Further refining of programmed states is illustrated in Figure 6 where the cell is simulated with 200 mV steps between source voltages, Vsl. This confirms that the HSPICE model of the bit cell works well in the dynamic state of fine programming or erase and read-out in the weak inversion region. The area highlighted with red box in Figure 6b indicates the operation window for neuromorphic computation where separation of the I-V curves is nearly constant.

V. CONCLUSION

The modeling of the dynamic states of a 180 nm split-gate memory cell with weak-inversion read-out had been successfully demonstrated. This work will enable additional exploration in the neuromorphic computing area.

REFERENCES

[1] F. Merrikh-Bayat, X. Guo, H. Om'mani, N. Do, K. Likharev, "Redesigning commercial floating-gate memory for analog computing applications," *IEEE ISCAS, 2015.*

[2] F. Merrikh-Bayat, X. Guo, M. Klachko, N. Do, K. Likharev, "Model-based high-precision tuning of NOR flash memory cells for analog computing applications," *74th Annual Device Research Conference, 2016.*

[3] BSIM3v3 model documentations available at http://www-device.eecs.berkeley.edu/bsim/?page=BSIM3

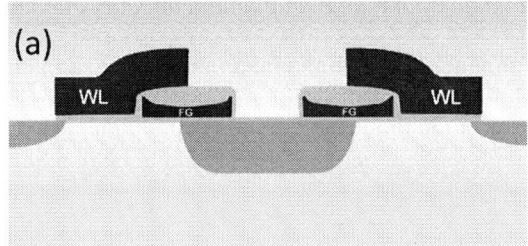

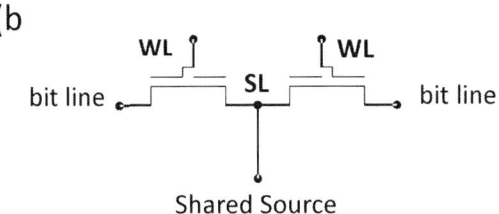

Figure 1. Cross section (a) and schematic representation (b) of the split-gate memory cell

	WL (SG)	Bit-line (BL)	Source (SL)
Erase	~11.5V	0V	0V
Program	~1.5V	~0.5V	~9.5V
Read	~2.5V	~1V	0V

Table 1. Memory cell operation in digital applications

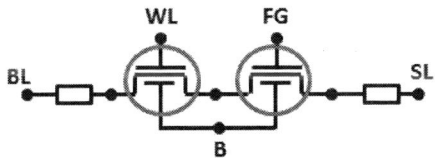

Figure 2. Equivalent circuit of the macro-model for split-gate memory cell

$$V_{fg} = (CRw_l * Vwl) + (CRsub * Vsub) + Qfg/Ct_{ot}$$

Where
CR_{wl} is the coupling ratio between WL and FG
CR_{sub} is the coupling ratio between FG and substrate
$C_{tot} = C_{wlfg} + C_{fgsub}$

and

$$Q_{fg} = Q_{fger}(\text{State}) + Q_{fgpr}(1\text{-State})$$

where
State defines the storage steady state of the memory ce

Figure 3. Capacitance coupling describes how floating gate voltage and charges can be modeled

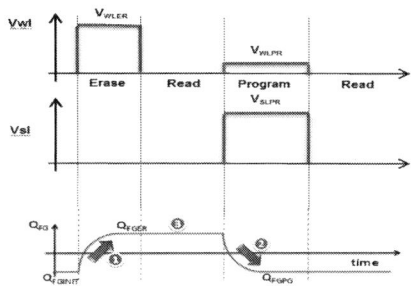

Figure 4. Description of charge/discharge cycle in transient and its associated bias conditions

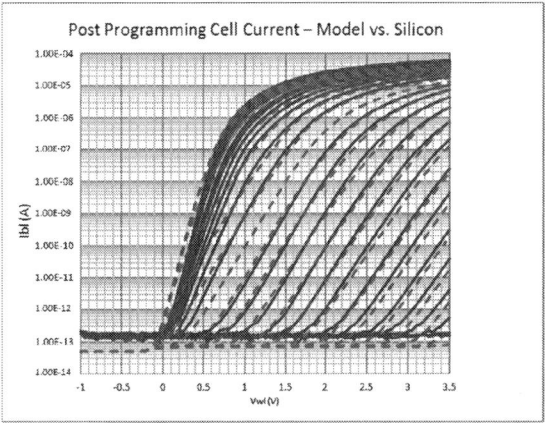

Figure 5. Fitting between measured I-V characteristics and Hspice model simulation

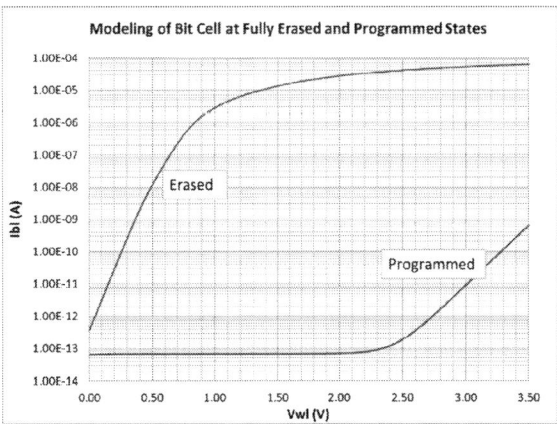

Figure 6a. Modeled I-V characteristics of programmed and erased steady states

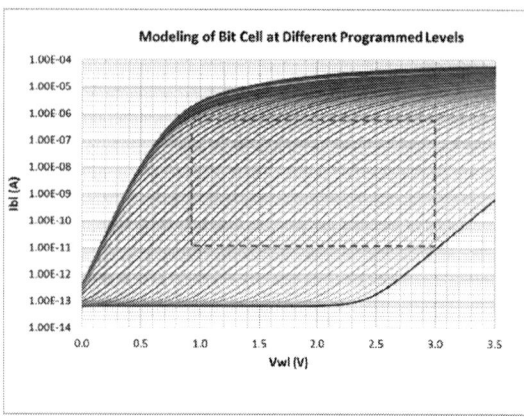

Figure 6b. Simulated I-V characteristics of the bit cell in different programmed levels and weak inversion read out.

Validation of the BSIM4 Irregular LOD SPICE Model by Characterization of Various Irregular LOD Test Structures

Bob Peddenpohl, Max Otrokov, Jeremy Wells
Cypress Semiconductor, Lexington, Kentucky
Email: ped@cypress.com

Abstract—Stress from FET Isolation regions affects the electrical behavior of transistors in modern technologies. Characterization and modeling of stress effects have primarily been done for transistor layouts that have rectangular (regular) shaped source and drains. However, circuits may include transistor layouts that have non-rectangular (irregular) shaped source and drains. This paper presents test structures to evaluate the effects caused by isolation stress on irregular source/drain transistor layouts, and shows that the standard way of calculating the effective distances for modeling stress effects is also reasonable for irregular source/drain transistor layouts.

Keywords—BSIM, LOD, STI Stress

I. INTRODUCTION

To enable accurate circuit simulations, layout dependent effects (LDE) have become critical components of Spice models [1]. These layout dependent effects do cause additional complications for the circuit designer because simulations during the schematic stage of the design might not always have layout information available. As a result, the schematic circuit designer must make estimates about layout information and not until the layout is completed can the designer determine the layout dependent effects. For many circuit simulations, it is critical that the designer simulate with the layout dependent effects. One example of the circuit impact is discussed in simulations of a current mirror ratio [1], and other examples are shown in [2][3].

One of the layout dependent effects is attributed to stress at the MOSFET Channel region caused by the Shallow Trench Isolation (STI) in modern processes. The BSIM4 model varies mobility and threshold related model parameters [4] based on the distance from the center of the MOSFET channel to the edge of the STI. This distance is commonly referred to as the LOD distance (**L**ength of **O**xide **D**efinition). In [3] the BSIM4 model equations are confirmed with data for both regular rectangular shaped LOD distances and some metrics are presented for irregular non-rectangular shaped LOD distances. This paper extends previous works on STI stress

effects [2][3] by focusing entirely on comparing results of irregular shaped LOD to regular shaped LOD. The measurements reported in this paper are from the thick oxide FETs on a 130nm process.

II. LOD TEST STRUCTURES

A. Regular LOD Test Structures

During model development, test structures with varying LOD are necessary to characterize how the FET performance varies with STI stress. Figure 1a shows an example of test structures that vary LOD in a "regular" rectangular distance by varying the distance SA and SB. The distance from the STI edge to the outside edge of the poly channel is referred to as SA (source side) and SB (drain side) in the BSIM4 model. The BSIM4 model predicts the FET performance to vary linearly with SA and SB using the following relationship [4]:

$$Stress = \frac{1}{LOD} = \frac{1}{SA_{eff} + 0.5*L} + \frac{1}{SB_{eff} + 0.5*L} \tag{1}$$

The justification of the BSIM4 LOD model is based on measurements, and also device simulations of the predicted stress along the diffusion and channel region [3][5]. Figure 1b shows the cross-section of the Figure 1a layout, and Figure 1c shows the predicted trend of stress along the diffusion and channel region.

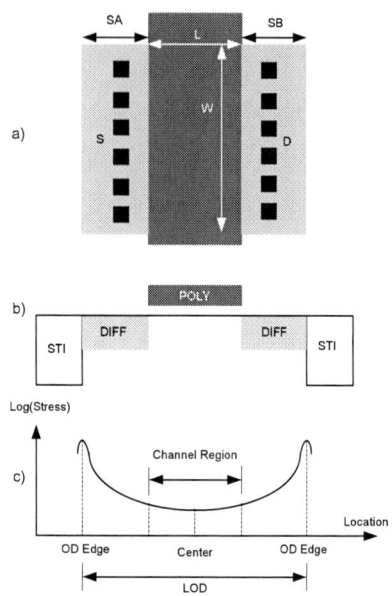

$$\frac{1}{SA_{eff} + 0.5 * L_{drawn}} = \sum_{i=1}^{n} \frac{SW_i}{W_{drawn}} * \frac{1}{SA_i + 0.5 * L_{drawn}}$$

$$\frac{1}{SB_{eff} + 0.5 * L_{drawn}} = \sum_{i=1}^{n} \frac{SW_i}{W_{drawn}} * \frac{1}{SB_i + 0.5 * L_{drawn}}$$

(2)

Figure 3 is a table of the regular and irregular shaped LOD structures that were created and measured. The irregular shaped LOD structures were designed to have an SAeff and SBeff as close as possible to a regular shaped LOD structure. The measurements of these irregular LOD structures on the same plots as regular LOD structures will be presented in the next section.

W_total (um)	SW1 (um)	SW2 (um)	SA1 (um)	SA2 (um)	SAeff (um)	Layout Type
5	5	N/A	5	N/A	5.000	Regular
5	5	N/A	1.11	N/A	1.110	Regular
5	4	1	1.645	0.265	1.076	Irregular
5	5	N/A	1.075	N/A	1.075	Regular
5	3	2	2.5	0.265	0.883	Irregular
5	3	2	1.645	0.265	0.746	Irregular
5	5	N/A	0.745	N/A	0.745	Regular
5	5	N/A	0.68	N/A	0.680	Regular
5	5	N/A	0.48	N/A	0.480	Regular
5	1	4	1.645	0.265	0.378	Irregular
5	5	N/A	0.375	N/A	0.375	Regular
5	5	N/A	0.35	N/A	0.350	Regular
5	5	N/A	0.265	N/A	0.265	Regular

Fig. 3. Table of layout test structures measured and the layout instance parameters (as defined in Figure 2) used to calculate the SAeff values. SA=SB so only the SA values are shown. (W_total = SW1 + SW2)

III. LOD MEASUREMENT RESULTS

A. Measurement Definition

On a 130nm process, the thick oxide FETs with a width of 5.0um were measured with the length ranging from 0.5um to 8.0um, for both regular and irregular NMOS and PMOS test structures. As shown in Equation 1, there is a linear dependence with 1/LOD. To account for the LOD effect, the BSIM4 model varies the modeling parameters for Vth, mobility, saturation velocity, body effect, and DIBL effect using the first-order model parameters kvth0, ku0, stk2, steta0 [4]. The graphs in the following sections will plot an important FET metric versus the 1/LOD distance and confirm that the measurements for both regular and irregular follow the linear trend that is implemented in the BSIM4 model.

B. Drain Current Measurement Results

There are two physical mechanisms that cause the FET characteristics to change based on the LOD distance. The first is a mechanical stress that affects the carrier mobility of the FET. With increased STI stress on a NMOS due to smaller SA and SB layouts, the Drain current decreases due to a decrease in electron mobility. With increased STI stress on a PMOS due to smaller SA and SB layouts, the Drain current increases due to an increase in hole mobility [6]. The saturation velocity variation with STI stress may also be an explanation for the drain current variation with stress [7]. The Saturation Drain current is plotted in Fig. 4 for the NMOS and Figure 5 for the PMOS, and the Linear Drain current is shown

Fig. 1. A) Layout view of a FET with Regular shaped LOD b) Simplified Cross-Section view and c) Trend of Expected Stress vs location on the FET

B. Irregular LOD Test Structures

Another consideration for test structure and model validation is to consider irregular shaped LOD as shown in Figure 2. During the design layout stage, it is possible to achieve area savings and also receive performance advantages (such as reduced junction capacitance) by sharing the source/drain diffusion regions of several FETs. If the FETs with shared diffusions have different channel widths, the FETs will have an irregular shaped LOD as shown in Figure 2. The area penalty for not allowing this type of irregular shaped LOD might be too severe, so this paper investigates how well the BSIM4 model predicts the LOD dependence of these irregular shaped LOD structures.

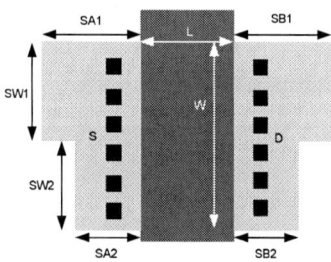

Fig. 2. Layout View of FET with Irregular shaped LOD.

The BSIM4 model [4] uses the following weighted average equation for calculating an effective SA and effective SB for irregular shaped MOSFETs:

in Figure 6 for the NMOS and Figure 7 for the PMOS. The Drain current trend with 1/LOD of the irregular structures follows the expected trend from Equation 2. The irregular structures follow a similar 1/LOD linear trend as the regular LOD structures.

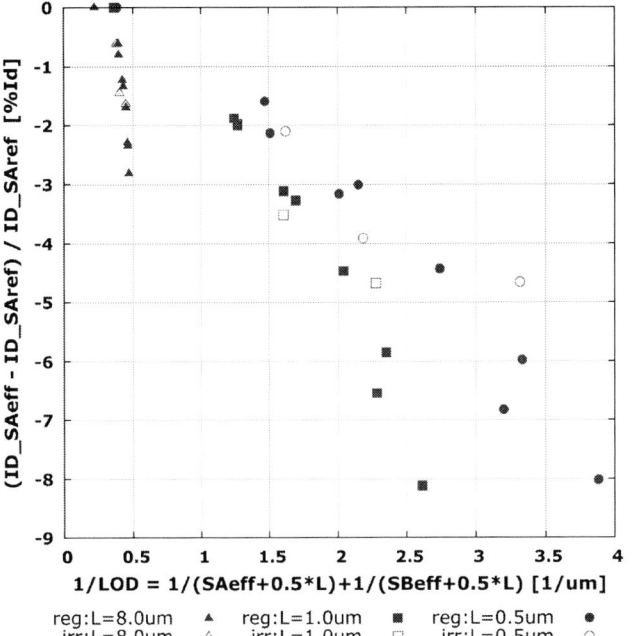

reg:L=8.0um ▲ reg:L=1.0um ▪ reg:L=0.5um ●
irr:L=8.0um △ irr:L=1.0um ▫ irr:L=0.5um ○

Fig. 4. NMOS Saturation Drain current (Vds=5.0V, Vgs=5.0V, Vbs=0.0) percent change between SAref=5.0um and other various regular and irregular SAeff values (SA=SB).

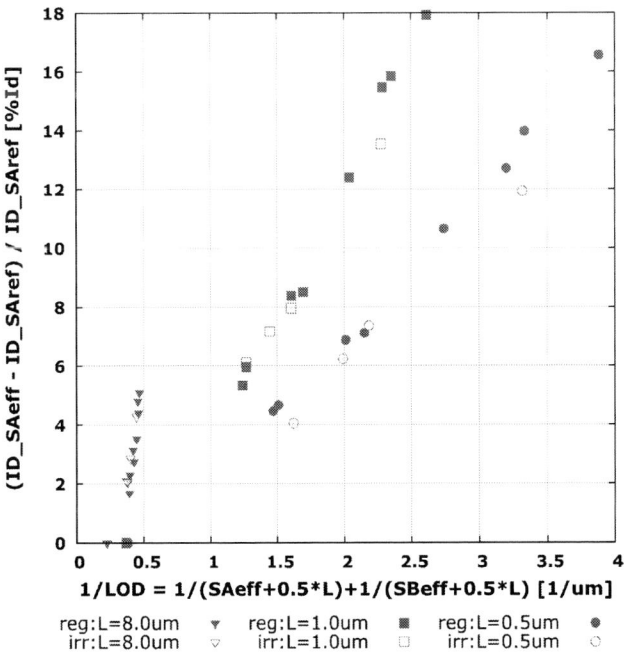

reg:L=8.0um ▼ reg:L=1.0um ▪ reg:L=0.5um ●
irr:L=8.0um ▽ irr:L=1.0um ▫ irr:L=0.5um ○

Fig. 5. PMOS Saturation Drain current (Vds=-5.0V, Vgs=-5.0V, Vbs=0.0) percent change between SAref=5.0um and other various regular and irregular SAeff values (SA=SB).

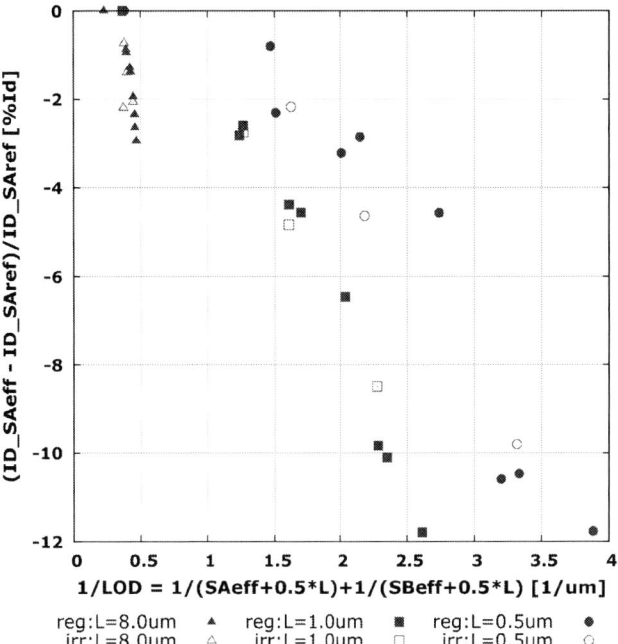

reg:L=8.0um ▲ reg:L=1.0um ▪ reg:L=0.5um ●
irr:L=8.0um △ irr:L=1.0um ▫ irr:L=0.5um ○

Fig. 6. NMOS Linear Drain current (Vds=0.1V, Vgs=5.0V, Vbs=0.0) percent change between SAref=5.0um and other various regular and irregular SAeff values (SA=SB).

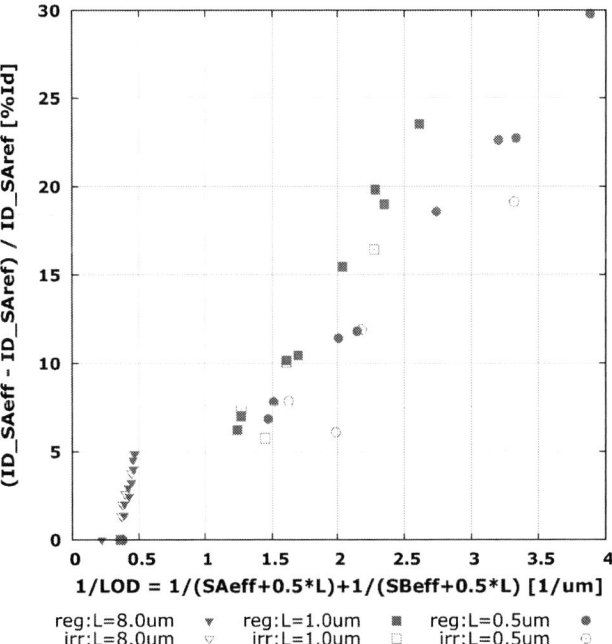

reg:L=8.0um ▼ reg:L=1.0um ▪ reg:L=0.5um ●
irr:L=8.0um ▽ irr:L=1.0um ▫ irr:L=0.5um ○

Fig. 7. PMOS Linear Drain current (Vds=-0.1V, Vgs=-5.0V, Vbs=0.0) percent change between SAref=5.0um and other various regular and irregular SAeff values (SA=SB).

C. Threshold Voltage Measurement Results

The other physical mechanism that causes the FET characteristics to vary with LOD distance is that the STI stress affects the diffusion [5]; therefore, causing the doping profile under the FET to change. This doping profile change causes the threshold voltage to vary with LOD distance.

The threshold voltage Vth using the peak-gm method at low Vds is shown in Figure 8 and Figure 9 for regular and irregular test structures of NMOS and PMOS. Figure 8 and Figure 9 show that the 1/LOD trend of the irregular LOD structures follows the same trend as the regular LOD structures for threshold voltage for both NMOS and PMOS. Measurements also were taken for the body effect and the DIBL effect, but those figures are not shown since there was not a significant LOD dependence for the body effect and DIBL effect in the process used for this paper.

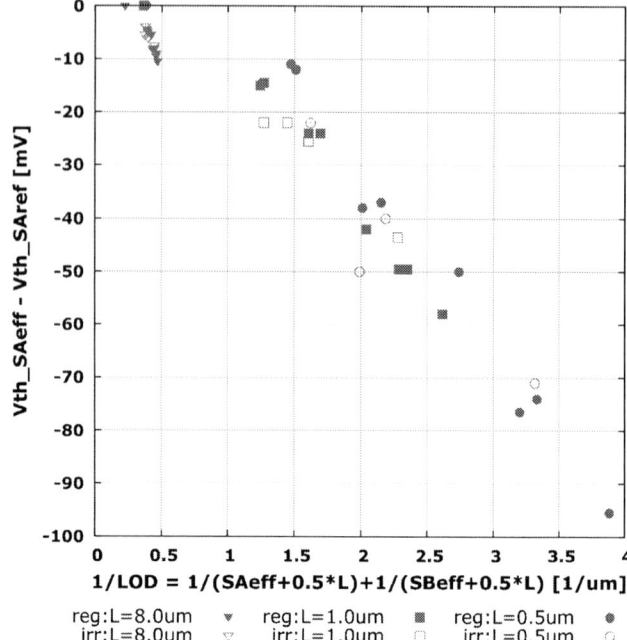

Fig. 9. PMOS peak-gm Vth (Vds=-0.1V, Vbs=0.0) differences between SAref=5.0um and other various regular and irregular SAeff values (SA=SB).

IV. CONCLUSIONS

This paper has confirmed that the BSIM4 model assumptions that irregular shaped LOD structures follow a weighted average along the transistor width are reasonably accurate. A review of the BSIM-BULK (formerly BSIM6) manual [8] shows the same equation for LOD, so it is expected that the BSIM-BULK LOD model will also be reasonably accurate for irregular shaped LOD structures. These irregular LOD model equations enable the layout designer to continue to take advantage of area and performance benefits of using the irregular shaped FETs and still maintain a reasonable amount of model accuracy.

V. REFERENCES

[1] P.G. Drennan, M.L. Kniffin, D.R. Locascio, "Implications of Proximity Effects for Analog Design", 2006 IEEE CICC, p169-176.

[2] R.A. Bianchi, G. Bouche, O. Roux-dit-Buisson, "Accurate Modeling of Trench Isolation Induced Mechanical Stress Effects on MOSFET Electrical Performance," 2002 IEEE IEDM, p 117-120

[3] K. W. Su, et al., "A Scaleable Model for STI Mechanical Stress Effect on Layout Dependence of MOS Electrical Characterization," 2003 IEEE CICC, p245-248

[4] BSIM4 Website from Berkeley: http://bsim.berkeley.edu/models/bsim4

[5] G. Scott, J. Lutze, M. Rubin, F. Nouri, M. Manley, "NMOS Drive Current Reduction Caused by Transistor Layout and Trench Isolation Induced Stress," 1999 IEEE IEDM, pg 827-830.

[6] Y.M. Sheu, K.Y.Y Doong, C.H. Lee, M.J. Chen, and C.H. Diaz, "Study on STI Mechanical Stress Induced Variations on Advanced CMOSFETs," 2003 IEEE ICMTS, Pg 3.205 – 3.208.

[7] A. Lochtefeld and D.A. Antoniadis, "Investigating the Relationship Between Electron Mobilty and Velocity in Deeply Scaled NMOS via Mechanical Stress," Dec 2001 IEEE TED, Vol 22, No12, p591-593.

[8] BSIM-BULK Website: http://bsim.berkeley.edu/models/bsimbulk/

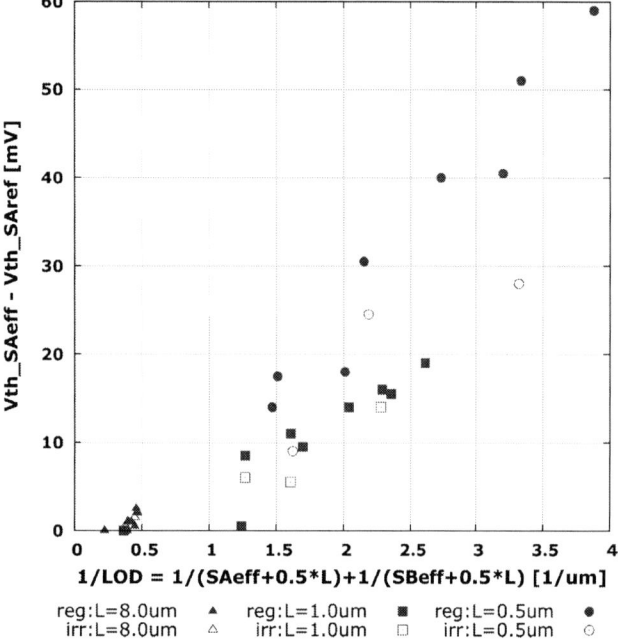

Fig. 8. NMOS peak-gm Vth (Vds=0.1V, Vbs=0.0) differences between SAref=5.0um and other various regular and irregular SAeff values (SA=SB).

2018 INTERNATIONAL CONFERENCE ON MICROELECTRONIC TEST STRUCTURES, MARCH 19-22, 2018, AUSTIN, TEXAS, USA

Efficient Parameter-Extraction of SPICE Compact Model through Automatic Differentiation

Michihiro Shintani[†], Masayuki Hiromoto[‡], and Takashi Sato[‡]

[†]Graduate School of Information Science,
Nara Institute of Science and Technology (NAIST), 8916-5 Takayama-cho, Ikoma 630-0192, Japan
[‡]Graduate School of Informatics, Kyoto University, Yoshida-hon-machi, Sakyo, Kyoto 606-8501, Japan
[†]shintani@is.naist.jp, [‡]paper@easter.kuee.kyoto-u.ac.jp

Abstract— A novel parameter extraction method for compact MOSFET models is proposed. The proposed method exploits automatic differentiation (AD) technique that is widely used in the training of artificial neural networks. In the AD technique, gradient of all the parameters of the MOSFET model is analytically calculated as a graph to reduce computational cost. On the basis of the calculated gradient, the model parameters are efficiently optimized. Through experiments using SPICE models, the parameter extraction using the proposed method achieved 7.01x speedup compared to that using the numerical-differentiation method.

I. INTRODUCTION

Prediction of circuit performance at the early design stage is important. Among others, an accurate formulation of compact MOSFET model and the careful extraction of their parameter set are crucial for obtaining reliable circuit simulation results. The compact MOSFET models, which have long been an intensive research topic, are composed of multiple nonlinear functions to capture various physical phenomena. In particular, surface-potential-based and charge-based compact models, which can represent the MOSFET behavior based on device physics, have been successfully applied for simulating various types of MOSFETs such as silicon and silicon carbide [1]–[7].

One of the most widely used parameter extraction methods is the gradient-based method [8], and is generally composed of two processes: the gradient calculation and parameter update. The two processes are repeated until a good agreement between model and measured characteristics are obtained, or the iteration limit is reached. The gradient is a set of partial derivatives for all model parameters in the model equation. In the parameter optimization, the gradient descent algorithm is applied to update the parameters using the obtained gradient. The gradient is a good indicator of which direction the parameters should be modified for attaining optimal parameters. However, the gradient decent method tends to require very long time until all the parameters are optimized because numerical differentiation (ND) of the parameters are repeatedly calculated, in which full evaluation of the model equation at least twice for a differentiation for each parameter. As explained later in more detail, times of the model evaluation during the parameter extraction is $2n$, where n is the number of the model parameters. The model evaluation is conducted quite a large number of times until the exit condition is satisfied. In our preliminary experiment using a simple device

model in [9], the gradient calculation occupies 99.8% of the parameter optimization time.

In this paper, we propose a novel model parameter extraction method through accelerating the computation of differentiation in the gradient-based method. We utilize automatic differentiation (AD) technique [10], which is commonly utilized in machine-learning communities to learn the weights of artificial neural networks through backpropagation [11], [12]. The AD technique enables to calculate the contribution of error between measurement and simulation in a single traversal of sensitivity graph, thus eliminating the evaluation of model equations during the gradient calculation. On the basis of the obtained gradient, the model parameters are optimized using the gradient descent technique. Through experiments, we demonstrate 7x speedup for a parameter extraction of eight parameters over to the ND method, which is close to the theoretical values.

The remainder of this paper is organized as follows. Section II provides the problem formulation of the ND. In Section III, the basic concept of AD and the AD-based parameter extraction method are described. The experiments using the simple compact model and power MOSFET model [6] are shown in Section IV to quantitatively evaluate the effectiveness of the proposed method. Finally, we will conclude the paper in Section V.

II. PARAMETER EXTRACTION BASED ON NUMERICAL DIFFERENTIATION

A. Gradient-based parameter extraction

We will first review the gradient-based parameter extraction, which is shown in Algorithm 1. In the Algorithm, a parameter extraction of the channel current model $f(\cdot)$ is considered as an example. Here, f is a function of gate-source voltage V_{gs} and drain-source voltage V_{ds}, using model parameters p in the current model equation.

The algorithm takes six inputs: the initial value of the parameter vector p, the vector of the small changes of each parameter δ, the measured current data I^{meas} and their corresponding bias voltages V, the target error E_{max}, and the maximum iteration N_{max}. The numbers of elements of vector p and vector δ are both n, which is equal to the number of model parameters. V is a matrix of the gate-source voltage V_{gs} and the drain-source voltage V_{ds} in the measurement of

978-1-5386-5072-1/18 $31.00 © 2018 IEEE

Algorithm 1 Gradient descent based parameter optimization

Require: $\boldsymbol{p} = (p_0, \ldots, p_{n-1})$, $\boldsymbol{\delta} = (\delta_0, \ldots, \delta_{n-1})$, $\boldsymbol{V} = ((V_{\mathrm{gs}_0}, V_{\mathrm{ds}_0}), \ldots, (V_{\mathrm{gs}_{m-1}}, V_{\mathrm{ds}_{m-1}}))$, $\boldsymbol{I}^{\mathrm{meas}} = (I_0^{\mathrm{meas}}, \ldots, I_{m-1}^{\mathrm{meas}})$, E_{max}, N_{max}
1: initialize $N = 0$
2: **while** ($N < N_{\mathrm{max}}$ or $E_{\mathrm{max}} < E$) **do**
3:     $\nabla E, E = \mathrm{gradient\_calc}(\boldsymbol{p}, \boldsymbol{\delta}, \boldsymbol{V}, \boldsymbol{I}^{\mathrm{meas}})$
4:     **for each** $p_i \in \boldsymbol{p}$ **do**
5:         Update p_i so as to reduce E based on $\frac{\partial E}{\partial p_i}$
6:     **end for**
7:     $N++$
8: **end while**
9: Return optimal parameter $\boldsymbol{p}$

Algorithm 2 ND-based gradient calculation

1: **function** $\mathrm{ND}(\boldsymbol{p}, \boldsymbol{\delta}, \boldsymbol{V}, \boldsymbol{I}^{\mathrm{meas}})$
2:     $e = 0$, $e_{\mathrm{delta}} = 0$
3:     **for each** $p_i \in \boldsymbol{p}$ **do**
4:         $\boldsymbol{p}' = p_i + \delta_i$
5:         **for each** $(V_{\mathrm{gs}_j}, V_{\mathrm{ds}_j}) \in \boldsymbol{V}$ **do**
6:             $I_{\mathrm{sim1}} = f(\boldsymbol{p}, (V_{\mathrm{gs}_j}, V_{\mathrm{ds}_j}))$
7:             $e = e + (I_j^{\mathrm{meas}} - I_{\mathrm{sim1}})^2$
8:             $I_{\mathrm{sim2}} = f(\boldsymbol{p}', (V_{\mathrm{gs}_j}, V_{\mathrm{ds}_j}))$
9:             $e_{\mathrm{delta}} = e_{\mathrm{delta}} + (I_j^{\mathrm{meas}} - I_{\mathrm{sim2}})^2$
10:     **end for**
11:     $E = \sqrt{\frac{e}{m}}$
12:     $E_{\mathrm{delta}} = \sqrt{\frac{e_{\mathrm{delta}}}{m}}$
13:     $\frac{\partial E}{\partial p_i} = \frac{E_{\mathrm{delta}} - E}{\delta_i}$
14:     **end for**
15:     Return $\nabla E, E$
16: **end function**

$\boldsymbol{I}^{\mathrm{meas}}$. The dimensions of the matrix $\boldsymbol{V}$ and $\boldsymbol{I}^{\mathrm{meas}}$ are both $m_{\mathrm{vd}} \times m_{\mathrm{vg}}$, where $m = m_{\mathrm{vd}} \times m_{\mathrm{vg}}$ is the total number of the measurement data and m_{vd} and m_{vg} are the number of measured voltages in V_{ds} and V_{gs} sweeps, respectively. The ND-based method is applied so that the cost function is reduced by changing the parameter $\boldsymbol{p}$. Here, as the cost function, the root mean square error (RMSE) between the simulation and measurement is widely used:

$$E = \sqrt{\frac{1}{m} \sum_m (I^{\mathrm{meas}} - I^{\mathrm{sim}})^2}. \tag{1}$$

In line 3 of Algorithm 1, through the function call of gradient_calc, the gradient of the cost function in terms of each parameter ∇E is obtained as follows:

$$\nabla E = \left(\frac{\partial E}{\partial p_0}, \ldots, \frac{\partial E}{\partial p_{n-1}} \right). \tag{2}$$

Then, in lines 4–6, by using the gradient, the parameters are updated to the direction that the cost function is lowered. With the updated model parameters, cost function and its gradient is again calculated. The gradient calculation and parameter update are alternatively repeated until either of the exit conditions in line 2 is satisfied — the cost function becomes lower than the target value or the iteration limit is reached.

B. Numerical differentiation

The gradient-based parameter extraction requires the calculation of the gradient as shown in line 3 of Algorithm 1. A simple way to calculate derivatives is ND. In the ND, one of the model parameters of interest is changed slightly for differentiation while the other model parameters and inputs are fixed. This process is versatile because it can be applied even when the model equation does not have a closed form. However, the ND may suffer from rounding error when the *slight change* is chosen to be too small. What is worse, the ND is computationally intensive because the entire model equation needs to be repeatedly evaluated for the combination of all

bias voltages and all parameters. Hence, the computational cost of the ND highly depends on the number of the model parameters and measurement points, and it tends to be very time consuming.

The ND-based gradient calculation, which is a function called as the function gradient_calc in Algorithm 1, is summarized in Algorithm 2. The ND approximates the derivative by the slope of two points, hence it requires evaluations of the model equation $f(\cdot)$ twice with respect to each model parameter, i.e., I_{sim1} and I_{sim2}. E and E_{delta} are the RMSEs for $\boldsymbol{p}$ and $\boldsymbol{p}'$, respectively. Using the errors, the partial derivative of E for each parameter is calculated as shown in line 13 in Algorithm 2.

In the ND-based method, $f(\cdot)$ is evaluated for $2mnN$ times, where N is the iteration count. Obviously, we can see the calculation time increases as n increases, i.e., the model equation becomes complex. In practical implementations, the calculation of ND can be further accelerated by avoiding calculation of E for all $\boldsymbol{p}$. It can be separately calculated and stored in advance to the parameter loop (line 3 in Algorithm 2). While it reduces the number of evaluations of $f(\cdot)$ from $2mnN$ to $(1+n)mM$, but that optimization does not alter computational complexity.

III. PARAMETER EXTRACTION BASED ON AUTOMATIC DIFFERENTIATION

The proposed parameter extraction is also based on the typical gradient-based extraction flow in Algorithm 1. By replacing the most time-consuming ND to the AD, parameter extraction is accelerated.

A. Automatic Differentiation

The basic concept of the AD is the decomposition of partial derivatives by the chain rule. The derivative $\mathrm{d}y/\mathrm{d}x$ of a composite function $y = g(h(x)) = g(w)$ is written by the chain rule as

$$\frac{\mathrm{d}y}{\mathrm{d}x} = \frac{\mathrm{d}y}{\mathrm{d}w}\frac{\mathrm{d}w}{\mathrm{d}x}. \quad (3)$$

The first and the second terms of Eq. (3) are individually calculated since $\mathrm{d}y/\mathrm{d}w = \mathrm{d}g(w)/\mathrm{d}w$ and $\mathrm{d}w/\mathrm{d}x = \mathrm{d}h(x)/\mathrm{d}x$. Then $\mathrm{d}y/\mathrm{d}x$ is derived by the product of the two terms. In the AD, the given expression is generally represented by using a computational graph.

The AD splits the above task into two modes: forward mode and backward mode. In the forward mode, the given expression is first decomposed into a set of primitive expressions, i.e., the simplest function, such as addition and multiplication functions. Through the forward mode, the model parameters and input values are propagated to obtain the value of the output. Then, in the backward mode, the output variable to be differentiated is first fixed, and the partial derivative value concerning each partial expression, e.g., $\mathrm{d}y/\mathrm{d}w$ and $\mathrm{d}w/\mathrm{d}x$ in the above example, is recursively calculated. To obtain the gradient through the AD, the calculations of the partial derivatives with respect to all the input parameters are conducted. Note that the forward and backward propagations in AD are respectively one time process, hence requiring no model evaluation for every parameter. The computational complexity of the forward propagation and the backward propagation are both roughly equal to that of a model evaluation.

B. Parameter Extraction

The AD-based parameter extraction method is described in Algorithm 3. which is also called at line 3 in Algorithm 1 as gradient_calc. In the AD, the parameter δ is not necessary and hence ignored. Lines 4 and 5 in Algorithm 3 correspond to the forward and backward modes, respectively. Without the for loop in p, gradient of the model equation is obtained. The calculation complexity hence roughly becomes $2mN$, which is independent of the number of parameters, n. With this, we can accelerate the gradient calculation. When compared with the conventional method, ND, the calculation efficiency is improved by n fold. This implies that as the number of the model parameters is increased, the acceleration achieved by the proposed method increases.

In the following, we explain the proposed parameter extraction method by using a simple example of the channel current equation f shown below [6] in Algorithm 3:

$$I_{\mathrm{ch}} = \frac{1}{1 + \mathbf{THETA} \cdot V_{\mathrm{gs}}}(1 + \mathbf{LAMBDA} \cdot V_{\mathrm{ds}}) \cdot \mathbf{K} \cdot I_{\mathrm{DD}}. \quad (4)$$

Here, I_{ch}, V_{gs}, and V_{ds} are the channel current, gate-source, and drain-source voltage, respectively. I_{DD} is an intermediate value obtained by the function of V_{gs}, V_{ds}, and surface potential at the metal-oxide-semiconductor (MOS) interface. The channel current equation considers the channel length modulation and mobility degradation. $\mathbf{K}$, $\mathbf{LAMBDA}$, and $\mathbf{THETA}$ are the model parameters that represent the scaling factor, the channel length modulation, and the channel mobility degradation, respectively.

Algorithm 3 AD-based gradient calculation

1: **function** AD(p, V, I^{meas})
2:     $e = 0$
3:     **for each** $(V_{\mathrm{gs}_j}, V_{\mathrm{ds}_j}) \in V$ **do**
4:         Calculate $I_{\mathrm{sim}} = f(p, (V_{\mathrm{gs}_j}, V_{\mathrm{ds}_j})$ and $e = e + (I_{\mathrm{meas}} - I_{\mathrm{sim}})^2$ through forward mode
5:         Calculate ∇E through backward mode
6:     **end for**
7:     $E = \sqrt{\frac{e}{m}}$
8:     Return $\nabla E, E$
9: **end function**

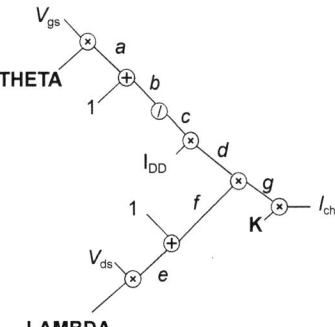

Fig. 1. Computational graph of Eq. (4) including five multiplication operations, two addition operations, and one division operation.

Figure 1 shows a computational graph representing Eq. (4). The variables on the leftmost leaves of the graph represent the inputs, such as $\mathbf{THETA}$ and V_{ds}. I_{ch} on the right is the output. Each vertex labeled as variables from a to g stands for intermediate values.

In the gradient calculation, first, the forward propagation is conducted, in which the current model equation is evaluated by traversing the graph to generate the internal vertexes at the bias voltage condition, V_{ds} and V_{gs}, as shown in Fig. 2. The values of the internal vertexes labeled as a to l are also calculated, e.g., $I_{\mathrm{ch}} = g \cdot \mathbf{K}$. In Fig. 2, the calculation of the RMSE of the simulated value against the measured value is added. Then, in the backward mode, the derivative of the output E is propagated backwards through the network in order to calculate the partial derivatives of E with respect to each model parameter.

Let us consider the backward mode for deriving the partial derivative $\frac{\partial E}{\partial \mathbf{K}}$. There are seven edges, E, n, l, j, h, I_{ch}, and $\mathbf{K}$, on the path from E to $\mathbf{K}$ in the graph. According to the chain rule, $\frac{\partial E}{\partial \mathbf{K}}$ is given as

$$\frac{\partial E}{\partial \mathbf{K}} = \frac{\partial E}{\partial E} \cdot \frac{\partial E}{\partial n} \cdot \frac{\partial n}{\partial l} \cdot \frac{\partial l}{\partial j} \cdot \frac{\partial j}{\partial h} \cdot \frac{\partial h}{\partial I_{\mathrm{ch}}} \cdot \frac{\partial I_{\mathrm{ch}}}{\partial \mathbf{K}}. \quad (5)$$

The derivatives of each edge with respect to the previous edge are easily obtained, e.g., $\frac{\partial n}{\partial l} = \frac{1}{2\sqrt{l}}$ since $n = \sqrt{l}$ and $\frac{\partial I_{\mathrm{ch}}}{\partial \mathbf{K}} = g$

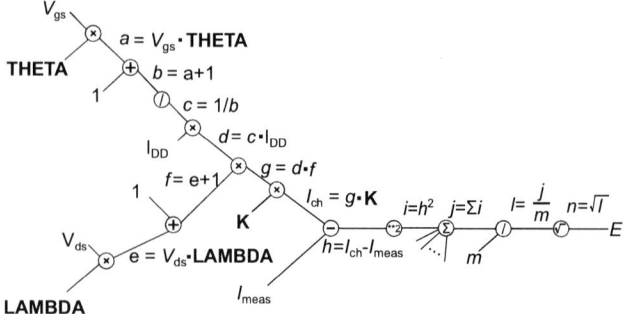

Fig. 2. Forward mode.

since $I_{\text{ch}} = g \cdot \mathbf{K}$. Substituting all the partial derivatives of the edges,

$$\frac{\partial E}{\partial \mathbf{K}} = \frac{1}{2\sqrt{l}} \cdot \frac{1}{m} \cdot 2h \cdot g \qquad (6)$$

is derived. The values of l, m, h, and g are already known by the forward mode. This calculation is conducted for all the partial derivatives of all the model parameters. The set of the partial derivatives is used as a gradient in the parameter update phase. Thus, by applying the chain rule, the computation of the gradient is conducted without approximation. The proposed method can calculate the gradient quickly by using the backward-propagation rule for each operation as shown in Fig 3. For example, the sum operation is used in the calculation of the RMSE. The backward mode of the sum operation is similarly calculated with the addition operation. Though it is not used in the example, the exponential operation, which is often used in the physics based compact model, is also available in the proposed method.

Note that the construction of the computational graph is required only once for a given MOSFET model. The constructed computational graph can be utilized as long as the same MOSFET model is used, so the construct time of the calculation graph is virtually negligible.

IV. EXPERIMENTAL RESULTS

The proposed method is applied to extract model parameters of two MOSFET models described in Subsection IV-A. In this experiment, the I-V curve of an SiC MOSFET [13] is used. The measurement of the I-V curve is conducted by a dedicated curve tracer [14]. As one of the termination conditions, we set the maximum number of the iterations to be 1,000. The parameter update is conducted by using AdaGrad [15], which is,

$$h_i = h_i + \left(\frac{\partial E}{\partial p_i}\right)^2 \quad \text{and} \qquad (7)$$

$$p_i = p_i - \eta \frac{1}{\sqrt{h_i}} \frac{\partial E}{\partial p_i}, \qquad (8)$$

where η is an update rate, which is optimized in the AdaGrad. The parameter update is performed to decrease the RMSE. In the AdaGrad, the parameter vector $\boldsymbol{h} = (h_0, ..., h_{n-1})$ is introduced as shown in Eq. (7). According to the gradient,

Fig. 3. Backward propagation of each operation. In this figure, $\partial L/\partial z$ propagates to the downstream.

$\eta \frac{1}{\sqrt{h_n}}$ is gradually reduced as shown in Eq. (7). In general, the initial values of all the elements of $\boldsymbol{h}$ are set to zeros.

All experiments were carried out on a Linux PC with Intel Xeon E5-2630 v2 2.60 GHz CPU using a single thread. The proposed method was implemented all with the Python programming language.

A. MOSFET Models

1) Threshold Voltage Based Model: The first model is the N-th-power-low MOSFET current model [9]. The number of the model parameters is eight. The drain current is calculated based on the threshold voltage of the MOSFET. The saturation voltage $V_{\text{ds,sat}}$ and the saturation current $I_{\text{d,sat}}$ are defined as

$$V_{\text{ds,sat}} = \mathbf{J} \left(V_{\text{gs}} - \mathbf{VTH}\right)^{\mathbf{M}}, \qquad (9)$$

$$I_{\text{d,sat}} = \mathbf{K} \left(V_{\text{gs}} - \mathbf{VTH}\right)^{\mathbf{N}}. \qquad (10)$$

$V_{\text{ds,mod}}$ is defined to replace V_{ds}, in order to represent the smooth transition between the linear and the saturation regions [6]:

$$V_{\text{ds,mod}} = \frac{V_{\text{ds}}}{\left[1 + \left(\frac{V_{\text{ds}}}{V_{\text{ds,sat}}}\right)^{\mathbf{DELTA}}\right]^{\frac{1}{\mathbf{DELTA}}}}. \qquad (11)$$

By using Eq. (11), $V_{\text{ds,mod}}$ is smoothly connected to $V_{\text{ds,sat}}$. The parameter $\mathbf{DELTA}$ controls the smoothness of the connection. The drain current I_{ds} is calculated while considering

TABLE I
MODEL PARAMETERS OF THE SURFACE POTENTIAL MODEL [6]

Parameter	Description
TOX	Oxide thickness [m]
VFBC	Flat-band voltage of channel region [V]
NA	Acceptor concentration [cm^{-3}]
K	Current gain factor [cm^2/V]
RD	Parasitic resistance at the drain side [Ω]
LAMBDA	Channel length modulation [1/V]
THETA	Channel mobility degradation [1/V]
DELTA	Smoothing parameter for gradual transition between linear and saturation regions [-]

the channel length modulation and mobility degradation [1], [2] as

$$I_{ds} = \mathbf{K}\left(2 - \frac{V_{ds,mod}}{V_{ds,sat}}\right)\frac{V_{ds,mod}}{V_{ds,sat}} \times$$
$$(1 + \mathbf{CLM}V_{ds})[1 + \mathbf{THETA}(V_{gs} - \mathbf{VTH})]. \quad (12)$$

2) Surface Potential Based Model: We also use the surface potential model for simulating the behavior of SiC power MOSFETs [6] in this experiment. It has eight model parameters for its current equation as shown in Table I. The current equation is represented as shown in Eq. (4). I_{DD} is derived by the surface potential. In the SiC power MOSFET model, iterative methods, such as Newton-Raphson method, are employed to obtain the surface potential. The iterative calculation cannot be represented by the computational graph. Hence, the partial derivatives of the parameters on the surface potential, i.e., **TOX**, **NA**, **VFBC**, and **DELTA** are obtained by the ND instead of the AD.

B. Results

The proposed AD-based parameter extraction is compared with the ND-based method. Figure 4 shows the RMSE of the two methods as a function of the CPU time for the N-th-power-law and surface potential based models, respectively. In Fig. 4a, the proposed method achieved 7x acceleration in parameter extraction, which is close to the ideal value of 8, which is the number of parameters. On the other hand, the result for the surface potential model shows only a 1.7x speedup. This is because not all the parameters in the surface potential based model was not represented in the computational graph of AD due to the existence of the loop in the surface potential calculation. As a result, the actual number of the model parameters considered in the AD is four in the eight parameters, the ideal value of the acceleration becomes 2x. The speedup is also close to the ideal value.

The measured and simulated I-V characteristics of the N-th-power-law model are shown in Fig. 5. The final values of the model parameters obtained by AD are used for drawing the graph. The MOSFET model accurately reproduces the I-V characteristics of the SiC MOSFET.

V. CONCLUSION

In this paper, we proposed a novel parameter extraction method of MOSFET model. The proposed method employs

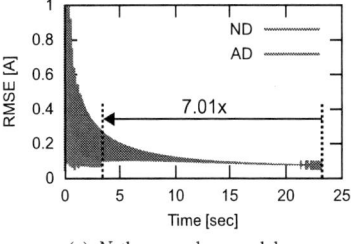

(a) N-th-power-law model

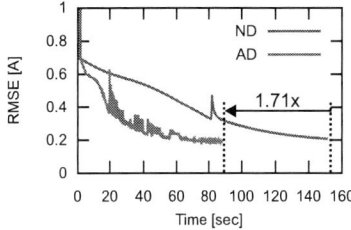

(b) Surface potential-based model

Fig. 4. RMSE as a function of the CPU time.

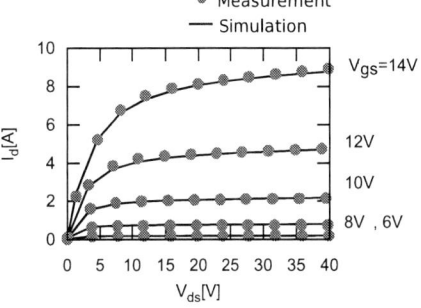

Fig. 5. Simulated and measured I-V curves.

an AD, which is commonly used in backpropagation of the artificial neural networks. The AD technique calculates the partial derivatives without the calculation of the numerical derivatives, and contributes to accelerate the CPU time of the parameter extraction. Experimental results show that the proposed method obtains the model parameters for the current equation of the MOSFET with 7x speedup compared to the conventional gradient descent method using two-point gradient approximations.

ACKNOWLEDGMENTS

This work was partially supported by JST Super Cluster Program and NEDO Cross-ministerial Strategic Innovation Promotion Program. This work is also a collaborative work with Sumitomo Electric Industries, Ltd.

REFERENCES

[1] M. Miura-Mattausch, N. Sadachika, D. Navarro, G. Suzuki, Y. Takeda, M. Miyake, T. Warabino, Y. Mizukane, R. Inagaki, T. Ezaki, H. J. Mattausch, T. Ohguro, T. Iizuka, M. Taguchi, S. Kumashiro, and S. Miyamoto, "HiSIM2: Advanced MOSFET model valid for RF circuit simulation," *IEEE Transactions on Electron Devices*, vol. 53, no. 9, pp. 1994–2007, 2006.

[2] G. Gildenblat, X. Li, W. Wu, H. Wang, A. Jha, R. van Langevelde, G. D. J. Smit, A. J. Scholten, and D. B. M. Klaassen, "PSP: An advanced surface-potential-based MOSFET model for circuit simulation," *IEEE Transactions on Electron Devices*, vol. 53, no. 9, pp. 1979–1993, 2006.

[3] W. Yao, G. Gildenblat, C. C. McAndrew, and A. Cassagnes, "SP-HV: A scalable surface-potential-based compact model for LDMOS transistors," *IEEE Transactions on Electron Devices*, vol. 59, no. 3, pp. 542–550, 2012.

[4] H. J. Mattausch, M. Miyake, T. Iizuka, H. Kikuchihara, and M. Miura-Mattausch, "The second-generation of HiSIM_HV compact models for high-voltage MOSFETs," *IEEE Transactions on Electron Devices*, vol. 60, no. 2, pp. 653–661, 2013.

[5] Y. S. Chauhan, S. Venugopalan, M.-A. Chalkiadaki, M. A. U. Karim, H. Agarwal, S. Khandelwal, N. Paydavosi, J. P. Duarte, C. C. Enz, A. M. Niknejad, and C. Hu, "BSIM6: Analog and RF compact model for bulk MOSFET," *IEEE Transactions on Electron Devices*, vol. 61, pp. 234–244, 2013.

[6] Y. Nakamura, M. Shintani, K. Oishi, T. Sato, and T. Hikihara, "A simulation model for SiC power MOSFET based on surface potential," in *Proceedings of International Conference on Simulation of Semiconductor Processes and Devices*, 2016, pp. 121–124.

[7] M. Shintani, Y. Nakamura, M. Hiromoto, T. Hikihara, and T. Sato, "Measurement and modeling of gatedrain capacitance of silicon carbide vertical double-diffused MOSFET," *Japanese Journal of Applied Physics*, vol. 56, no. 04CR07, pp. 04CR07-1–04CR07-5, 2017.

[8] Q. Zhou, W. Yao, W. Wu, X. Li, Z. Zhu, and G. Gildenblat, "Parameter extraction for the PSP MOSFET model by the combination of genetic and Levenberg-Marquardt algorithms," in *Proceedings of International Conference on Microelectronic Test Structures*, 2009, pp. 137–142.

[9] T. Sakurai and A. R. Newton, "A simple MOSFET model for circuit analysis," *IEEE Transactions on Electron Devices*, vol. 38, no. 4, pp. 887–894, 1991.

[10] M. Bartholomew-Biggs, S. Brown, B. Christianson, and L. Dixon, "Automatic dierentiation of algorithms," *JCAM*, vol. 124, no. 1, pp. 171–190, 2000.

[11] A. G. Baydin, B. A. Pearlmutter, and A. A. Radul, "Automatic differentiation in machine learning: a survey," *The computing Research Repository (CoRR)*, vol. abs/1502.05767, 2015.

[12] Y. LeCun, Y. Bengio, and G. Hinton, "Deep learning," *Nature*, vol. 521, 2015.

[13] ROHM Co., Ltd., *SCT2450KE N-channel SiC power MOSFET Datasheet*, 2013.

[14] Y. Nakamura, M. Shintani, T. Sato, and T. Hikihara, "A high power curve tracer for characterizing full operational range of SiC power transistors," in *Proceedings of International Conference on Microelectronic Test Structures*, 2016, pp. 90–94.

[15] J. Duchi, E. Hazan, and Y. Singer, "Adaptive subgradient methods for online learning and stochastic optimization," *The Journal of Machine Learning Research*, vol. 12, pp. 2121–2159, 2011.

Session 3

Reliability

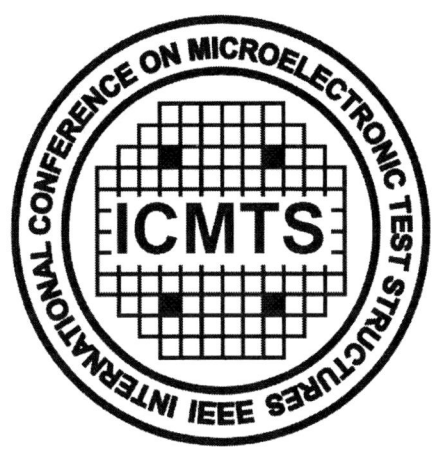

March 20, 2018

14:10-14:50

Chair:

Anthony J. WALTON, University of Edinburgh, Scotland

978-1-5386-5072-1/18 $31.00 © 2018 IEEE

Test Structure Design for Model-Based Electromigration

Ertugrul Demircan, Mehul D. Shroff, Hsun-Cheng Lee

NXP Semiconductors, Austin, TX 78735, USA

Email: ertugrul.demircan@nxp.com

Abstract— As VLSI technology features are pushed to the limit with every generation and with the introduction of new materials and increased current densities to satisfy performance demands, failure risk due to Electromigraton (EM) is ever-increasing. In this paper, we present experimental results using a novel set of test structures to validate a new model-based EM risk assessment approach. In this method, EM risk can be assessed for any interconnect geometry through an exact solution of the fundamental stress equations. This approach eliminates the need for complex look-up tables for different geometries and can be implemented in CAD tools very easily as we demonstrate on real design examples.

Keywords— *VLSI, CMOS, SOI, Reliability, Electromigration, EM.*

I. INTRODUCTION

Electromigration (EM) is the failure mechanism caused by diffusing metal atoms under the influence of electrical current flowing through interconnects [1,2]. It leads to creation of voids or hillocks at cathode and anode locations respectively, and it is a significant reliability challenge in advanced semiconductor technologies. With technology scaling, introduction of new materials, and increased current densities to satisfy performance demands, EM failure risk is ever-increasing. Furthermore, with the transition to Fully-Depleted SOI (FDSOI) and finFETs, EM failure risk can be further exacerbated by the heat dissipation from the device to the interconnects.

Several new methods have been recently proposed [3-8]. A novel model-based methodology where EM risk can be assessed for any interconnect geometry through an exact solution of the fundamental stress equations has been proposed recently [6]. In this paper, we present test structure design and experimental confirmation of this approach and demonstrate design enablement with real design examples.

II. THEORETICAL BACKGROUND

The mechanical stress caused by the electron-induced backflow of metal ions is given by the following equations:

$$\frac{\partial \sigma}{\partial x} = Cj \tag{1}$$

$$\frac{\partial u}{\partial x} = \alpha \sigma \tag{2}$$

$$\sum_j u_{ij} w_{ij} = 0 \tag{3}$$

Here, σ is the EM stress, $C = Z^* e\rho/\Omega$, Z^* is the effective charge and Ω is the atomic volume of the underlying metal atoms, e is the electron charge and ρ is the bulk resistivity, u_{ij} is the displacement at j^{th} node on a branch between nodes i and j, $M = 1/\alpha$ is the Young's modulus of the interconnect material, and w_{ij} is the width of the branch. The interconnect segment is defined as the contiguous shape on a single metal layer comprised of all branches. The nodes, such as the vias and pins and width changes that may cause a change in current density, are the branching points.

In the simple case where the interconnect segment is a straight line, the solution implies that the stress is inversely proportional to the length of the line and hence the critical current that causes EM failure is also inversely proportional to the line length. This is the well-known Blech effect [2].

Numerical solutions to the EM stress equations (EMSE) are not difficult to implement [3], but they are CPU-intensive and require additional run time to complete circuit simulation. Thus,

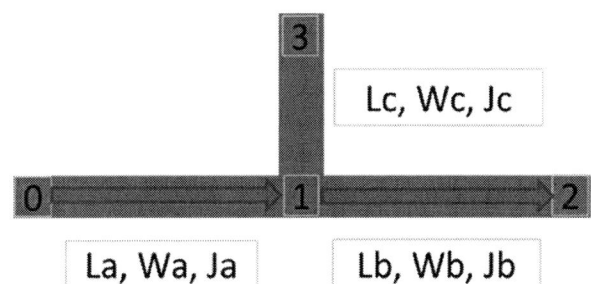

Figure 1: Illustration of T-junction interconnect with directed graph inserted to indicate electron flow (L = Length, W = Width, J = current density).

the generally accepted method of EM failure risk determination involves a rule-based algorithm where the current density limit on an interconnect segment is determined by a look-up table of geometric properties like the length and width.

However, the Blech length of an interconnect segment is not well-defined except in the simple case of a straight line, hence, an approximation is usually applied as the longest distance between any two end points. This approximation has two drawbacks: 1) It does not take into account the current direction in branches, and 2) It ignores the effects of some branches on the stress distribution. These two exceptions may cause false fails but more importantly may lead to missing some actual fails.

In order to avoid these problems, we approach the problem from a more fundamental point of view. For an interconnect network of N nodes, equation (1) provides (N-1) equations and (N) unknown node stresses. Hence, we need one more independent equation from the other two EMSE.

We first start by rewriting the fundamental equations for an interconnect graph with current density j_{ij} and length l_{ij} of the branches and σ_j stress at a node:

$$\sigma_i - \sigma_j = C j_{ij} l_{ij}, \tag{4}$$

$$u_{ji-j} - u_{ji-i} = \alpha l_{ij} \frac{\sigma_j + \sigma_i}{2}. \tag{5}$$

Multiplying (5) by w_{ij} on both sides and summing over all branches and using (3) leads to a single equation:

$$\sum_\alpha a_\alpha \overline{\sigma}_\alpha = 0. \tag{6}$$

Here, a_α is the area and $\overline{\sigma}_\alpha$ is the average stress of the branch α. Then it is possible to show that we can rewrite this equation in terms of node stress:

$$\sum_j a_j \sigma_j = 0, \tag{7}$$

where a_j is the total area of branches connected to node j.

We also note that the branch areas satisfy the identity for total segment area A:

$$\sum_j a_j = 2A. \tag{8}$$

On the other hand, after using Ohm's Law to convert to node potentials, (4) can be written as:

$$\sigma_j = \sigma_i + \frac{Z^* e}{\Omega} V_{ij}, \tag{9}$$

where V_{ij} is the potential difference between the nodes i and j.

Then substituting (9) into (7) and defining the EM potential of the segment as:

$$V_E = \frac{1}{2} \sum_j V_j \frac{a_j}{A}, \tag{10}$$

we arrive at the expression for stress at any node i:

$$\sigma_i = \frac{Z^* e}{\Omega} (V_E - V_i). \tag{11}$$

Alternatively, the EM potential can be written in terms of the branch average voltages $\overline{V}_\alpha$ and branch areas a_α as:

$$V_E = \frac{1}{A} \sum_\alpha \overline{V}_\alpha a_\alpha. \tag{12}$$

In the continuum limit this leads to the expression:

$$V_E = \frac{1}{A} \iint V(x, y) dx dy. \tag{13}$$

This equation has a very simple physical interpretation: EM stress at any node is proportional to the difference between the node potential and the EM potential for the interconnect segment. Therefore, EM stress determination is reduced to a problem of analyzing the node potentials and combining them with geometric information from the interconnect segment. Since node potentials are readily available after circuit simulations, there is no need for an additional numerical solution. It is also important to note that we avoid the complications of current loops leading to an incomplete set of equations which requires special handling during a numerical solution of EMSE [4].

The main implication of this Voltage-Based EM (VBEM) method is that EM risk can be represented with the EM potential for any interconnect topology. This new formulation removes the need to define a "Blech length" which is not well defined for complex interconnect topologies and the need for inaccurate look up tables, and implicitly takes care of the current direction.

An important realization from the VBEM method is that it predicts an impact on EM lifetime from segments that do not carry a DC current (i.e., a stub). A simple but very common structure that can illustrate this effect is the T-Line structure shown in Fig. 1.

EM risk can be assessed by comparing this stress value to the critical stress that leads to EM failures, which can be determined empirically from silicon measurements on typical structures. Defining the critical stress $\sigma_C = (Z^* e/\Omega) V_C$ we can state the passing criteria for EM as:

$$V_i > V_E - V_C. \tag{14}$$

Without any loss of generality, we can define the EM potential relative to any node on the interconnect segment. We can set $V_i = 0$ for any node and calculate stress values since equation (11) is invariant under this transformation.

Also, if the lowest potential node (cathode) satisfies the pass criteria in (14) then all other nodes will also pass. Hence, we can rewrite (14) in a more compact form as:

$$V_{E0} < V_C . \tag{15}$$

Here V_{E0} is the EM potential of the interconnect segment calculated relative to lowest cathode potential on it.

III. EXPERIMENTAL RESULTS

To confirm the predictions of the VBEM method for the effect of a stub, we have built various T-Line test structures in a 90nm technology and measured their Mean-Time-To-Failure, which is then correlated to the VBEM-calculated stress.

The structure geometries are given in Table I in terms of the parameters defined in Fig. 1. The actual test structure layouts are shown in Fig. 2. Structures A and B have a main branch length of 10μm and a stub length of 5μm. In structure A, the stub is placed close to the center of the main line and in structure B, it is placed at one end. In structures C to F, the main branch length is 20μm and stub lengths are 5μm and 10μm. In structure C, the stub is close to center of the main line as in structure B and in others it is at the end.

TABLE I
TEST STRUCTURE DIMENSIONS (μm)

Case	La	Lb	Lc	Wa	Wb	Wc
A	6	4	5	0.14	0.14	0.28
B	10	0	5	0.14	0.14	0.28
C	11	9	10	0.14	0.14	0.14
D	20	0	10	0.14	0.14	0.14
E	20	0	5	0.28	0.28	0.14
F	20	0	10	0.28	0.28	0.14

EM package-level tests using ceramic dual-inlaid packages were performed using a current of 1.35mA and at a stress temperature of 325°C. The metal resistance for each device under test was monitored using a modular integrated reliability analyzer (MIRA) system. Resistance measurements vs. time were taken during specific time intervals and plotted to measure the device resistance with an increase of 20% as the fail criterion.

Measurements were performed for both the "Forward" bias condition where the stub is closer to the anode and for the "Reverse" bias condition where the stub is closer to the cathode. Based on predictions of the VBEM methodology, we expect the Forward and Reverse bias conditions to have significant differences in EM risk when the stub is closer to the end of the line segment. Similarly, when the stub is close to the middle of the line segment, we expect a smaller difference in EM risk.

The Mean-Time-To-Fail (MTTF) data is presented in Fig. 3. Structures A and C both have the stub close to the center and show much smaller changes in MTTF, whereas all other structures with stubs at the ends show significant increase when the stub is close to the anode (Reverse bias).

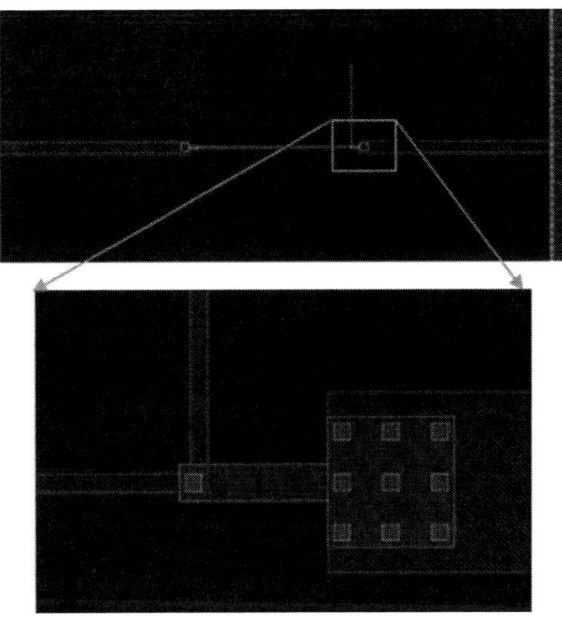

Figure 2: Test structure layout for Structure D shown in Table I. The top image shows placement between pads and the lower one shows detail at via connection. The lines on Metal 1 are in red and Metal 2 are on blue while vias are on green. The jumpers on connector leads from the pads are set to make sure the leads do not fail before the main structure during testing.

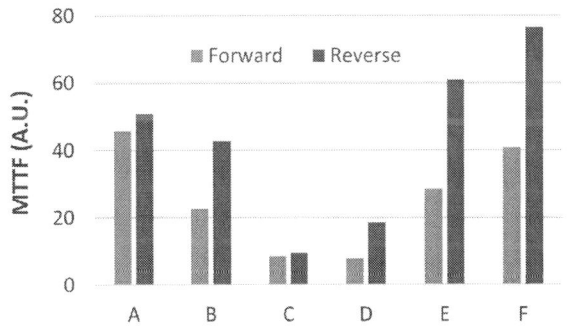

Figure 3: Comparison of MTTF for different current directions. In forward direction, the cathode is at node "0" and in reverse direction it is at node "2" in relation to Fig. 1. In both cases, $J_c = 0$.

In Fig. 4, the MTTF data for the various structures is plotted against the maximum stress calculated from VBEM method and clearly indicates a strong correlation, with both Forward and Reverse bias conditions following the same trend. This data can be used to set an EM stress limit which can be used for any interconnect topology for a given product life time requirement.

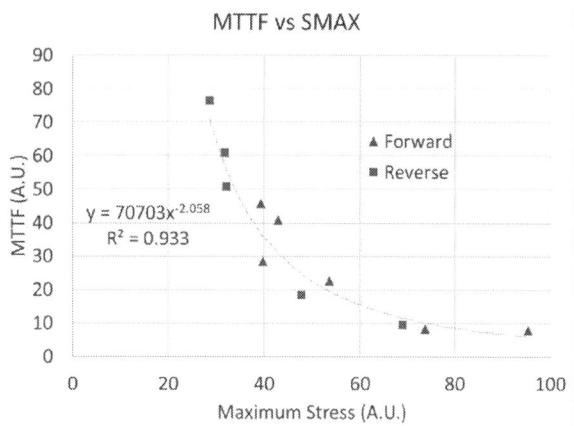

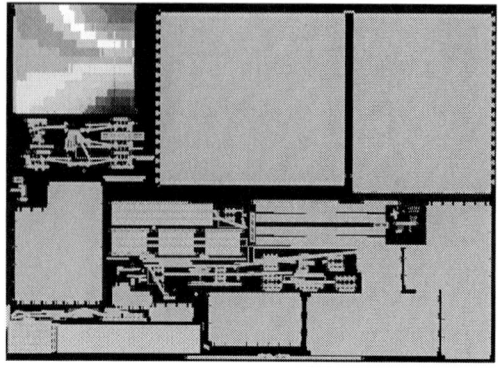

Figure 4: Correlation of measured MTTF with maximum EM stress (SMAX) at cathode calculated from VBEM method. Structure B Reverse bias data is excluded as an outlier.

Figure 6: Color map of VBEM-calculated EM stress for an analog block M1 layer. Higher stress regions are clearly visible in the upper left corner sub-block while the majority of the design has a vanishing EM stress.

IV. DESIGN FLOW IMPLEMENTATION

Our in-house spice simulator was enhanced to use the VBEM methodology. In Fig. 5, we present the M1 layer EM stress color map for a standard cell layout on an advanced technology node obtained from VBEM simulations. The gradual increase in stress (from blue to red) along the U-shaped interconnect follows the expected current flow between the NMOS and PMOS devices on the lower and upper parts of the design. In a typical commercial EM analysis tool, the segment shown in the red oval would be highlighted as the highest EM risk since it would have the maximum current density. However, from the VBEM color map, the highest stress occurs at the cathode end of the interconnect segment.

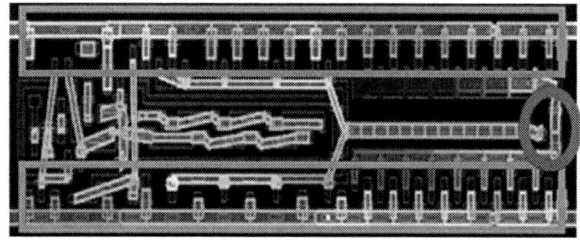

Figure 5: Color map of VBEM calculated EM stress for a standard cell M1 layer. The red (blue) regions indicate high (low) EM stress. PMOS and NMOS regions are shown in green and blue boxes respectively.

We also applied the VBEM flow on a larger analog block to test the efficiency of the implementation. This is an analogy IP block in an advanced technology node as in the previous example. The calculated EM stress distribution is as shown in Fig. 6. The CPU cost for adding the EM stress during the spice simulation for this test case was on the order of several minutes, which was less than 1% overall increase. On the other hand, the same test case required over 9 hours additionally for EM analysis using a commercial EDA EM tool.

V. CONCLUSIONS

In this paper, we showed experimental confirmation of the VBEM method using a novel test structure and presented design examples that attest to its efficiency and accuracy. Our results show that we can use the VBEM method to determine EM stress for product life time requirements.

The closed form expression for node stress in terms of node voltages allows us to predict MTTF for each interconnect segment. We can establish a clear EM risk assessment and pass/fail criteria based experimental MTTF data as shown in Fig. 4. The data shows that below a certain stress value the MTTF is effectively infinite, i.e., the interconnect is immortal. Hence, we can use this stress value to determine the critical stress limit used in (15). With the use of this critical stress limit, we can filter the interconnect segments in the design as low EM risk factors.

The interconnect segments that are over the critical stress value will eventually fail if given enough time to operate. The exact MTTF for these interconnects will depend on the geometry and other material properties. We are currently developing a time-dependent formulation of VBEM and a fast numerical solution method to analyze interconnect segments in this regime [9].

The empirical relation shown in Fig. 4 also allows chip level statistical EM budgeting. The MTTF distribution can be characterized over a large sample size and then can be used to calculate the chip level EM risk value.

REFERENCES

[1] J. R. Black, "Mass transport of Al by momentum exchange with conduction electrons," *Proc. IEEE Int. Rel. Phys. Symp., 1967, pp. 148-159.*

[2] I. A. Blech, "Electromigration in thin aluminum films on titanium nitride," *J. Appl. Phys., vol. 47, no. 4, pp. 1203–1208, 1976.*

[3] M. Gall, et. al., "Chip-Level Electromigration Reliability for Cu Interconnects," *AIP Conference Proceedings, Vol. 741, 2004*

[4] H. Haznedar, M. Gall, V. Zolotov, P. S. Ku, C. Oh, R. Panda, "Impact of Stress-Induced Backflow on Full-Chip Electromigration Risk

978-1-5386-5072-1/18 $31.00 © 2018 IEEE

Assessment," *IEEE Trans. On Comp. Aided Design, Vol. 25, no. 6, pp. 1038-1046*

[5] M. H. Lin, A. Oates, "An electromigration failure distribution model for short-length conductors incorporating passive sinks/reservoirs," *IEEE Trans. On Device and Materials Reliability, Vol. 13, no. 1, pp. 322-326, 2013*

[6] E. Demircan, M. Shroff, "Model based method for electromigration stress determination in interconnects," *in Proceedings of the International Reliability Physics Symposium (IRPS), 2014, pp.IT.5.1-IT.5.6.*

[7] Z. Sun, E. Demircan, M. Shroff, T. Kim, X. Huang, S. Tan "Voltage-Based Electromigration Immortality Check for General Multi-Branch

Interconnects," *in Proceedings of the International Conference on Computer Aided Design (ICCAD), 2016, 9A.4.*

[8] P. Gibson, M. Hogan, V. Sukharev, "Electromigration analysis of full-chip integrated circuits with hydrostatic stress," in Proceedings of the International Reliability Physics Symposium (IRPS), 2014, pp.IT.2.1-IT.2.7

[9] C. Cook, Z. Sun, E. Demircan, M. Shroff, S. Tan, "Fast Electromigration Stress Evolution Analysis for Interconnect Trees using Krylov Subspace Method," *IEEE Trans. On VLSI Systems*, in press.

Electrostatic Test Structures for Transmission Line Pulse and Human Body Model Testing at Wafer Level

Robert Ashton[1], Stephen Fairbanks[2], Adam Bergen[1], Evan Grund[3]

[1]Minotaur Labs, Mesa, Arizona, USA
Email: rashton@minotaurlabs.com, abergen@minotaurlabs.com
Telephone: +1-480-283-6393, +1 480 279 2036
[2]SRF Technologies, Mesa, Arizona, USA
Email: stephen@srftechnologies.com
[3]Grund Technical Solutions, Milpitas, California, USA
Email: evan@grundtech.com

Abstract— New two two-pin ESD testers are capable of doing both Transmission Line Pulse (TLP) and Human Body Model (HBM) testing at wafer level. These systems facilitate using test structures to link fundamental circuit element parameters measured with TLP and expected HBM results on final products.

Keywords—ESD, HBM, TLP

I. INTRODUCTION

Electrostatic discharge (ESD) is a reliability threat to all microelectronic devices. Even when electronic systems are manufactured in an environment designed to reduce the incidence of ESD, all integrated circuits (ICs) must be able to withstand some level of ESD stress. All ICs have specially designed circuits included to protect sensitive parts of the circuit from ESD threats they may experience in manufacture and assembly. Transmission Line Pulse (TLP) has become the standard tool for studying the properties of IC technologies at the current levels (usually up to 10 A) and time domain (several to 100s of ns) of ESD events [1].

Once the final IC has been designed and manufactured it must be tested for ESD robustness to ensure that the ESD design robustness goals have been met. Human Body Model (HBM) is the oldest and most widely performed ESD reliability test for integrated circuits, and is included in almost all integrated circuit qualification plans [2]. TLP testing is often done at wafer level, often on test structures, while most HBM testing is done using matrix based HBM test systems on packaged integrated circuits.

Recently new test systems, utilizing wafer prober technology, have been introduced which can do both TLP and HBM on both packages and wafers. These new testers, often called two-pin testers, have several useful characteristics. One is very low tester parasitics. Capacitive and inductive parasitics, which can be significant in matrix based HBM testers, have been found to create false failures on some integrated circuits. [3,4] A second feature is the ability to do both HBM and TLP at wafer level. Before the development of two-pin testers it was difficult to deliver in specification HBM waveforms to a wafer. The third advantage of the two-pin

tester is the ability to measure the current and voltage during an HBM event. The combination of these features in two pin testers improves the usefulness of ESD test structures, providing additional data and options for ESD designers not present when HBM testing was restricted to packaged devices. This paper will discuss the basics of TLP and HBM testing and present TLP and HBM measurements on test structures in a 180 nm high voltage CMOS technology, including DMOS and bipolar structures. The measurements and discussion will demonstrate how TLP and HBM measurement on test structures complement each other and provide more complete information for ESD protection design.

II. HBM AND TLP TESTING

The HBM test is usually represented as a 100 pF capacitor, precharged to a selected voltage, and then discharged through a 1500 ohm resistor to the device under test (DUT), see Fig. 1. The joint JEDEC/ESDA HBM test standard JS-001 [2] defines the stress as a current waveform with requirements through a short and a 500-ohm resistor. The required waveform properties are listed in Table 1.

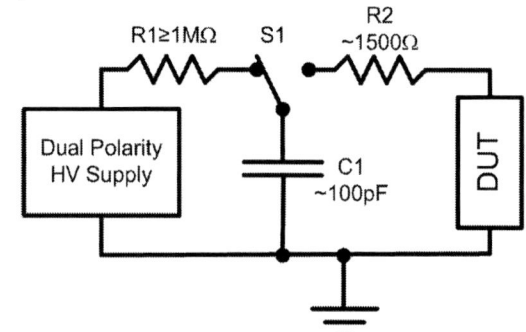

Fig. 1 Schematic of the HBM test

A simplified TLP tester schematic is shown in Fig. 2. A length of coaxial cable is charged to a predetermined voltage and discharged through the device under test. This creates a known amplitude square wave stress pulse whose length, typically 100 ns, as determined by the length of the coax

978-1-5386-5072-1/18 $31.00 © 2018 IEEE

cable. Voltage across the device and current through the device are measured during the TLP pulse. By stressing at progressively higher charging voltages, current versus voltage curves can be developed point-by-point. Fig. 3 shows a comparison of the current for a 2000 V HBM pulse with a 100 ns long, 1 A, TLP pulse. The level of stress to a device by induced heating is very similar between the two pulses. TLP has the advantage of creating accurate current versus voltage curves from which one can extract parameters such as breakdown voltage, trigger currents, and dynamic resistance. These parameters are useful for basic understanding as well as circuit modeling.

TABLE 1 Nominal HBM waveform parameters, †500 Ω only specified at 1000 V and 4000 V

Voltage (V)	Ipeak Short (A/kV)	Ipeak 500 Ω (A/kV)	Rise t Short (ns)	Rise t 500 Ω (ns)	Decay t Short (ns)	Ring Current Short
50 to 8000	0.667	0.46†	2 - 10	5 − 25†	150	< 15 % Peak I

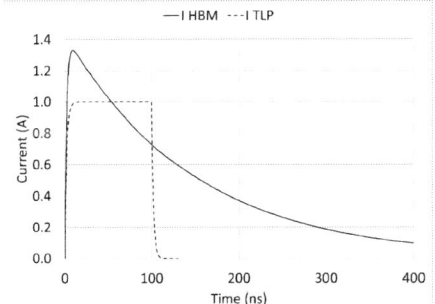

Fig. 2 Simplified TLP schematic (elements to remove unwanted reflections are not shown)

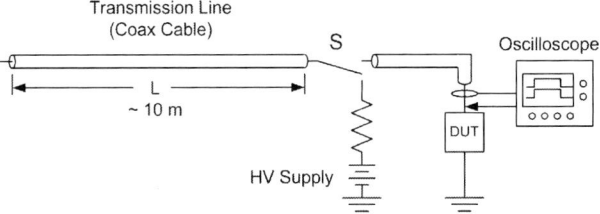

Fig. 3 Comparison of a 2000 VHBM pulse with a 1 A 100 ns long TLP pulse

III. TEST STRUCTURES FOR ESD TESTING

Test structures for ESD can be individual circuit elements, sub circuits containing the elements of an IO buffer, full IO buffers or IO buffers arranged in a full pad ring, but without core circuit elements. The test structures in this paper concentrate on individual circuit elements to show how fundamental ESD parameters, measured with TLP, relate to performance under HBM stress. Examples will be seen where the correlation is good, as well as examples where the difference between TLP and HBM stress become evident and analysis of the voltage and current during HBM events can give valuable information.

Data will be presented on 5 circuit elements and a full protection design, all in a 180 nm high voltage, 25 V, bipolar, CMOS, LDMOS (BCDMOS) technology. The 5 circuit elements consist of a high voltage N-Well diode, a 25 V nLDMOS transistor, 2 SCR designs and a pnp structure.

IV. TEST SYSTEM

All measurements were made on wafer with a Pure Pulse system from Grund Technical Solutions which can perform two-pin testing of wafers and packages with HBM, Machine Model (MM) and TLP stress. The wafer probers are mounted on computer driven micro-manipulators on an automated probe station.

V. RESULTS

A. N-Well Diode:

Diodes are frequently used as part of ESD protection strategies, frequently in forward bias where they are used to steer current away from sensitive elements and toward protection structures. Fig. 4 shows TLP current versus voltage and leakage data for the 40 um wide p-NWELL diode in forward bias. The first observation is that the diode survives 100 ns TLP pulses up to 2 A before leakage develops, revealing junction damage. Fig. 4 also shows a linear fit to the IV curve between 0.1 A and 1 A, yielding a dynamic resistance of 2.2 ohms and a voltage intercept of 1.35 V. These values can be used to predict voltage drops across the diode during an ESD event with currents up to 1 A. For higher currents a more detailed model could be used based on these measurements.

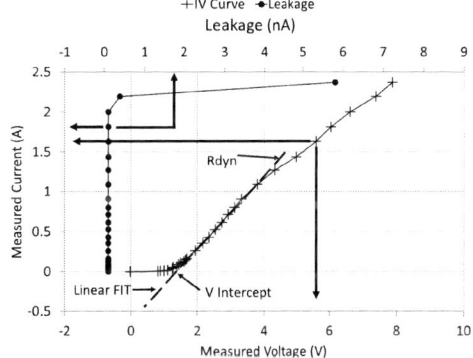

Fig. 4 TLP measurements of a p-NWell diode in forward bias. + marks are TLP IV points. ● marks are leakage measurements made after each TLP pulse. The arrows in this figure, and in the remaining TLP IV measurements, indicate the axes used for plotting each set of points. The y value for the leakage is the TLP pulse current for the pulse proceeding the leakage measurement.

The p-NWell diode was also tested using an HBM stress, without damage, to the system maximum of 4 kV. For many structures it has been found that for each 1 A of 100 ns TLP stress which the structure can survive correlates to 2000 V of HBM robustness. [5] The TLP and HBM measurements on the p-NWell diode are consistent with this correlation.

B. nLDMOS

N channel lateral diffusion transistors serve as drive transistors in high voltage technologies and must be able to survive ESD stress on their own, or be protected by another

2018 INTERNATIONAL CONFERENCE ON MICROELECTRONIC TEST STRUCTURES, MARCH 19-22, 2018, AUSTIN, TEXAS, USA

element. The grounded gate nLDMOS transistor test structure was measured with both TLP and HBM. The TLP IV curve and leakage measurements are shown in Fig. 5. The TLP IV data shows that the transistor breaks down above 48 V, well above the 25 V working voltage of the technology. On the first TLP pulse above breakdown the device jumps from 20 mA to over 1 A and has high leakage after the 1 A pulse, indicating device damage.

The jump in TLP pulse current from 20 mA to over 1 A is not, however, purely a feature of the nLDMOS device, whether damaged or not. The jump in current is an interaction of the TLP system's 50-ohm source impedance and properties of the nLDMOS device. This is complicated by the fact that nMOS devices often have "snapback" characteristics when the parasitic npn is activated and the voltage can drop to lower values without damage. There may be a section of the nLDMOS's IV curve which can carry current without damage, but is invisible because of the TLP' 50-ohm load line. A potential, but invisible, section of IV curve below the load line in shown in Fig. 5.

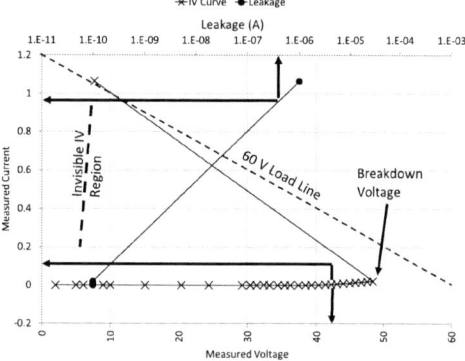

Fig. 5 TLP IV and leakage measurements on the nLDMOS transistor test structure as well as the load line for the final TLP pulse

Each TLP pulse has its own load line. The load lines connect the voltage the TLP system would deliver across an open (zero current), and the current the TLP pulse would deliver through a short (zero voltage). All TLP voltage and current measurements for each pulse must lie on the pulse's load line. The slope of all load lines on the IV is always 1/(50 ohms), but each pulse has a different load line, shifted from the others. The final TLP pulse in Fig. 5 was at 60 V and this load line is shown in the figure. The measured voltage and current for that pulse lies on the load line, as it must. Unfortunately, the TLP current in excess of 1 A damaged the device.

Insight on the potential snapback region can be gained from HBM measurements, since the load line for HBM is determined by the series 1500 ohms resistance, and therefore has a lower slope. Due to the lower slope HBM measurements may be able to "see" the snapback region invisible to TLP. (Note that the load line in HBM is continually changing during each pulse because the HBM pulse is continually changing.)

The HBM data on the nLDMOS showed minor increase in leakage to the transistor at 250 V and 500 V and high leakage

after a stress of 750 V. This is well below the usual design targets of 1 or 2 kV HBM. It is therefore necessary that protection structures be added to the IC design to protect the nLDMOS drivers. The TLP data, however, gives the valuable information that the nLDMOS has a breakdown voltage at or above 48 V. The necessary protection device needs to have a trigger voltage below 48 V to protect the nLDMOS transistor, but above the 25 V working voltage for the technology to prevent unintended operation.

Because the two-pin tester measures both voltage and current during each HBM pulse we can look for the presence of a snapback region. Fig. 6 and Fig. 7 show the current and voltage as a function of time for the three HBM pulses. The current looks superficially like the expected exponential decay but drops to zero late in the pulse, rather than exponentially approaching zero as current through a short or resistor would. After an initial voltage spike the voltage drops to below 5 volts before increasing to a higher voltage at the same time as the current drops to zero. This is the snapback region which could not be seen in the TLP data. The increase in voltage late in the pulse is when the current gets too low to sustain the snapback.

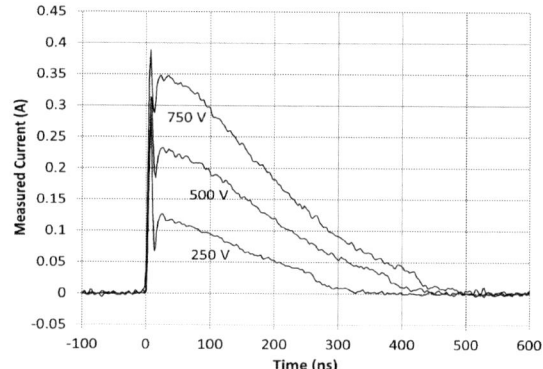

Fig. 6 Current versus time for the three HBM pulse on the nLDMOS

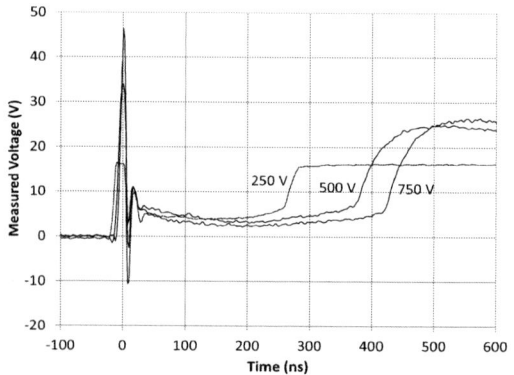

Fig. 7 Voltage versus time for the three HBM pulses on the nLDMOS

It is also possible to construct an IV curve from the HBM data. Fig. 8 shows an IV curve derived from the 750 V HBM pulse. For each 2 ns time interval from 32 ns to 600 ns, a current versus voltage point is plotted, forming an IV point-by-point. This creates an IV curve as the current decays. This

978-1-5386-5072-1/18 $31.00 © 2018 IEEE

2018 INTERNATIONAL CONFERENCE ON MICROELECTRONIC TEST STRUCTURES, MARCH 19-22, 2018, AUSTIN, TEXAS, USA

clearly shows the snapback region and how the low voltage state is lost when the current falls below 40 mA.

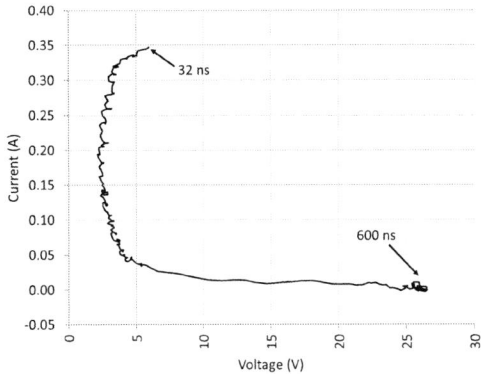

Fig. 8 IV curve of the nLDMOS device derived by plotting current versus voltage time point by time point starting at 32. ns into the pulse and ending at 600 ns

C. SCRs

Silicon Controlled Rectifiers, which consist of an npnp structure, can be very effective ESD protection devices. SCRs can have a high breakdown voltage, but when triggered they snapback to low voltages. With the low snapback voltage SCRs can carry considerable current, while dissipating relatively low power. This makes them very efficient ESD protection structures.

Fig. 9 shows TLP data for SCR Design 1. The device has an initial breakdown of 43 V, deep snapback to below 5 V and carries up to 4 A of current without damage. This predicts 8 kV of HBM robustness, which was is consistent with its passing the maximum 4 kV level of the HBM stress. Protection devices with deep snapback, however, need to be used with caution. There are two issues. Will the device present a latch-up risk during normal operation, and will the protection device turn off prematurely as the HBM current decreases?

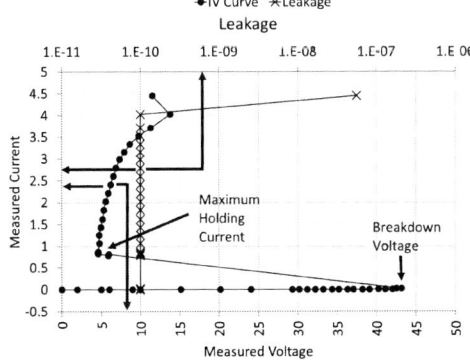

Fig. 9 TLP measurements of SCR Design 1. The device shows deep snapback and a failure at 4 A.

If the holding current for the snapback state is less than what the system power supply can supply there is danger of latch-up or oscillation. The data in Fig. 9 suggests a holding current of about 0.8 A. As discussed in the section on the nLDMOS, the properties of the TLP system can mask part of

the snapback region due to the 50-ohm load line. The TLP measurement of 0.8 A for the holding current is therefore a maximum value. The true holding current could be much less. The source impedance of the HBM source is 1500 ohms and can give better insight into the true holding current of the snapback state.

Fig. 10 and Fig. 11 for SCR Design 1 show current and voltage during HBM stress.

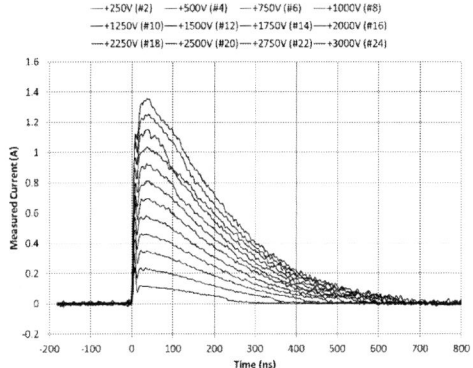

Fig. 10 Current versus time measurements on select HBM pulses for SCR Design 1

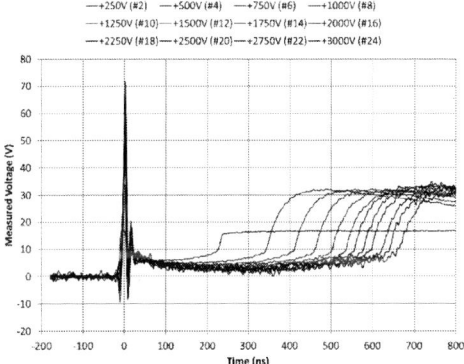

Fig. 11 Voltage versus time measurements on select HBM pulses of Field Oxide Device

Fig. 10 shows the expected HBM current waveforms, except that at some point the current becomes zero, rather than asymptotically approaching zero. Fig. 11 for voltage shows an interesting story. After an initial voltage spike, the SCR turns on, clamped the voltage to below 5 V, as predicted by the TLP curve, Fig. 9. Later in the pulse the voltage rises again, when there is insufficient current to maintain the SCR action, but the HBM 100 pF capacitor still has 30 V remaining from the initial charge. This is consistent with the concern raised above, that at some point the SCR can no longer sustain the low voltage state. Any device being protected by this SCR needs to be able to either survive a relatively long exposure to about 30 V or be able to handle the current from the charge remaining on the 100 pF HBM capacitor. A question remains, what is the holding current for SCR Design 1? Further insight can be found by looking in greater detail at the HBM current waveforms, as shown in Fig. 12. This is the same data as shown in Fig. 10, but with an expanded current axis. At

978-1-5386-5072-1/18 $31.00 © 2018 IEEE   51

roughly 38 mA all of the HBM current pulses break from their exponential decay and drop more rapidly to zero current. This drop in current coincides with the increase in voltage seen in the voltage versus time curves in Fig. 11. This suggests that the holding current for the SCR is 38 mA or less.

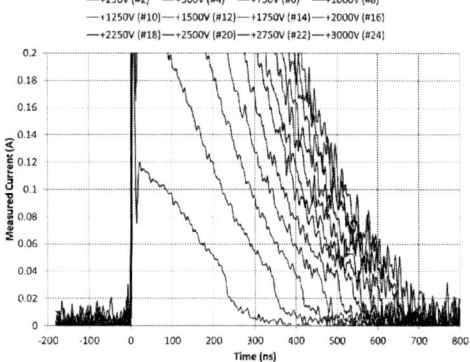

Fig. 12 HBM current versus time for SCR Design 1 with an expanded current scale

TLP results for SCR Design 2 are shown in Fig. 13. This SCR did not work as well as SCR Design 1. The breakdown voltage is about 48 V, followed by immediate damage to the device. The HBM data showed passing results at 250 V, minor damage at 500 V and 750 V, more damage at 1000 V and extensive damage at 1250 V. The use of a test structure to test this design saved a great deal of time and silicon area by not committing to a full protection strategy using this layout.

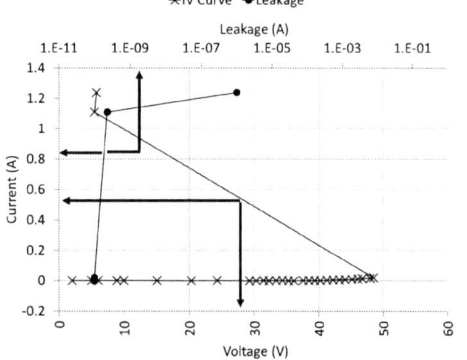

Fig. 13 TLP measurements of SCR Design 2

D. pnp

The final single element test structure is the pnp with the n base tied to one of the p diffusions. The TLP results are shown in Fig. 14. The TLP IV curve looks very similar to a reverse bias diode with a 31 V breakdown. The pnp structure, however, gives a lower dynamic resistance than would be seen for a similar sized diode. A linear fit to the pnp data was performed for the data between 10 mA and 100 mA, yielding a voltage intercept of 31.2 V and a dynamic resistance of 46.3 ohms. The 31.2 V breakdown is a good choice for protection of the nLDMOS device discussed earlier. 31.2 V is well above the expected 25 V working voltage for the transistor and well below the transistor's 48 V breakdown. The onset of damage at about 0.1 A for the pnp is not sufficient to provide adequate protection. The benefit of the pnp structure is that, since it has

no snapback, multiple copies can be placed in parallel to improve the overall current carrying capability as well as reduced dynamic resistance. Using parallel devices, which have deep snapback such as SCRs, is often unsuccessful because once one finger of a multi finger device goes into snapback other fingers are prevented from going into snapback and the single finger must carry all of the current.

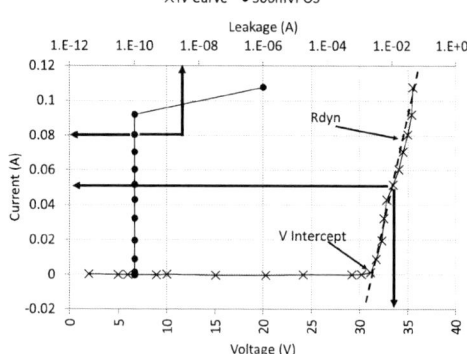

Fig. 14 TLP measurement of pnp device. The dashed line is a linear fit to the data over the range of 10 to 100 mA

E. Full Protection Design:

TLP data for a full protection design is shown in Fig. 15. The protection design consists of 26 parallel replications of the pnp device, discussed in section III.D, in series with 5 parallel replications of the p-NWELL diode, discussed in section III.A. 26 parallel pnp devices would be expected to carry up to 2.39 A of 100 ns long TLP current based on a single pnp being able to carry 92 mA. Similarly, 5 p-NWELL diodes in parallel should carry 10 A of 100 ns TLP current based on the 2 A capability of a single diode. The pnp is therefore the weaker element of the chain and should give a better prediction of the current carrying capability of the full protection device. The data in Fig. 15 shows the full protection device carrying 2.12 A of 100 ns TLP current without damage. This is slightly less than the predicted 2.39 A of current, but it is likely that individual elements in a closely packed array will not be able to carry as much current as an isolated device, due to lower ability to dissipate heat.

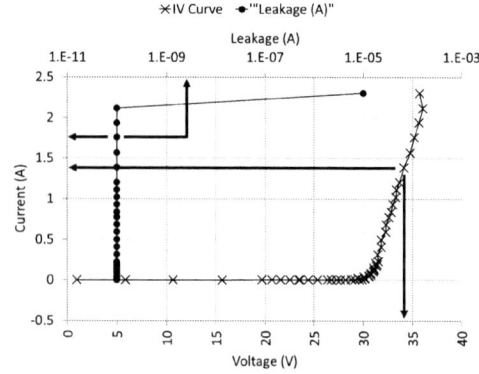

Fig. 15 TLP on full ESD protection structure using pnp devices

A linear fit of the data between 0.1 A and 2 A yields a voltage intercept of 30.8 and a dynamic resistance of 2.46 ohms. The HBM data showed passing results at 3800 V and failure at 4000 V. This is consistent with the earlier rule of thumb of 2 kV of HBM robustness for 1 A of survival for a 100 ns TLP pulse.

An advantage of the two-pin HBM system is that it is sometimes possible to pinpoint the exact time of failure during HBM testing by observing the measured voltage versus time curve. This is illustrated in Fig. 16 which shows the voltage captures for 3800 V, which did not cause damage, and for 4000 V, which resulted in damage. After an initial voltage spike, the voltage is clamped at about 35 V. For the 3800 V pulse the voltage remains clamped for the full duration of the measured waveform. At 4000 V the time of damage can be seen at 226 ns, when the voltage across the protection device drops from 35 V to about 3.5 V.

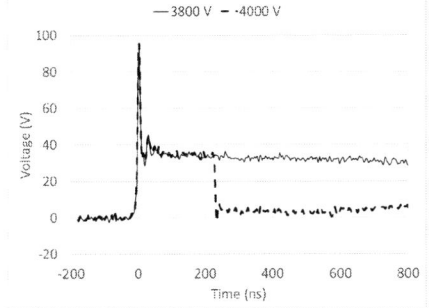

Fig. 16 Voltage capture on the 3800 V and 4000 V HBM pulses on the full protection design using the pnp structure

VI. CONCLUSIONS

This paper has presented TLP and HBM measurements on ESD test structures using a two-pin tester. The two-pin tester's unique ability to perform both measurements at wafer level, and its ability to measure voltage and current during HBM testing improves the usefulness of ESD test structures and can significantly shorten the design cycle.

TLP is usually performed at wafer level and provides high quality IV curves in which each data point is measured in the time domain of an ESD event. TLP IV curves yield important design parameters such as breakdown voltages and dynamic resistance. TLP data is also used to create model files, valid for the high currents and shorts times of ESD events.

As shown, TLP usually provides a good prediction of HBM pass/fail levels. 50-ohm TLP has limitation however, especially when measuring devices with snapback. Measuring voltage and current during each HBM pulse, supplements the TLP data and provides insight beyond the pass/fail data available from most HBM testers. HBM waveform measurements provide information on device performance during an exponential current decay, not available from TLP with its abrupt decay in current. HBM, with its 1500-ohm source impedance, can explore snapback behavior at lower currents than 50-ohm TLP. Additionally, two-pin testing provides data without the parasitic effects present in matrix based HBM test systems, and can be performed without expensive and time-consuming packaging and test board development.

REFERENCES

[1] T. Maloney, N. Khurana, "Transmission line pulsing techniques for circuit modeling of ESD phenomena", proc. 8th EOS/ESD Symposium, p 49-54, 1985

[2] ANSI/ESDA/JEDEC JS-001-2017, "Electrostatic discharge sensitivity testing human body model (HBM)"

[3] M. Chain et al., HBM tester parasitic effects on high pin count devices with multiple power and ground pins, EOS/ESD Symposium, 2006, pp 146-151.

[4] H. Kunz, R. Steinhoff, C. Duvvury, G. Boselli, L. Ting, "The effect of high pin-count tester parasitics on transiently triggered ESD clamps", EOS/ESD Symposium, 2004, pp 187-198

[5] P. Salome, C. Leroux, P. Crevel, J. Chante, "Investigations on the thermal behavior of interconnects under ESD transients using a simplified thermal RC network", EOS/ESD Symposium, 1998, pp 187-198.

978-1-5386-5072-1/18 $31.00 © 2018 IEEE

Session 4

Materials Characterization

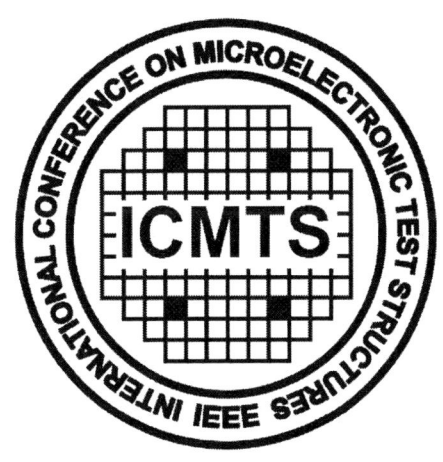

March 20, 2018

15:40-17:20

Co-Chairs:

Tatsuya OHGURO, Toshiba, Japan

Chadwin YOUNG, University of Texas at Dallas, USA

978-1-5386-5072-1/18 $31.00 © 2018 IEEE

2018 INTERNATIONAL CONFERENCE ON MICROELECTRONIC TEST STRUCTURES, MARCH 19-22, 2018, AUSTIN, TEXAS, USA

Reliability analysis of the metal–graphene contact resistance extracted by the Transfer Length Method

Stefano Venica*, Francesco Driussi*, Amit Gahoi[†], Satender Kataria[†],
Pierpaolo Palestri*, Max C. Lemme[†‡] and Luca Selmi*[§]

Abstract—The transfer Length Method is a well–estab experimental technique to characterize the contact resista semiconductor devices. However, its dependability is ques for metal–graphene contacts. We investigate in–depth the st cal error of the extracted contact resistance values and we strategies to limit such error and to obtain reliable result method has been successfully applied to samples with diff contact metals.

I. INTRODUCTION

Two–dimensional (2D) materials are regarded as pror candidates to boost the performance of RF analog elect [1]. As an example, graphene–based devices, such a Graphene Field Effect Transistors (GFETs), nowadays very high cut–off frequency (f_T), comparable with those of state–of–the–art RF transistors [2]. However, the maximum oscillation frequency (f_{max}) of GFETs is largely limited by the Metal–Graphene (M–G) contact resistance (R_C) at the source/drain of devices [1], [3]. Furthermore, also the performance of other device concepts based on the 2D technology is hampered by the detrimental effect of the metal–2D material contact resistance [4], [5].

Despite the improvements to the quality of the M–G contacts achieved recently [6], the physics underlying the conduction through the M–G contact is not completely understood. This comprehension is mandatory to devise strategies to produce significant breakthrough in the engineering of the M–G junction. Therefore, reliable investigation procedures and experimental techniques are required to assess the nature of the M–G junction [7].

Transfer Length Method (TLM) is an established technique to measure R_C in semiconductors [7]. However, the reliability of the TLM applied to graphene based devices is often questioned, since R_C data for M–G contacts reported in the literature are characterized by large error bars [8] and, in several instances, by negative R_C values [9]–[13]. In this paper, we report an in–depth analysis of the statistical errors

* Universitá degli Studi di Udine, Dipartimento Politecnico di Ingegneria e Architettura (DPIA), via delle Scienze 206, 33100 Udine, Italy (email: venica.stefano@spes.uniud.it).
† Chair of Electronic Devices, Faculty of Electrical Engineering and Information Technology, RWTH Aachen University, Otto-Blumenthal-Str. 25, 52074 Aachen, Germany.
‡ AMO GmbH, Advanced Microelectronic Center Aachen, Otto–Blumenthal–Str. 25, 52074 Aachen, Germany.
§ now with Universitá degli Studi di Modena e Reggio Emilia, Dipartimento di Ingegneria "Enzo Ferrari" (DIEF), via Pietro Vivarelli 10, 41125 Modena, Italy.

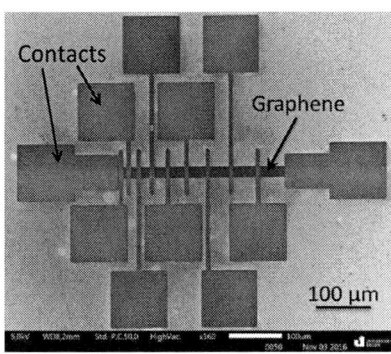

Fig. 1. SEM image of the TLM structure. The space between contacts (L_{CH}) varies from 5 μm to 50 μm (from left to right) with a 5 μm increase between adjacent channels. For this structure the graphene width (W) is 20 μm. The graphene ribbon and metal pads are indicated by arrows.

of extracted R_C values and, by accounting for the GFET specificity, we devise strategies to limit such errors and to achieve reliable M–G contact resistance values.

II. DEVICES AND EXPERIMENTS

TLM test structures based on CVD graphene and consisting of a series of back–gated GFETs with channel length (L_{CH}) ranging from 5 to 50 μm have been fabricated (Fig. 1). Boron–doped (p–type) silicon substrates were used for TLM fabrication. Wafers were thermally oxidized to achieve an 85 nm SiO_2 back oxide. After the CVD graphene transfer, nickel (Ni), copper (Cu) or gold (Au) were used to make the electrical contacts. More details concerning the device fabrication can be found in [14]. DC characterization was carried out at Lakeshore Cryotronics using a Keithley SCS4200 parameter analyzer in an ultra–high vacuum chamber at 300 K.

III. RESULTS AND DISCUSSION

Figure 2(a) reports typical source–drain currents (I_{DS}) measured on GFETs with Ni contacts and different L_{CH}, that vary with the applied back–gate bias (V_{BG}). The minimum conduction point, i.e. Dirac Point (DP), is clearly visible. For this sample, the DP of GFETs is located at negative back–gate voltages ($V_{DP} < 0$ V), with just a slight dependence on L_{CH}. This can be due to some unintentional graphene doping/interfacial impurities and/or to residual air/humidity that shift the DP position with respect to $V_{BG} = 0$ V [15]. Transfer characteristics in Fig. 2(a) are used to calculate the total resistance R_T^{exp} of the devices, which is contributed by

978-1-5386-5072-1/18 $31.00 © 2018 IEEE

2018 INTERNATIONAL CONFERENCE ON MICROELECTRONIC TEST STRUCTURES, MARCH 19-22, 2018, AUSTIN, TEXAS, USA

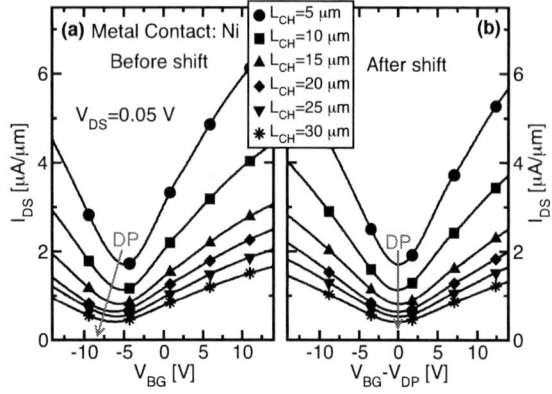

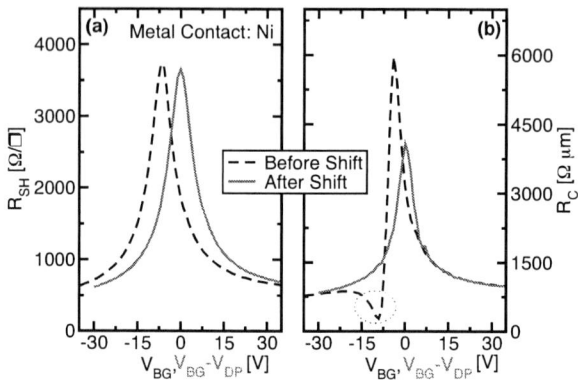

Fig. 2. (a) Typical current measured through the GFETs of the TLM structure with W=20 μm and Ni contacts. The experiments are limited to the shortest six devices ($L_{CH} \leq 30\ \mu$m, Fig. 1). DP at $V_{DP}<0$ is observed in the whole TLM, with a value slightly depending on L_{CH}. (b) Transfer characteristics shifted ~~with respect to the DP position~~

Fig. 4. Sheet resistance (a) and contact resistance (b) as a function of V_{BG} (dashed lines) extracted from the linear fittings in Fig. 3. Near the DP, R_C shows an evident drop (highlighted area). When considering the shifted transfer charateristics in Fig. 2(b), the trend of R_C strongly changes and the dip disappears.

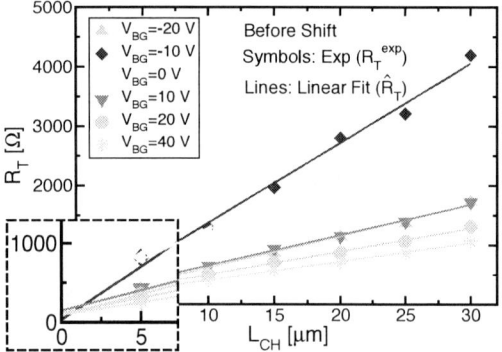

Fig. 3. Measured total resistance R_T^{exp} (symbols) versus L_{CH} at different V_{BG} obtained from the data in Fig. 2(a). By linear regression (lines), we extract R_{SH} and R_C at different V_{BG}. The zoom near the origin of the axis highlights the intercept value ($2R_C$).

the channel resistance, R_C and the resistance of the metal contact (generally neglected) [7]. In particular, by assuming uniform CVD graphene sheet resistance R_{SH} in the whole TLM structure, R_T^{exp} should linearly depend on L_{CH}, as:

$$R_T^{exp} = \frac{V_{DS}}{I_{DS}} = \frac{R_{SH}}{W}L_{CH} + 2R_C. \quad (1)$$

This seems confirmed by the measured R_T^{exp} (symbols) versus L_{CH} data in Fig. 3. By linear regression (lines), we extract R_{SH} and R_C from the slope and the intercept with the y–axes of the linear fit, respectively (Eq. 1). Note that, by looking to the zoomed area near the origin of the axis, the $2R_C$ value appears to be non–monotonic with $|V_{BG}|$ and rapidly drops for $V_{BG}=-10$ V (blue line).

Figure 4 (dashed lines) reports the extracted R_{SH} (a) and R_C (b) versus V_{BG} as obtained from Fig. 3. R_{SH} shows the trend typical of graphene channels with large resistivity near the DP. Although a high correlation between R_{SH} and R_C has been demonstrated in [16], the extracted R_C in Fig. 4(b,

dashed line) shows a clear dip at $V_{BG} \simeq -10$ V (not present for R_{SH}), suggesting issues in the extraction technique.

To assess the reliability of the R_{SH} and R_C extraction through the TLM, we analyzed the statistical errors by using Eqs. 2–4 to calculate the regression coefficient (r^2) and the R_{SH} and R_C errors (ε) [17]:

$$r^2 = 1 - \frac{\sum_i^N (R_T^{exp} - \widehat{R}_{T_i})^2}{\sum_i^N (R_T^{exp} - \overline{R}_T)^2} \quad (2)$$

$$\varepsilon(R_{SH}) = \pm\ W\sqrt{\frac{\sigma^2}{S_{xx}}} \quad (3)$$

$$\varepsilon(R_C) = \pm\ \frac{1}{2}\sqrt{\sigma^2\left[\frac{1}{N} + \frac{\overline{L_{CH}}}{S_{xx}}\right]}, \quad (4)$$

where:

$$\sigma^2 = \frac{\sum_i^N (R_{T_i}^{exp} - \widehat{R}_{T_i})}{N-2} \quad (5)$$

$$S_{xx} = \sum_i^N (L_{CH_i} - \overline{L_{CH}})^2 \quad (6)$$

$$\widehat{R}_{T_i} = \frac{R_{SH}}{W}L_{CH_i} + 2R_C. \quad (7)$$

Figure 5 reports r^2 for the data in Fig. 4 (dashed lines); although close to the unity, r^2 reduces near the DP. Figure 6 (dashed lines) instead shows the relative errors for R_{SH} (a) and R_C (b); while the R_{SH} error is quite small, reassuring for the reliability of the extracted R_{SH}, the R_C error explodes exactly where the dip is visible in the R_C curve of Fig. 4(b) (near the DP).

We assumed that this huge error is due to the small V_{DP} dependence on L_{CH} visible in Fig. 2(a), thus we shifted the I_{DS}–V_{BG} curves of each device by the corresponding V_{DP} value; in other words, we plot I_{DS} versus ($V_{BG} - V_{DP}$) [Fig. 2(b)] to compensate the slightly different V_{DP} in the measured GFETs, allowing the comparison of R_T in much

978-1-5386-5072-1/18 $31.00 © 2018 IEEE   58

2018 INTERNATIONAL CONFERENCE ON MICROELECTRONIC TEST STRUCTURES, MARCH 19-22, 2018, AUSTIN, TEXAS, USA

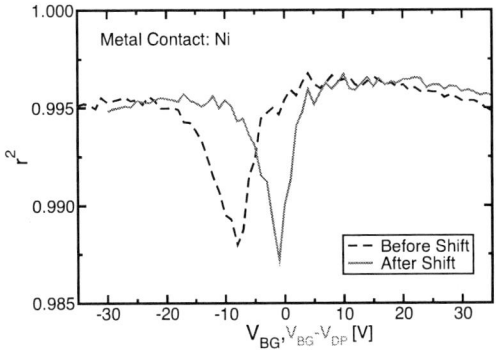

Fig. 5. Linear regression coefficient r^2 versus V_{BG} of R_T-L_{CH} data calculated considering the transfer characteristics in Fig. 2(a) (dashed line) or in Fig. 2(b) (solid line). In both cases the quality of the fit decreases in pro:

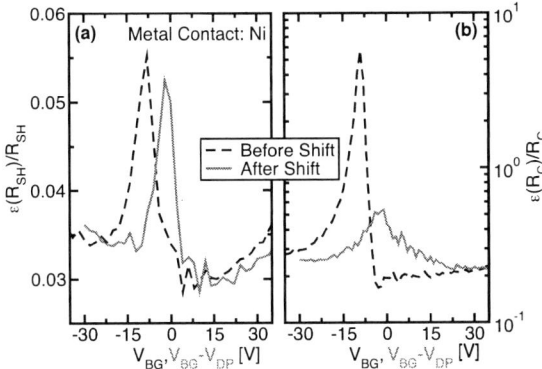

Fig. 6. Relative error on R_{SH} (a) and on R_C (b) as a function of V_{BG} calculated by using Eqs.3–7 and the data in Fig. 2(a) for the R_T calculation (dashed lines). Although the error is always limited for R_{SH}, it becomes huge near the DP for R_C. When considering the transfer characteristics versus $(V_{BG} - V_{DP})$ [Fig. 2(b)], the error is much less near the DP (solid lines).

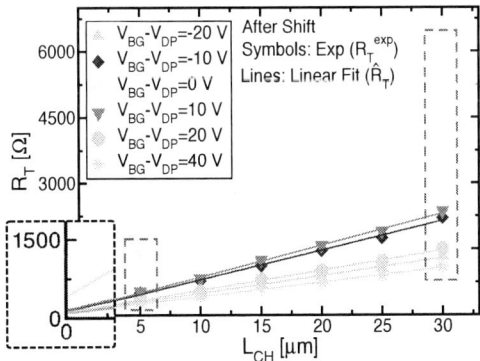

Fig. 7. Experimental total resistance R_T^{exp} (symbols) as a function of L_{CH} at different V_{BG} biases calculated with the shifted transfer characteristics reported in Fig. 2(b). The shortest and longest devices (5 and 30 μm) are potentially responsible for the dip in the regression coefficient in proximity of the Dirac Point seen in Fig. 5.

more similar charge carrier concentration in the graphene channels of the different GFETs. By using now these shifted I_{DS} curves, we repeated the R_{SH} and R_C extraction.

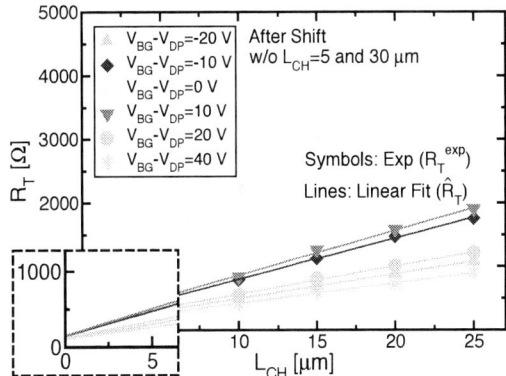

Fig. 8. Experimental total resistance R_T^{exp} (symbols) as a function of L_{CH} at different V_{BG} bias for the shifted transfer characteristics reported in Fig. 2(b) excluding the outlier devices. The linear fitting of data is much better than in Fig

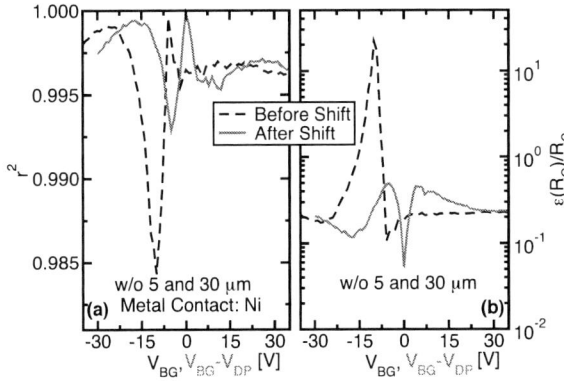

Fig. 9. Linear regression coefficient (a) and R_C relative error (b) as a function of V_{BG} excluding the outlier devices. The comparison between results considering the original (dashed lines) and the shifted transfer characteristics (solid lines) is also reported. An improvement in r^2 and a drastic reduction of the R_C relative error are achieved only for the shifted characteristics.

Figure 7 reports the R_T^{exp} values obtained after the shift. The curves do not show anomalous behaviors; in particular, the intercept seems now monotonic in $|V_{BG}|$, differently from Fig. 3. R_{SH} and R_C extracted from the shifted curves are compared in Fig. 4 (solid lines) with the values extracted without the $I_{DS}-V_{BG}$ shift (dashed lines). While R_{SH} is pretty similar, R_C is very different and the dip near the DP disappears. Moreover, although the r^2 value is not very different in the two cases (Fig. 5) and the R_{SH} error remains small [Fig. 6(a), solid line], the new R_C extraction is characterized by a much smaller relative error [Fig. 6(b), solid line], making the obtained R_C values much more reliable.

Furthermore, by carefully observing the new R_T^{exp} in Fig. 7, we note that the shortest and the longest measured GFETs seem to slightly deviate from the linear behavior of the other devices. Thus, it is possible to improve the extraction procedure by excluding these outliers; the linear fit of the remaining devices becomes better as shown in Fig. 8, with a significant improvement of both r^2 and of the R_C error (see Fig. 9, solid

978-1-5386-5072-1/18 $31.00 © 2018 IEEE   59

2018 INTERNATIONAL CONFERENCE ON MICROELECTRONIC TEST STRUCTURES, MARCH 19-22, 2018, AUSTIN, TEXAS, USA

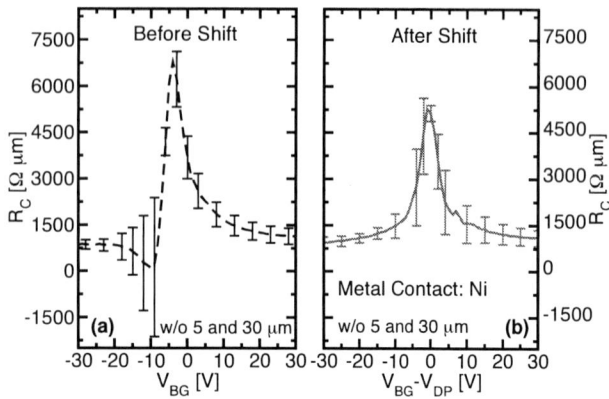

Fig. 10. Contact resistance as a function of V_{BG} with the corresponding error bars $[\pm\varepsilon(R_C)]$ obtained excluding the outliers. Data are extracted considering the original (a) and the shifted transfer characteristics (b). The latter can be

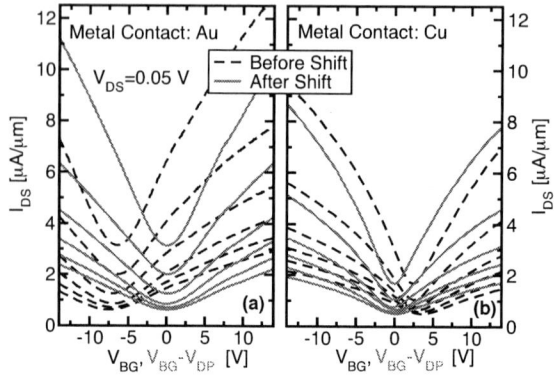

Fig. 12. Transfer characteristic of the GFETs belonging to the TLM structure for a samples with Au (a) and Cu (b) contacts (W is 20 and 30 μm, respectively). The dashed lines are the original I_{DS}–V_{BG} curves, while the solid curves are the transfer characteristics shifted with rispect to the DP po:

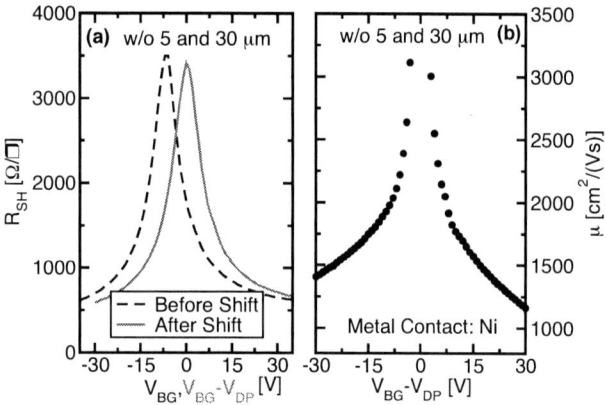

Fig. 11. Sheet resistance (a) and corresponding majority carrier mobility (b) versus V_{BG} extracted excluding the outlier devices. Mobility is calculated by means of the model reported in [18].

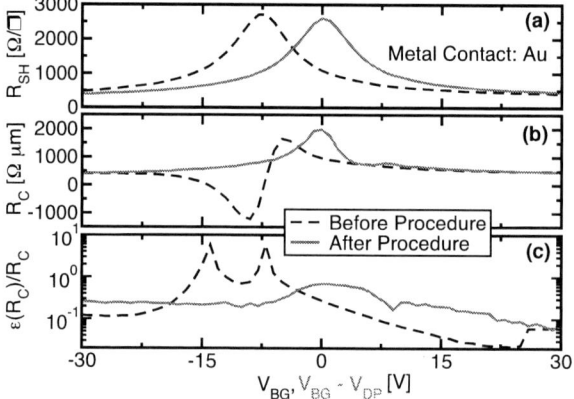

Fig. 13. Sheet resistance (a), contact resistance (b) and R_C relative error (c) for the samples with Au contact extracted from the raw data (dashed lines) and after the procedure described in this work (solid lines), namely the shift with respect to the DP and the exclusion of the outliers.

lines). We also verified that the outliers are not responsible of the huge error in the R_C (i.e. the dip near the DP) extracted from the non–shifted transfer characteristics; indeed, if we extract R_C by excluding the outliers from the original I_{DS}–V_{BG} curves, R_C shows again the dip [Fig. 10(a)] and the error is still huge near the DP (Fig. 9, dashed line). Hence, the DP variability is mainly responsible for the large R_C errors, making the shift of the I_{DS} characteristics mandatory to obtain reliable R_C.

With the developed procedure, we extract reliable R_C values [Fig. 10(b)] from the TLM structure of the investigated Ni–graphene contact. Furthermore, the R_{SH} obtained in Fig. 11(a) allows to estimate the carrier mobility in the graphene [Fig. 11(b)], calculated as $\mu=(\rho R_{SH})^{-1}$, where ρ is the charge density due to electrons/holes calculated with the model in [18]. Note also the very similar shape of R_{SH}–V_{BG} in Fig. 11(a) extracted before (dashed lines) and after (solid lines) applying the procedure of the present work, due to the limited relative error of R_{SH} already obtained using the

original transfer characteristics [Fig. 6(a)].

As a further proof of the validity of the procedure described above, we applied the latter also to data of TLM structures with Au and Cu contacts. In particular, Fig. 12 shows the transfer characteristics of GFETs belonging to TLM structures (again GFETs with L_{CH} ranging from 5 to 30 μm have been measured) with Au (a) and Cu (b) metal contacts. In both cases, the measured I_{DS}–V_{BG} (dashed lines) show a residual DP shift with respect to $V_{BG}=0$: V_{DP} is negative in the Au case (similarly to Ni contacts), while it is positive in the Cu case, reflecting different M–G interactions in the two structures and/or interfacial impurities at the M–G interface.

By calculating R_T^{exp} by means of Eq. 1 and extracting the corresponding R_{SH} and R_C from the raw data, the contact resistance in the Au case shows anomalous negative values in proximity of the DP [$V_{DP}\approx-8$ V, black dashed lines, Fig. 13(b)]. However, in this V_{BG} range, the error associated to R_C is huge and unacceptable [dashed lines, Fig. 13(c)].

978-1-5386-5072-1/18 $31.00 © 2018 IEEE

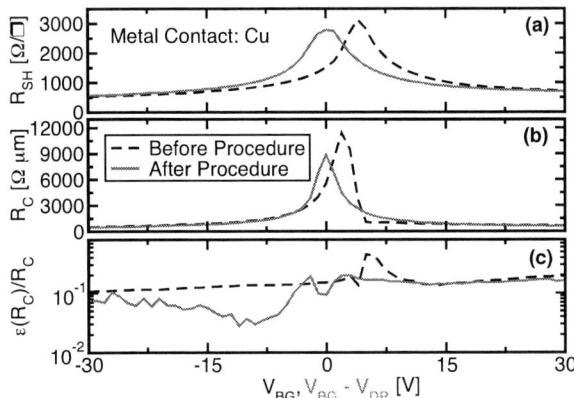

Fig. 14. Sheet resistance (a), contact resistance (b) and R_C relative error (c) for the samples with Cu contact before (dashed lines) and after the procedure described in this work (solid lines), namely the shift with respect to the DP and the exclusion of the outliers. Note the much higher contact resistance in corrispondence of the DP with respect to Ni and Au samples.

Therefore, we applied the procedure developed so far to improve the reliability of the extracted values. While the newly extracted R_{SH} is similar as before [Fig. 13(a)], the shift of the transfer characteristics and the exclusion of the outliers largely reduce the R_C relative error and R_C is positive in the whole range of V_{BG} [see Figs. 13(b) and (c), solid lines], making evident that the negative resistance values obtained from the raw data are just an artifact of the extraction procedure and not the result of peculiar physical mechanisms [10].

Also the R_{SH} and R_C extracted from the raw data for Cu contacts are affected by errors [Fig. 14(c)]. In Fig. 14(b) (dashed lines), R_C shows a largely asymmetric peak around the DP. By considering, instead, the shifted transfer characteristics with respect to the DP position and excluding from the extraction procedure the outlier devices, R_C recovers an almost symmetric behavior with respect to $V_{BG}-V_{DP}=0$.

As for the Ni sample, errors in the R_{SH} extraction are less critical as evident from Figs. 13(a) and 14(a), showing similar R_{SH} values before and after the $I_{DS}-V_{BG}$ shift. Nevertheless, after the application of the procedure to limit the errors, the R_{SH} obtained for the Ni, Au and Cu samples are quite close each others [compare Figs. 11(a), 13(a) and 14(a)], confirming the graphene uniformity among the samples and pointing out that the different R_C values extracted for the three contacts are directly correlated to the used metals and not to the graphene quality.

IV. CONCLUSION

We have shown how the extraction of the contact resistance between metal and graphene from TLM may lead to substantial errors. These latters are due to slight variations through the TLM array due to graphene transfer and fabrication procedure and/or to residual air/humidity that induce a V_{DP} shift along the TLM. This hampers the charge density uniformity among the GFETs and, thus, the correct extraction of R_{SH} and R_C through Eq. 1.

However, we also demonstrated that the combination of a shift of the $I_{DS}-V_{DS}$ characteristics (aimed to compensate the DP variability) and the exclusion of outliers from the linear regression of R_T data leads to an improvement of the regression coefficient and to a drastic reduction of the error related to the extracted values in the present study. This has been verified also by characterizing TLM structures with different metal contacts.

The combined procedure above represents a guideline to extract reliable M–G contact resistance values through TLM structures and to avoid artifacts in the R_C versus V_{BG} curve. Furthermore, this analysis is synergist to the assessment of models and techniques able to de–convolve the different contributions to the M–G contact resistance [7], [19], thus helping in the better understanding of the physics governing the conduction through the M–G junction.

ACKNOWLEDGMENT

The work was partially funded by the EU through the FP7-ICT STREP Project "GRADE" under the grant 317839 via the IU.NET consortium and through an ERC starting grant (InteGraDe, 307311), as well as the German Research Foundation is gratefully acknowledged.

REFERENCES

[1] Y. Wu, X. Zou, M. Sun, Z. Cao, X. Wang, S. Huo, J. Zhou, Y. Yang, X. Yu, Y. Kong, G. Yu, L. Liao and T. Chen, "200 GHz Maximum Oscillation Frequency in CVD Graphene Radio Frequency Transistors", *ACS Applied Materials & Interfaces*, vol. 8, no. 39, pp. 25645–25649, Sep. 2016, DOI: 10.1021/acsami.6b05791.

[2] F. Schwierz, J. Pezoldt and R. Granzner, "Two–Dimensional Materials and Their Prospects in Transistor Electronics", *Nanoscale*, vol. 7, pp. 8261–8283, Apr. 2015, DOI: 10.1039/C5NR01052G.

[3] F. Giubileo and A. Di Bartolomeo, "The Role of Contact Resistance in Graphene Field–Effect Devices" *Progress in Surface Science*, vol. 92, no. 3, pp. 143–175, Aug. 2017, DOI: 10.1016/j.progsurf.2017.05.002.

[4] S. Venica, F. Driussi, P. Palestri, D. Esseni, S. Vaziri and L. Selmi, "Simulation of DC and RF Performance of the Graphene Base Transistor", *IEEE Transactions on Electron Devices*, vol. 61, no. 7, pp. 2570–2576, Jun. 2014, DOI: 10.1109/TED.2014.2325613.

[5] Y.–W. Lan, C. M. Torres, X. Zhu, H. Qasem, J. R. Adleman, M. B. Lerner, S.–H. Tsai, Y. Shi, L.–J. Li, W.–K. Yeh and K. L. Wang, "Dual–Mode Operation of 2D Material–Base Hot Electron Transistors", *Scientific Reports*, vol. 6, p. 32503, Sep. 2016, DOI: 10.1038/srep32503.

[6] A. Meersha, H. B. Variar, K. Bhardwaj, A. Mishra, S. Raghavan, N. Bhat and M. Shrivastava, "Record Low Metal – (CVD) Graphene Contact Resistance Using Atomic Orbital Overlap Engineering", *2016 IEEE International Electron Devices Meeting (IEDM)*, pp. 5.3.1–5.3.4, 2016, DOI: 10.1109/IEDM.2016.7838352.

[7] S. Venica, F. Driussi, A. Gahoi, V. Passi, P. Palestri, M. C. Lemme and L. Selmi, "Detailed Characterization and Critical Discussion of Series Resistance in Graphene–Metal Contacts", *2017 International Conference of Microelectronic Test Structures (ICMTS)*, pp. 27–31, 2017, DOI: 10.1109/ICMTS.2017.7954259.

[8] F. Xia, V. Perebeinos, Y.–M. Lin, Y. Wu and P. Avouris, "The Origins and Limits of Metal-Graphene Junction Resistance", *Nature Nanotechnology*, vol. 6, pp. 179–184, Feb. 2011, DOI: 10.1038/nnano.2011.6.

[9] P. Blake, R. Yang, S. V. Morozov, F. Schedin, L. A. Ponomarenko, A. A. Zhukov, R. R. Nair, I. V. Grigorieva, K. S. Novoselov and A. K. Geim, "Influence of Metal Contacts and Charge Inhomogeneity on Transport Properties of Graphene Near the Neutrality Point", *Solid State Communications*, vol. 149, no. 27, pp. 1068 1071, Jul. 2009, DOI: 10.1016/j.ssc.2009.02.039.

[10] R. Nouchi, T. Saito and K. Tanigaki, "Observation of Negative Contact Resistances in Graphene Field–Effect Transistors", *Journal of Applied Physics*, vol. 111, no. 8, p. 084314, Apr. 2012, DOI: 10.1063/1.4705367.

[11] T. Chari, R. Ribeiro–Palau, C. R. Dean and K. Shepard, "Resistivity of Rotated GraphiteGraphene Contacts", *Nano Letters*, vol. 16, pp. 4477–4482, Jul. 2016, DOI: 10.1021/acs.nanolett.6b01657.

[12] N. F. W. Thissen, and R. H. J. Vervuurt, A. J. M. Mackus, J. J. L. Mulders, J.–W. Weber, W. M. M. Kessels and A. A Bol, "Graphene Devices with Bottom–Up Contacts by Area–Selective Atomic Layer Deposition", *2D Materials*, vol. 4, no. 2, p. 025046, Feb. 2017, DOI:10.1088/2053-1583/aa636a.

[13] W. Wang, M. Muruganathan, J. Kulothungan and H. Mizuta, "Study of Dynamic Contacts for Graphene Nano–Electromechanical Switches", *Japanese Journal of Applied Physics*, vol. 56, no. 4S, p. 04CK05, Mar. 2017, DOI: 10.7567/JJAP.56.04CK05.

[14] A. Gahoi, S. Wagner, A. Bablich, S. Kataria, V. Passi and M. C. Lemme, "Contact Resistance Study of Various Metal Electrodes with CVD Graphene", *Solid–State Electronics*, vol. 125, pp. 234–239, Nov. 2016, DOI: 10.1016/j.sse.2016.07.008.

[15] A. D. Smith, K. Elgammal, F. Niklaus, A. Delin, A. C. Fischer, S. Vaziri, F. Forsberg, M. Rasander, H. Hugosson, L. Bergqvist, S. Schroder, S. Kataria, M. Östling, M. C. Lemme, "Resistive Graphene Humidity Sensors with Rapid and Direct Electrical Readout", *Nanoscale*, vol. 7, pp. 19099–19109, Oct. 2015, DOI: 10.1039/C5NR06038A.

[16] T. Cusati, G. Fiori, A. Gahoi, V. Passi, M. C. Lemme, A. Fortunelli and G. Iannaccone, "Electrical Properties of Graphene–Metal Contacts", *Scientific Reports*, vol. 7, p. 5109, Jul. 2017, DOI:10.1038/s41598-017-05069-7.

[17] D. C. Montgomery and G. C. Runger, "Applied Statistics and Probability for Engineers", 3rd Edition, *John Wiley & Sons, Inc.*, 2003.

[18] S. Venica, M. Zanato, F. Driussi, P. Palestri and L. Selmi, "Modeling Electrostatic Doping and Series Resistance in Graphene-FETs", *2016 International Conference on Simulation of Semiconductor Processes and Devices (SISPAD)*, pp. 357-360, 2016, DOI: 10.1109/SISPAD.2016.7605220.

[19] S. Venica, F. Driussi, G. Amit, P. Palestri, M. C. Lemme and L. Selmi, "On the adequacy of the Transmission Line Model to describe the graphene–metal contact resistance", *accepted for IEEE Transaction on Electron Devices*.

Test Structures for Seed Layer Optimisation of Electroplated Ferromagnetic Films

C.M. Mackenzie Dover, A.W.S. Ross, S. Smith, J.G. Terry, A.R. Mount and A.J. Walton

School of Engineering, University of Edinburgh, EH9 3FF, UK

Email: C.Dover@ed.ac.uk

Abstract— This paper presents a full wafer test structure, designed to quantify the effect of seed layer thickness and conductivity on the plating uniformity of patterned electroplated structures. The test structure enables the effect of IR drop on the electroplated film to be evaluated and provides information to help facilitate the optimisation of seed layer thickness.

Keywords— *Seed layer, test structures, electroplated ferromagnetic films, nickel, integrated magnetic technologies, optimisation, characterisation.*

I. INTRODUCTION

The electroplating method is perhaps the most widespread manufacturing technique in microelectronics/ microsystems for producing thick metallic films (tens of microns) e.g. Cu and NiFe. The properties of these films have been evaluated using test structures, for example in the characterisation of the magnetic properties of electroplated NiFe [1]. Generic test structures suitable for characterising elements of the electroplating process and the material film properties include Greek crosses (sheet resistance) [2,3], split cross bridges (line width) [4] and stress test structures [5-7]. Such structures have also been used to characterise the plating process on single chip devices [8]. However, to the authors' knowledge no test structures have been reported for characterising full wafer electroplating. This paper details a test structure that can help characterise the effect of seed layer thickness on subsequent electroplated layers.

II. BACKGROUND

The electroplating process typically involves the blanket deposition of a seed layer upon which photoresist is coated and patterned to define a mould for the following bottom-up plating process [9] as shown in Figure 1.

Once the electroplating has been completed, the photoresist is stripped, followed by the removal of the seed layer which has been used to electrically connect all the metal structures. This latter step is routinely

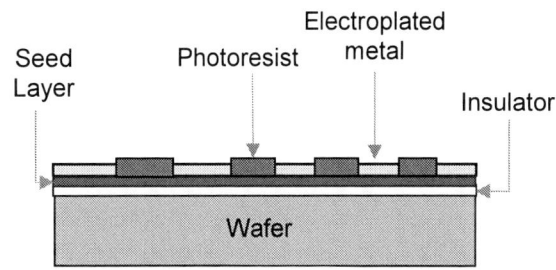

Figure 1: Bottom up plating using a photoresist mask (note thicknesses not to scale).

performed using a wet-etch dip. The ideal situation is to use a seed layer with a good etch selectivity with respect to the electroplated film. In this case the electroplated film is not affected by the seed layer removal and any undercut can be minimised by reducing the seed layer thickness. However, this option is not available for very thin electroplated structures, especially for magnetic materials, as conductive seed layers such as Cu are less than desirable as they provide an unwanted conduction path for eddy currents [10].

The effect of seed layer undercut, or the etch attacking the electroplated film, when removing the seed layer is not an issue for large structures. However, this is not the case for structures with small dimensions. In this situation the undercut can release the structure from the substrate. Additionally, a thick seed layer of the structure material will result in the electroplated film also being attacked while the seed layer is being removed. Hence, for small structures there is a real need to minimise seed layer thickness.

For large structures (>10 µm) a seed layer of 100nm Ti and 300nm Cu has been typically employed for both Ni and NiFe electroplated films [1]. However, for micron sized mesas this at best results in an excessive eddy current path or in the worst case release of the mesas during the seed removal step. Clearly, for films with these dimensions the seed layer needs to be as thin as possible and made of the same material as the electroplated film. Unfortunately, thin seed layers can result in IR drops and non-uniform thicknesses, particularly in the instance of sparsely patterned structures. This paper presents the first test structure specifically designed to characterise such processes.

Typically, electroplating systems are set up to deliver uniform blanket depositions, which minimises any non-uniformity related to current crowding resulting from the pattern. In this regime any IR drop across the wafer is rapidly reduced as the plating proceeds to increase the effective thickness of the seed layer and will have a negligible effect on uniformity and final thickness of the deposited films. For example, with a blanket layer electroplated using a seed layer of 30 nm of Ti and 100 nm of Ni it is possible to achieve a uniformity of 3-5% on 75mm wafers. However, for patterned films only the exposed area is electroplated as the thin seed layer carries the current to the plating areas. Hence, depending on the pattern and the current distribution in the seed layer, there is potential for different voltages to result in local, exposed areas. Clearly, for any test structure to examine such a process it should provide a more pattern independent characterisation of the seed layer performance.

III. DESIGN CONSIDERATIONS

Plating uniformity is governed by a large number of parameters including pattern uniformity (current crowding) and the physical layout of the plating bath, as well as the circulation of the plating electrolyte. Minimisation of non-uniformity from these sources requires the bath geometries to first be optimised for uniform blanket plating. This provides a benchmark for plating on patterned surfaces. Next, a test structure is required, which provides a uniform pattern density (ratio of plated to non-plated area).

Figure 2 shows the mask layout architecture, where the area ratio (plated to non-plated) is 1:1 and the radial resistances of the seed areas between each plated annulus are designed to be the same value by adjusting the annular width so the radial resistance R1 between E1 and E2 (the electroplated regions) is the same as R2 (between E2 and E3) and so on[†].

During electroplating there will be an IR voltage drop (resulting from the seed layer) between each of the adjacent electroplating regions (E1 to E5). The test structure will therefore enable the effect of the seed layer

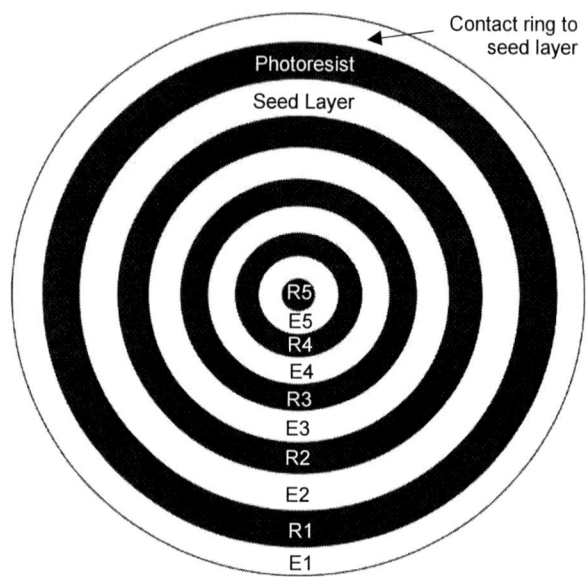

Figure 2: Schematic view of mask indicating distinct electroplating (E1-E5) and resistive (R1-R5) areas

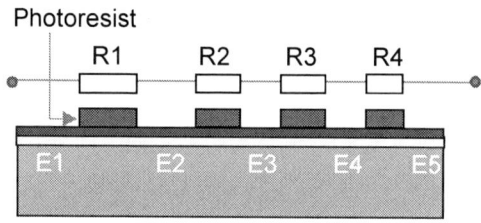

Figure 3. Radial cross-section schematic of the electroplating annulus structure showing the seed layer radial resistance. Note the electroplated areas (E1- E5) form the probe pads for measuring R1-R5, (LHS wafer edge, RHS centre, oxide – yellow, seed layer – blue)

thickness to be quantified in a pattern independent mode. Table I shows the design table where each electroplated annulus is adjacent to a resistive (photoresist covered) area of equal size and provides constant annular resistance values.

Figure 3 shows a schematic cross-section along the radius of the wafer. Assuming the local electroplating thickness is proportional to the current distribution, this design enables the current flow in each of the plated regions to be estimated from the plating thickness and the total plating current.

IV. FABRICATION AND MEASUREMENTS

A. 75mm Wafers

The initial design in figure 2 was first evaluated on 75mm wafers using a custom jig with three contacts on

---

[†] This can be achieved by ensuring the width to length ratio (w/l) of each resistor are identical.

TABLE I. MASK DESIGN PARAMETERS FOR THE 75MM WAFER DESIGN

Ring	Outer radius (cm)	Inner radius (cm)	E/R area (cm²)	Length l (cm)	Width w (cm)	Radius r	w/r
E1	3.750	3.250	10.996	21.991	0.500	3.500	0.143
R1	3.250	2.660	10.954	18.567	0.590	2.955	0.200
E2	2.660	1.894	10.956	14.306	0.766	2.277	0.336
R2	1.894	1.549	3.728	10.817	0.345	1.722	0.200
E3	1.549	1.101	3.728	8.327	0.448	1.325	0.338
R3	1.101	0.901	1.262	6.289	0.201	1.001	0.200
E4	0.901	0.639	1.263	4.837	0.262	0.770	0.340
R4	0.639	0.522	0.425	3.649	0.117	0.581	0.200
E5	0.522	0.370	0.425	2.804	0.152	0.446	0.341
R5	0.370	0.302	0.144	2.113	0.068	0.336	0.201
E6	0.302						

TABLE II. AVERAGE SEED LAYER RESISTANCE VALUES FOR SEED LAYER AREAS R1 TO R5 (INDICATED IN FIGS 2 AND 3)

Resistor number	5 nm Al / 5 nm Ni (Ω)	50 nm Al / 50 nm Ni (Ω)
R1 (E1 to E2)	6.67	0.5
R2 (E2 to E3)	5.91	0.5
R3 (E3 to E4)	6.64	0.5
R4 (E4 to E5)	7.61	0.6

the outer electrode, which was exposed to the electrolyte. Seed layers of 100 nm (50 nm Al/ 50 nm Ni) and 10 nm (5 nm Al/ 5 nm Ni) were initially deposited on 75mm silicon wafers using magnetron sputtering. Photoresist was then coated and UV exposed using the photomask layout shown in figure 2 to define the annular pattern on the wafer. The 35 litre plating bath uniformity was first confirmed using a blanket deposition and a current density of 5 mA cm⁻². This current density was then used to electroplate the annulus structures for identical times and the thickness and resistance between

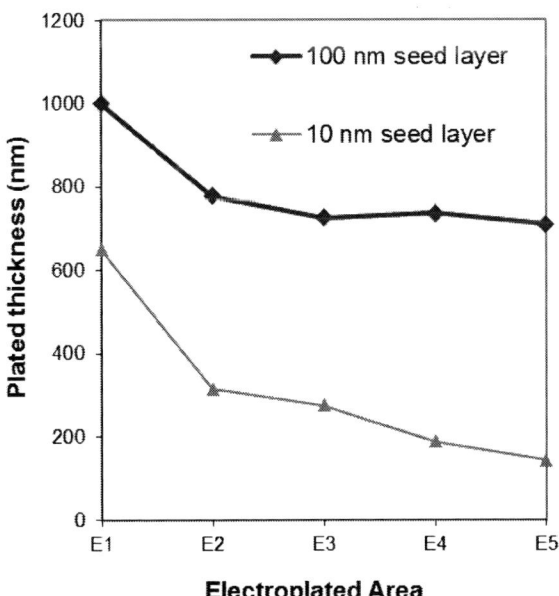

Figure 4: Plot of the electroplated thickness of Ni for each of the open annuli (E1-E5) displayed in figure 2.

adjacent E1 to E5 regions measured. If the electroplated layers result in significant thickness non-uniformity this suggests a pronounced IR voltage drop in the seed layer distributing the electroplating current. Table II shows the measured resistance values for R1 to R5 for two different seed layer thicknesses.

Figure 4 shows a plot of thickness of the electroplated regions E1 to E5 for the two seed layers in table I. The 10 nm seed shows evidence of an IR drop towards the centre of the wafer whereas the thicker 100 nm seed layer displays considerably greater radial uniformity. It should be noted that E1 is open to the solution and is electrically connected at three positions using 2.5 mm diameter copper wire. In this jig an area of approximately 78 mm² of copper is exposed to the electrolyte per contact. This copper is therefore also electroplated and adds to the total plating current. Clearly this connection arrangement is not ideal.

The large E1 plating thickness for both wafers is the result of current crowding at the wafer edge, which could be significantly reduced by using a dummy ring around the wafer periphery, which is discussed in the following section. In addition there are also benefits ensuring the outer contact ring is not exposed to the electrolyte.

B. 150mm Wafers

For the 150mm wafers the uniformity of the deposited seed layers was first determined as the Balzer sputter tool was known to be non-uniform for 150mm wafers (it was originally configured for 75mm wafers). For this measurement photo-resist was patterned as shown in layout in figure 5. The seed layer thickness was then measured after photoresist removal, with figure 5 also showing the vertical non-uniformity across the wafer to be greatest.

Once the seed layer was sputtered the 150mm wafers were electroplated using the same bath as that used for the 75mm wafers. However in this case the wafers were mounted in an AMMT lollipop jig. This jig uses the outer annulus of the wafer as the seed layer connection, which is electrically isolated from the electrolyte using an O-ring as shown in the cross section presented in figure 6. It is important that the metal contact ring thickness on the wafer periphery sufficient to provide a good electrical connection to the contact pins on the jig as well as ensure the ring sets an equipotential on the outer ring during the electroplating process. To achieve this the seed layers were first deposited, and photoresist coated and patterned to expose the outer electrode contact ring. A lift-off process was then used to pattern Al on the seed layers on outer electrode to provide good

2018 INTERNATIONAL CONFERENCE ON MICROELECTRONIC TEST STRUCTURES, MARCH 19-22, 2018, AUSTIN, TEXAS, USA

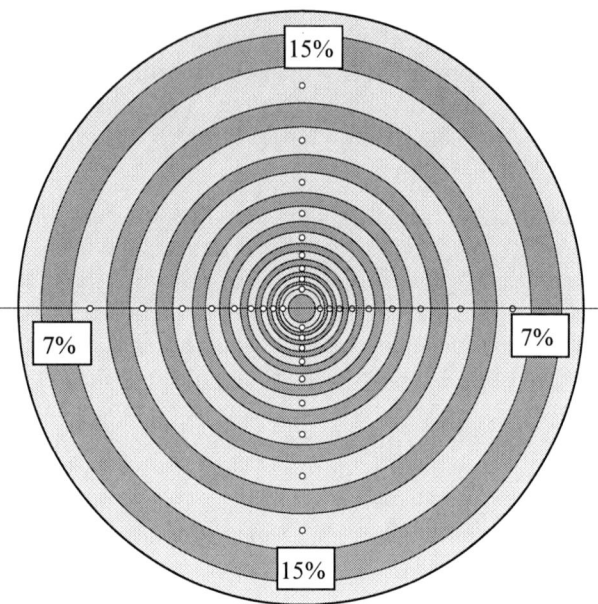

Figure 5. Mask layout for the measurement of seed layer deposition uniformity together with radial uniformity

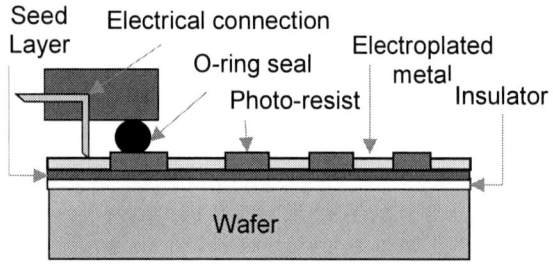

Figure 6. Schematic cross-section along the wafer radius showing the electrical connection of the test structure assembled in the AMMT jig

electrical contact to the jig. Figure 7 shows the assembled plating jig with an added current thief [11] to aid plating uniformity.

TABLE IV. RESISTANCE OF SEED LAYER RESISTORS MEASURED AFTER ELECTROPLATING ELECTRODES.

Resistor No.	50Al/50Ni (Ω)	10Al/10Ni (Ω)	5Al/5 Ni (Ω)
R9	2.2	2.6	4.4
R8	3.1	2.6	5.7
R7	2.5	4.0	6.8
R6	2.7	2.3	7.5
R5	4.3	3.4	7.8
R4	5.0	3.0	7.9
R3	4.4	3.0	8.3
R2	4.4	2.4	8.4
R1	2.4	3.5	8.1

The mask was designed using the previously described approach. Table III shows the design parameters for the mask layout calculated by using the Excel spread sheet

TABLE III   MASK DESIGN PARAMETERS FOR THE 150MM WAFER DESIGN

Ring	Outer radius (cm)	Inner radius (cm)	E/R area (cm²)	Length l (cm)	Width w (cm)	Radius r	w/r
E1	7.500	6.910	26.710	45.270	0.590	7.205	0.082
R1	6.910	6.110	32.723	40.904	0.800	6.510	0.123
E2	6.110	5.187	32.733	35.492	0.923	5.649	0.163
R2	5.187	4.586	18.457	30.705	0.601	4.887	0.123
E3	4.586	3.893	18.466	26.638	0.694	4.240	0.164
R3	3.893	3.441	10.405	23.040	0.452	3.667	0.123
E4	3.441	2.920	10.406	19.984	0.521	3.181	0.164
R4	2.920	2.581	5.852	17.283	0.339	2.751	0.123
E5	2.581	2.190	5.857	14.991	0.391	2.386	0.164
R5	2.190	1.936	3.294	12.964	0.254	2.063	0.123
E6	1.936	1.643	3.296	11.243	0.294	1.789	0.164
R7	1.643	1.452	1.858	9.720	0.191	1.547	0.123
E7	1.452	1.230	1.859	8.425	0.221	1.341	0.165
R8	1.230	1.087	1.042	7.281	0.143	1.159	0.123
E8	1.087	0.922	1.042	6.311	0.166	1.005	0.165
R8	0.922	0.814	0.587	5.453	0.108	0.868	0.123
E9	0.814	0.689	0.589	4.722	0.125	0.752	0.166
R9	0.689	0.608	0.329	4.076	0.081	0.649	0.123
E10	0.608	0.515	0.329	3.529	0.094	0.562	0.167
E11	0.515	0.364					
R11	0.364						

(with iterative macros) designed to meet the design criteria of a 1:1 electrode to non-plated area ratio and constant annular resistance values. The mask layout is shown in figure 8.

Table IV shows the measured seed layer resistance for the resistors R1 to R9. For each seed layer thickness these values should be the same and the variation observed indicates the non-uniformity of the metal deposition equipment used. The resistance calculations also assume that the effect of any resistance of the electroplated electrodes can be neglected.

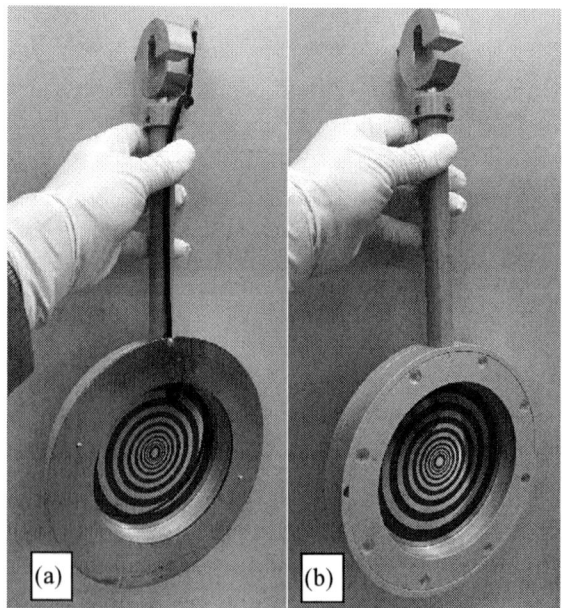

Figure 7. Electroplating jig used for plating 150mm wafers, which are located in the cavity. (a) With current thief, (b) Without current thief.

978-1-5386-5072-1/18 $31.00 © 2018 IEEE

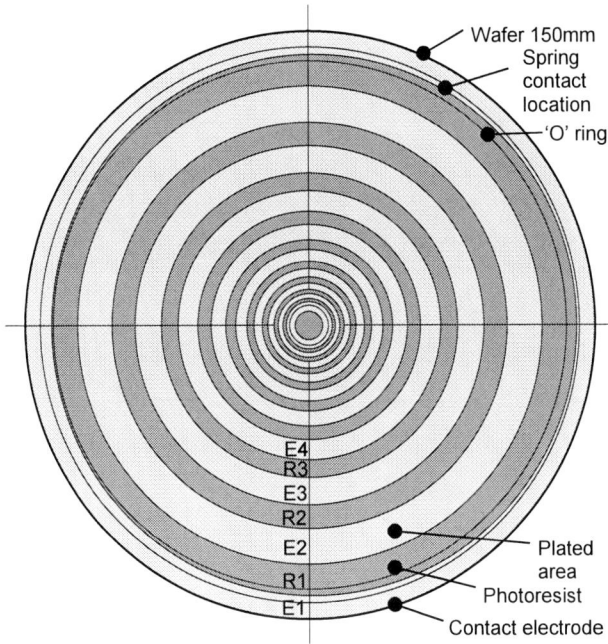

Figure 8. Mask layout for the electroplating of the 150mm wafer showing the position of the O ring and jig contacts

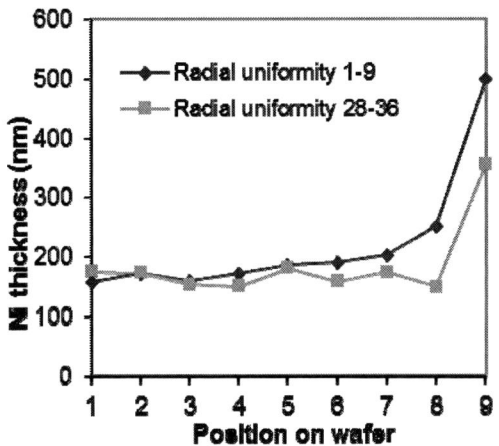

Figure 9. Radial uniformity measurements for 5 nm Al / 5 nm Ni seed layer. (1-9 centre-east. 28-36 centre-north)

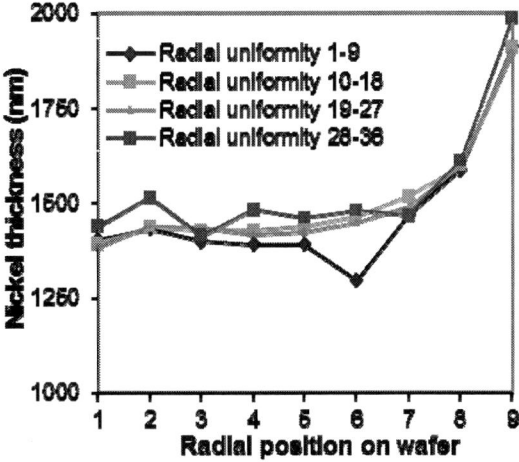

Figure 10. Radial uniformity measurements for 10 nm Al / 10 nm Ni seed layer. (1-9 centre-east, 10-18 centre-west, 19-17 centre-south, 28-36 centre-north)

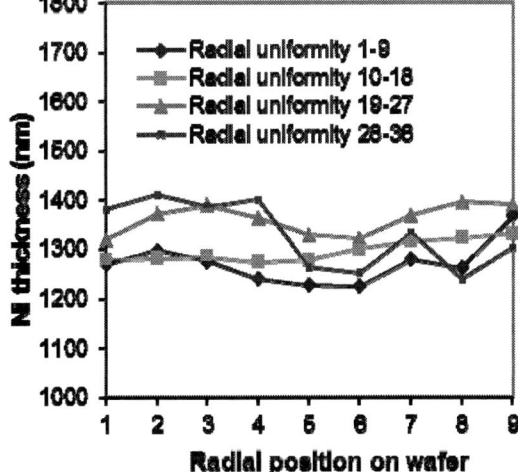

Figure 11. Radial uniformity measurements for 50 nm Al / 50nm Ni seed layer. (1-9 centre-east, 10-18 centre-west, 19-17 centre-south, 28-36 centre-north)

Figures 9, 10 and 11 show the measured radial thickness of the Ni electroplated electrodes on three different thicknesses of Al/Ni seed layers with a total plating current of 0.65mA (Note: a single power supply was used with the wafer and thief currents being individually recorded). These graphs show that the thickness of the seed layer does affect the uniformity of plating with the 5/5 and 10/10 exhibiting thicker electroplated films on the wafer periphery. It should be noted that the 5/5 seed layer plating was much slower with wafer plating currents of 0.068mA being forced compared with 0.21 mA and 0.16 mA for the 10/10 and 50/50 seed respectively.

These results indicate that it is possible to successfully electroplate on patterned moulds using very thin seed layers. However it should be noted that the thickness (conductivity) of the seed layers significantly affects the plating rates and radial uniformity, especially for the 5/5 seed.

V DISCUSION AND CONCLUSIONS

A test structure has been presented, which has been specifically designed to quantify the IR drop across a wafer during electroplating in a more pattern independent manner. The characterisation of thin seed layers is made possible by the proposed test structure and is particularly useful for optimisation of structures that require a minimised seed layer thickness; where the seed layer hinders device performance (e.g. eddy currents), or where wet-etching can cause undercutting which can potentially result in the unwanted release of the plated structures. The next step is to evaluate the effect of dummy structures to aid distribution of currents for thin seed layers with sparse patterns.

ACKNOWLEDGMENT

This work has been supported in part by EPSRC (EP/K034510/1).

The data reported in this paper is available from https://datashare.is.ed.ac.uk/handle/10283/758

REFERENCES

[1] Sirotkin, E., Smith, S., Walker, R., Terry, J. G. & Walton, A.J., "Test structures for the wafer mapping and correlation of electrical, mechanical and high frequency magnetic properties of electroplated ferromagnetic alloy films", IEEE International Conference on Microelectronic Test Structures (ICMTS), pp. 182-187, 2015.

[2] M. G. Buehler, W. R. Thurber, "An experimental study of various cross sheet resistor test structures", J. Electrochemical Soc - Solid State Technology, vol. 125, no. 4, pp. 645-650, April 1978.

[3] M.I. Newsam, A.J. Walton, M. Fallon; "Numerical Analysis of the Effect of Geometry on the Performance of the Greek Cross Structure", IEEE International Conference on Microelectronic Test Structures, pp. 247-252, March 1996.

[4] M. Buehler, S. Grant and W. Thurber, "Bridge and van der Pauw Sheet Resistors for Characterising the Line Width of Conducting Layers", J. Electrochem. Soc.: Solid-State Science and Technology, Vol. 125, pp. 650-654, April 1978.

[5] G. Schiavone, J. Murray, S. Smith S, M.P.Y. Desmulliez,, A.R. Mount, A.J Walton, "A wafer mapping technique for residual stress in surface micromachined films", Journal of Micromechanics & Microengineering., Vol 26, No 9, pp 095013, 2016

[6] S. Smith, N.L. Brockie, J. Murray, C.J. Wilson, A.B. Horsfall, J.G. Terry, J.T.M. Stevenson, A.R. Mount, A.J. Walton, "Analysis of the Performance of a Micromechanical Test Structure to Measure Stress in Thick Electroplated Metal Films", IEEE International Conference on Microelectronic Test Structures, 22-25 March 2010, Hiroshima, Japan, pp 80-85.

[7] S. Smith, N. Brockie, J.G. Terry, N. Wang, A.B. Horsfall, A.J. Walton, "Application of a Micromechanical Test Structure to the Measurement of Stress in an Electroplated Permalloy Film", IEEE International Conference on Microelectronic Test Structures, p. 75-80, 2009.

[8] J. Murray, R. Perry, J. G. Terry, S. Smith, A. R. Mount and A. J. Walton, "Chip level characterisation studies of Ni and NiFe electrochemical deposition using test structures," 2016 International Conference on Microelectronic Test Structures (ICMTS), Yokohama, 2016, pp. 178-183.

[9] J. Murray, G. Schiavone, S. Smith, J. Terry, AR. Mount, AJ. Walton, "Characterisation of Electroplated NiFe Films using Test Structures and Wafer Mapped Measurements," 24th International Conference on Microelectronic Test Structures, Amsterdam, pp 63 – 68, 4th – 7th April 2011.

[10] R. Walker, E. Sirotkin, J.G. Terry, S. Smith, M.P.Y. Desmulliez, A.J. Walton, "Effect of seed layers on the performance of planar spiral microinductors", IEEE International Conference on Microelectronic Test Structures, 2014, 2014. pp 135-140.

[11] Ang, J, "Method for electroplating metal films including use a cathode ring insulator ring and thief ring", US Patent 574401

Test structures without metal contacts for DC measurement of 2D-materials deposited on silicon

L. K. Nanver[1], X. Liu[1], and T. Knezevic[2]

[1]MESA+ Institute for Nanotechnology, University of Twente, Enschede, Netherlands
Email: l.k.nanver@utwente.nl

[2]University of Zagreb, Faculty of Electrical Engineering and Computing, Micro and Nano Electronics Laboratory, Croatia

Abstract—A set of ring-shaped test structures is presented for electrical characterization of 2D as-deposited layers on Si that electrically interact with the substrate. The test method is illustrated by investigation of 3 different nm-thin layers that are expected to form an interfacial layer of negative fixed charge. A test procedure is described that gives a low turnaround time and non-destructive way of evaluating different deposition methods in terms of diode characteristics, interface conductance, and electron carrier injection into the deposited layer.

Keywords—aluminum, atomic layer deposition, boron, chemical vapor deposition, electron injection, interface charge

I. INTRODUCTION

The interest in 2D-materials for fabricating new electrical and optoelectrical devices has taken flight over the last years. Very often their electrical manipulation is enabled by deposition on a semiconductor crystal, preferably Si because successful integration with Si is the fast-route to commercialization of any such new material. The methods of deposition on Si are numerous, ranging from defoliation followed by manual deposition to advanced Si processing techniques such as chemical vapor deposition (CVD) and atomic layer deposition (ALD). After deposition, further processing is often undesirable because it can destroy the virgin properties that need to be understood and exploited. In this paper a set of ring-shaped test structures is presented for enabling the characterization of several properties of as-deposited layers that electrically interact with the Si.

The use of the test structures is illustrated here by looking at 3 different types of nm-thin layers composed of boron and/or aluminum and they are expected to form an interfacial layer of negative fixed charge, Q_f, when deposited on Si. The first type, CVD layers of pure boron (PureB), are known to attract an almost complete monolayer of holes to the interface, which can be used to form low-saturation current p^+n diodes [1]. Similarly, the second type, ALD layers of Al_2O_3, are commonly used for passivating p-type surfaces in c-Si solar cells because a significant Q_f is formed. The third type, deposited by exposing the Si surface to both B and Al precursors, was the result of a preliminary experiment. In [1, 2] the interfacial holes in PureB devices were monitored by measuring the sheet resistance, R_{sh}, along the interface. The measurement was performed after the processing of the Deposition Under Test (DUT), by directly contacting implanted regions instead of adding a metal interconnect layer, the processing of which could disrupt the DUT. As opposed to this, commonly used test structures for investigating new materials such as graphene or nanowires often make use of metal patterning before depositing the DUT. However, for materials deposited in equipment that needs to be front-end CMOS compatible, such as the layers studied here, contamination issues mainly do not allow this option.

The measurements of the R_{sh} presented in [2] of PureB layers deposited at 400°C were used to monitor the integrity of the interface as a function of pre- and post-deposition processing steps. For fine differences, however, we show here that the exact substrate doping, N_{sub}, can have an undesirably large influence on the R_{sh}. Normally, C-V profiling could be used to monitor and then compensate for variations in N_{sub} but without metal contacts the contact resistance to the probes is so high that the extraction of the doping profile becomes unreliable.

In this paper, to supplement the R_{sh} measurements, another straightforward *I-V* measurement technique is presented. It is used to obtain more information on the quality of the DUT while also eliminating the influence of variable N_{sub}. By using a differential measurement the electron injection into the DUT could be estimated. The lower the electron injection, the better the blocking power of the DUT which for many applications is the property that needs to be optimized.

The 2D layers studied in this paper are high-ohmic to insulating and it is their interaction with the Si that is of interest. This is in contrast to many other studies of 2D materials, where it is the electrical properties of the layer itself, such as an exceptionally high conductivity, that are studied. For this, other types of structures are needed for isolating patterned regions of the 2D material. Examples from the literature include several bulk micromachined Si structures such as Greek-cross sheet-resistance test structures [3] and field-effect devices [4]. In these structures the DUT is separated from the Si by an insulating oxide layer to avoid contact with the Si.

II. TEST STRUCTURES

The process flow used to fabricate the test structures presented in this paper was originally developed to allow an accurate determination of the sheet resistance, R_{sh}, of the interfacial hole layer in PureB diodes [1]. For this, it was important that electrical I-V measurements could be performed directly after the PureB deposition. The simple 2 mask process flow is shown schematically in Fig. 1. The starting substrates are (100) 1-10 Ωcm n-type Si wafers that are first thermally oxidized to an oxide thickness of 235 nm. Contact pads for the probe-needles are formed by B^+ implants annealed at 950°C for 20 min in argon gas so that a 0.5-μm-deep junction is achieved with a surface doping of 10^{19} cm^{-3}. Windows to the Si are then opened to give access to the p^+-regions and they define where the DUT will be deposited.

Sets of ring-shaped structures were designed as illustrated in Fig. 2. Each ring structure has a fixed radius r_g =164 μm so the total perimeter, P = 2061 μm, is always the same despite the different width, L, of the rings. Sets of each 5 ring structures are designed as shown in Fig. 3 with L = 10 μm, 20 μm, 40 μm, 100 μm, and 200 μm. Five of such sets were designed. Sets S2 and S3 have a p^+-guard ring covering the whole perimeter region. This option is included for 2 reasons: for the first, any perimeter leakage due to a deficient DUT

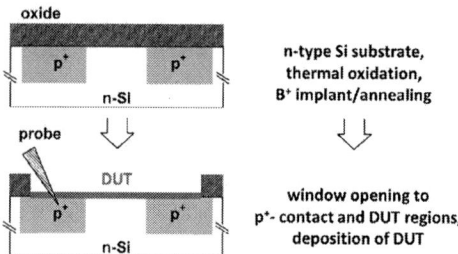

Fig. 1. Schematic process flow for the fabrication of p^+ contact regions and the DUT-only regions.

coverage at the edge of the oxide window is then eliminated, and, secondly, if the R_{sh} is very high, the low-ohmic p^+-guard ring will nevertheless ensure a rotational symmetric current flow from the inner to outer p^+ ring. In set S1 the p^+-guard ring is absent. In the sets S4 and S5, the p^+-region extends under the whole ring that in S4 is also covered with the DUT.

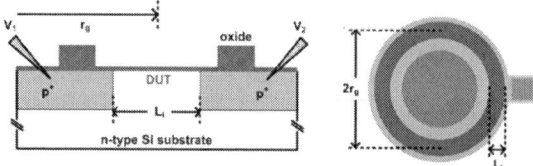

Fig. 2. Schematic cross section (left) and layout (right) of the ring-shaped measurement structures. In each set the perimeter of the DUT-only region is constant and equal to $4\pi r_g$ while the width of the ring has varying values L_i.

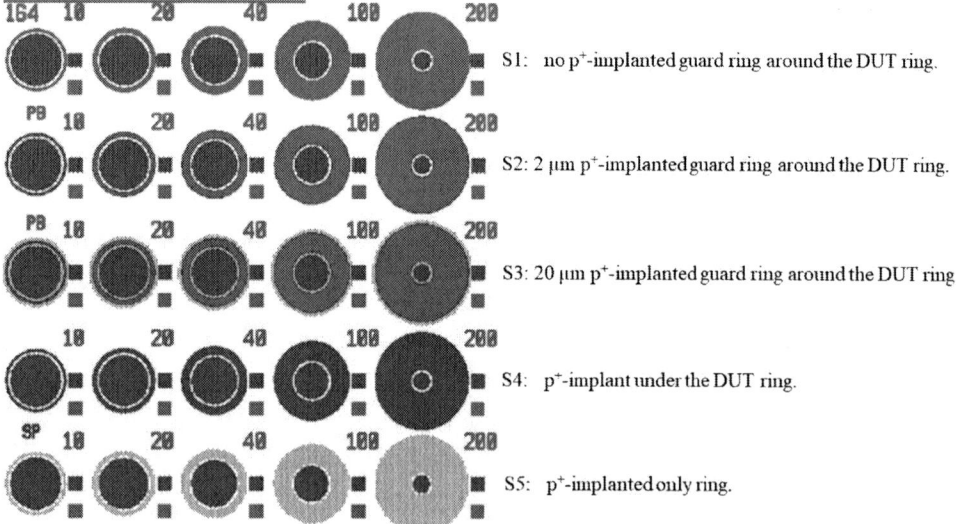

S1: no p^+-implanted guard ring around the DUT ring.

S2: 2 μm p^+-implanted guard ring around the DUT ring.

S3: 20 μm p^+-implanted guard ring around the DUT ring.

S4: p^+-implant under the DUT ring.

S5: p^+-implanted only ring.

Fig. 3. (1) Mask layout of a set of the ring-shaped test structures. The radius to the center of the ring is r_g = 164 μm and in each of the 5 rows the ring length, L_i, is 10 μm, 20 μm, 40 μm, 100 μm, and 200 μm. The deposition-under-test (DUT) region is shown in red and the p^+-implanted regions are grey as also indicated in Figs. 1 and 2. A p^+-implanted guard ring is placed around the DUT rings in rows S2 and S3. In row S4 a p^+-implant is placed under the whole DUT ring and in row S5 the ring only comprises the p^+-implant. For JFET-like operation, the drain contact is the p^+ contact in the center and the source is the small p^+ pad contacting the outer DUT ring.

III. EXPERIMENTAL DEPOSITED LAYERS

The use of the test structures is illustrated here by looking at 3 different types of nm-thin layers that are expected to form an interfacial layer of negative fixed charge. They are all deposited in a Picosun ALD system. A HF dip followed by rinsing/drying is used to remove native oxide from the Si windows immediately before loading the samples for deposition. The main deposition conditions are described in following subsections but it is beyond the scope of this paper go into all details about the depositions and their properties.

A. CVD layers of pure boron (PureB)

The PureB layers were deposited in CVD mode from diborane in an argon or nitrogen atmosphere at temperatures from 300°C to 450°C with thicknesses varying from 2 nm to 10 nm. Similar layers were in the past deposited in the 400°C to 700°C temperature range in a commercial CVD reactor and these have been studied extensively [5]. The driving PureB application for these layers has been low-dark-current photodiode detectors with high sensitivity for radiation that only penetrates a few nm into the Si. Especially their high stability and robustness have made them the detector of choice for detecting soft x-rays [5], vacuum-ultraviolet light [6], and electrons with energies below 1 keV. For the latter the bare pure B layer can be implemented as light-entrance window [7]. The overall results indicated that an interfacial hole layer was induced with $Q_f > 10^{14}$ cm^{-2}, even for the 400°C deposition where no B doping of the Si can be expected [1]. At this temperature the measured R_{sh} was ~ 35 kΩ/sq which was accorded entirely to the interfacial hole layer since the pure B layer itself has high resistivity, > 500 Ωcm [1].

B. ALD layer of Al_2O_3

A layer of Al_2O_3, 10 nm thick, was deposited in ALD mode at 250°C. This type of layer is commonly used as a passivation layer on p-type surfaces in c-Si solar cells. The distinguishing properties for this application have been found to be a field-effect passivation induced by negative fixed charges at the interface combined with a low interface defect density. Values of Q_f have been reported to be from 10^{12} cm^{-2} - 10^{13} cm^{-2} [8]. The Al_2O_3 is an insulator, which means that the Q_f can be determined by C-V measurements but this would require metallization to form metal-insulator-silicon (MIS) structures which is not available in our structures. Similar to the PureB detectors, the application of the Al_2O_3-to-Si interfacial hole layer in a photodiode detector configuration has also been demonstrated in [9], in which it was applied to enhance the performance of Si nanowire/quantum-dot light-emitting diodes.

C. Experimental Al-B CVD layers

The layers that we refer to here as Al-B layers were grown by first exposing the Si surface to trimethylaluminum (TMA) and then diborane, or cycles of both. Apart from the electrical characterization presented in this paper, little is known about these layers.

IV. MEASUREMENT TECHNIQUES

The characterization of the layers using the ring-structures includes 3 types of basic measurements as shown schematically in Fig. 4 and described in the following subsections. All measurements were performed at room temperature. First the influence of the substrate doping on the sheet resistance is estimated by simple simulations.

A. Influence of the substrate doping

When measuring the conductance along interface with the DUT, the substrate n-doping can have an impact in two ways:

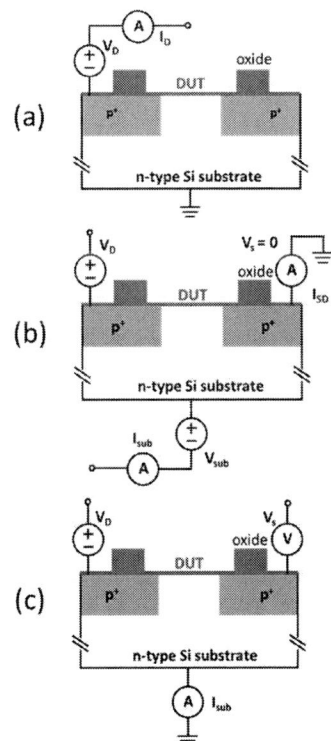

Fig. 4. Schematics of the measurement configuration for (a) diode I-V measurements, (b) voltage dependent JFET-like source-to-drain measurements, and (c) currentless voltage measurements.

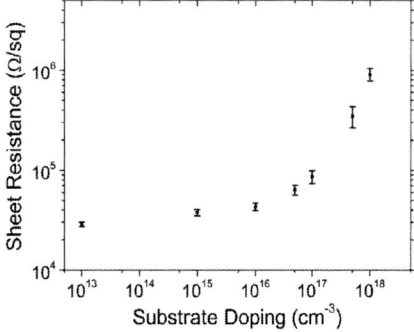

Fig. 5. Measured sheet resistance of the 400°C PureB deposition described in [1] as a function of the n-doping of the substrate. The exact doping of the intrinsic substrate marked 10^{13} cm^{-3} is uncertain.

978-1-5386-5072-1/18 $31.00 © 2018 IEEE

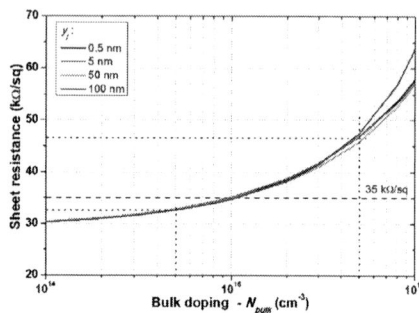

Fig. 6. Sentaurus Device simulation [10] of the R_{sh} of a p-type B-doped surface region with a Gaussian profile as a function of the n-doping concentration of the bulk Si substrate. The p-type region is designed to have a R_{sh} of 35 kΩ/sq for $N_{sub} = 10^{15}$ cm^{-3} and a varied thickness of 0.5 nm to 100 nm. The R_{sh} variation from 32 kΩ/sq to 47 kΩ/sq in the 1-10 Ωcm resistivity range of the Si substrates is indicated.

there will be a direct compensation of the holes from the n-type background doping and the depletion of holes at the junction will depend on N_{sub}. Both effects increase the R_{sh} as the N_{sub} increases because the effective number of mobile holes decreases. In addition, scattering effects, particularly at the interface, will lower the mobility. This is particularly true for nm-thin hole layers that are largely confined to the interface. The electric field created by the interfacial acceptor layer that binds the holes to the interface will also decrease the mobility similar to the way a vertical electrical field attenuates the inversion layer mobility in MOS devices [11]. The holes at a distance from the interface will have the highest mobility and will dominate the lateral conductance. For large changes in N_{sub} the increase in R_{sh} was measured as shown in Fig. 5 for the 400°C PureB depositions discussed in [1].

In the experiments presented here, the applied Si wafers have a specified spread in substrate doping from 10^{14} cm^{-3} to 10^{15} cm^{-3}. The effect on R_{sh} even in this range was experimentally found to be so large from wafer to wafer that the measured R_{sh} could not always be used to optimize the processing of the layers. To get an impression of the effect that the varying substrate doping can have on the sheet resistance, the situation is approximated by simulating boron-doped p-regions on the Si surface. In Fig. 5 an example is shown where the p-region R_{sh} is fixed to 35 kΩ/sq for $N_{sub} = 10^{15}$ cm^{-2} and the thickness is varied from 0.5 nm to 100 nm. For all these configurations the R_{sh} varies from about 32 kΩ/sq to 47 kΩ/sq for the given spread in wafer doping.

B. I-V diode measurements

As shown in Fig. 4a, the ring-shaped test structures can be biased as diodes to the substrate. An example of such a measurement is shown in Fig. 7 for devices where both the DUT diodes with and without a p$^+$-guard ring show signs of leakage. This means that both perimeter and bulk leakage is present which in this case was explained by a known contamination of the Si surface before the deposition. In contrast, the S4 and S5 devices where the whole ring has a p$^+$-implantation show ideal behavior. This will always be the case if the DUT does not deteriorate the oxide passivation or create

a conductive layer on the surface of the oxide which would connect neighboring structures. The depositions studied in the following do not have these shortcomings and mainly, when p$^+$ guard rings are used, near-ideal characteristics are achieved. In this situation, the diode current in the p$^+$-implanted S4 and S5 devices is dominated by hole injection, I_h, into the substrate. This is governed by the Gummel number of the n-substrate which is inversely proportional to $N_{sub} \times W_{sub}$, where $W_{sub} = 525$ µm. This is about 10 to 100 times lower than the Gummel number of the p$^+$-region, depending on the exact value of N_{sub}. Correspondingly, the electron injection into the p$^+$-region, I_{ep+}, is about 10 to 100 times lower than I_h. The Gummel number of the DUT will depend on the properties of the deposited material itself and the properties of the interface.

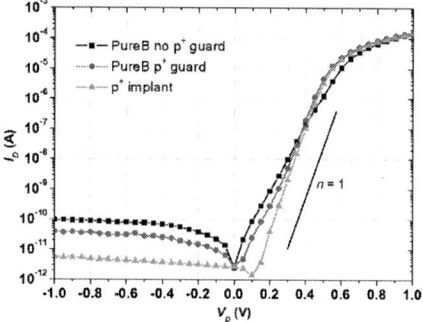

Fig. 7: I-V diode characteristics of $L = 200$ µm ring-structures with a 450°C PureB DUT deposited on a contaminated Si surface; measured in the same die on the sets S1, S3 and S5.

C. Voltage dependent resistance measurements

With the measurement configuration shown in Fig. 4b, JFET-like measurements can be used to measure the S-D resistance for all L and the voltage dependent R_{sh} can be extracted. As compared to Van Der Pauw structures the advantage of using the ring structures is that the perimeter can be guarded. Any leakage to the substrate will invariantly lower the R_{sh}. Therefore, if the leakage is significant it should always be taken into account [12].

The extraction of the R_{sh} using differential measurements to eliminate the series resistance associated with the contacts is described in [13]. Since all the ring structures are designed here with a fixed total perimeter length, only two variables are important for the extraction. These are the resistance extracted from the I-V measurement, R_{mi}, and the ring width L_i, where the indices $i = 1,...,5$ refer to each specific test structure in the given set of n structures. As described in [13], the sheet resistance R_{sh} can be calculated as:

$$R_{sh} = \frac{R_{ij}}{\alpha_{ij}} = \frac{R_{mj} - R_{mi}}{\alpha_{ij}} \qquad (1)$$

where R_{ij} is the differential resistance between 2 ring structures and

$$\alpha_{ij} = \frac{1}{2\pi} \ln\left(\frac{(r_g - \frac{1}{2}L_i) \cdot (r_g + \frac{1}{2}L_j)}{(r_g + \frac{1}{2}L_i) \cdot (r_g - \frac{1}{2}L_j)} \right) \qquad (2)$$

Table I: Currents and breakdown voltage measured on ring-structures with $r_g = 200$ µm and the extracted I_{eDUT} and R_{sh}.

Deposition	Deposition temperature (°C)	I_h at V=0.3V (nA)	$I_h + I_{eDUT}$ at V=0.3V (nA)	I_{eDUT} at V=0.3V (nA)	R_{sh} for $V_{sub}=0$ (kΩ)	R_{sh} for $V_{sub}=-10V$ (kΩ)	R_{sh} for $V_{sub}=-40V$ (kΩ)	Breakdown voltage (V)
PureB	300	2.1	15.3	13.2	104	-	-	-10
PureB	350	3.1	10.3	7.2	75	1328	-	-20
PureB	400	2.1	8.1	6.0	58	80	-	-30
PureB	450	2.2	3.5	1.3	30	44	85	-50
Al-B	350	2.3	10.0	7.7	80	263	2640	< -100
Al$_2$O$_3$	250	4.8	5.9	1.1	37	43	63	-60

D. Currentless voltage measurement

In the measurement configuration shown in Fig. 4c, possible limitations from current crowding can be identified. If there is significant current crowding due to a very high R_{sh}, the monitored voltage, V_S, will be lower than the applied voltage, V_D. Current crowding will also be enhanced if there is a high leakage current, I_{sub}, to the substrate. Therefore, if $V_S < V_D$, this indicates that, for example, in the diode measurements the entire length of the DUT may not be biased at the applied voltage.

V. RESULTS

For the interpretation of the measurement results we make use of the possibility to easily do differential measurements. In principle, for the diode measurements the fact that the total perimeter of the DUT rings is fixed means that the contribution of the perimeter can be directly subtracted and the laterally uniform diode current J_{dA} per unit area can be extracted from the relationship

$$I_d = J_{dA} A + J_{dP} P \qquad (3)$$

where I_d is the measured diode current, J_{dP} is the perimeter current per unit area, and $A = 2\pi L \times r_g$ is the ring area. However, just like for the extraction of R_{sh}, the curvature of the ring structures must be taken into account. The lateral extend of the spreading of the I_h into the substrate is of the same order as the thickness of the wafer of 525 µm. Since the inner radius of the rings designed here is smaller than this, the inner perimeter will have an J_{dP} value that decreases as L increases.

For the extraction of I_{eDUT}, which is the parameter of interest here, it is not necessary to use the relationship (3). For diodes with the same L, the measured diode currents on S3, I_{dDUT}, and S5, I_{dp+}, structures are related as

$$I_{dDUT} = I_h + I_{eDUT} \qquad (4)$$

$$I_{dp+} = I_h + I_{ep+} \qquad (5)$$

Therefore, if I_{ep+} can be neglected

$$I_{eDUT} = I_{dDUT} - I_{dp+} \qquad (6)$$

For the DUT layers studied here the I_{eDUT} is in fact large enough to be determined by these measurements. An example of I-V measurements showing this are displayed in Fig. 8. At a forward voltage $V_D = 0.3$ V, the characteristics are ideal, and the current level is high enough to be measured accurately

while being low enough to avoid attenuation from the high series resistance of the metal-less probing.

All in all, there is a compromise to be made in the radius of the ring structures. The larger the r_g and the L, the higher the measured currents, which increases the accuracy when differential extractions need to be made. Moreover, the larger the inner radius of the rings, the smaller the influence on the radial current spreading into the substrate. However, the larger the structures, the higher probability that imperfections in the processing will cause leakage currents that inhibit the extraction of I_{eDUT}. For the devices studied here, it was found that the number of near-ideal ring diodes with L = 100 µm and 200 µm was high enough to allow the desired extractions.

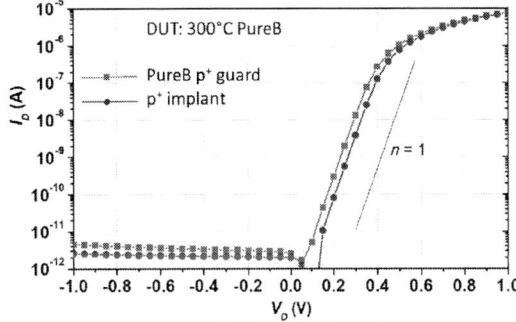

Fig. 8. I-V diode characteristics of $L = 200$ µm ring-structures with a 300°C PureB DUT, measured in the same die on the sets S3 and S5.

The values extracted for I_{eDUT}, R_{sh} and the breakdown voltage of the diodes to the substrate, are listed in Table I for the 3 different types of depositions. For all the studied layers, the behavior was as expected for structures where a significant interfacial hole layer is present. For example, R_{sh} was measurable and increased with increasing reverse V_{sub}. For the PureB depositions, decreasing the deposition temperature has the clear effect of increasing I_{eDUT} and decreasing the breakdown voltage. This is consistent with a degradation of the interface properties and/or a reduction in the concentration of holes at the interface. The former could be due to a deficient coverage of the Si. The 300°C -350°C PureB and the Al-B layer have a high R_{sh} that increases rapidly with V_{sub} into the MΩ/sq range, indicating a low hole density at the interface. This is shown in Fig. 9 for the Al-B layer. In contrast, the Al$_2$O$_3$ layer and the 450°C PureB layer have a low R_{sh} and low I_{eDUT}, indicating a much higher concentration of interfacial holes than in the other cases. Accordingly, the JFET measurements do not show pinch-off before breakdown to the

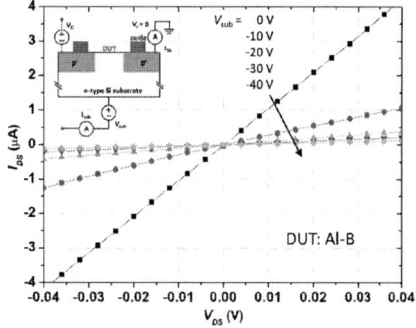

Fig. 9. Source-drain current measured as shown in Fig. 4b for a Al-B DUT on an $L = 100$ μm ring-structure in set S3.

substrate is reached. Even though these two types of layers are very different, the one being insulating and the other conductive but high-ohmic, the similarity of the results indicates that interfacial hole layer is dominating the electrical behavior.

As opposed to the other layers, the Al-B layer displays a JFET output characteristic with low pinch-off combined with high breakdown to the substrate. This is seen in Fig. 10 where the measurement configuration of Fig. 4c was applied. A clear current crowding as the R_{sh} increases is seen at $V_D = -20$ V. In the other cases, a crowding effect was only witnessed when there was a significant increase in the I_{sub}. An example is shown in Fig. 11 for a PureB deposition at 350°C.

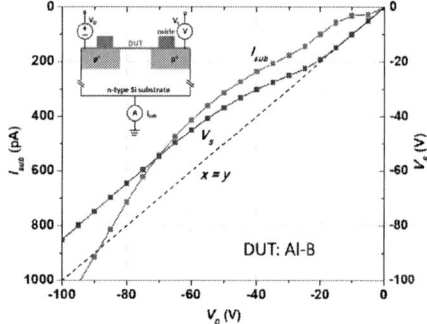

Fig. 10. Substrate current and source voltage as a function of drain voltage for a Al-B DUT on an $L = 200$ μm ring-structure in set S3.

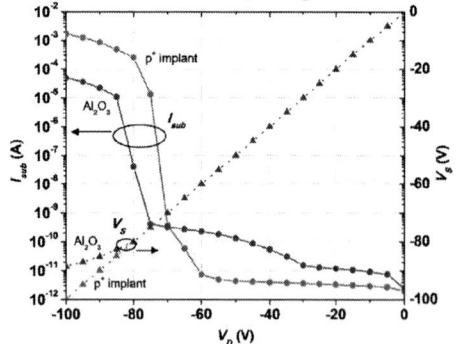

Fig. 11. Substrate current and source voltage as a function of drain voltage for an Al_2O_3 DUT on an $L = 200$ μm ring-structure in sets S3 and S5.

VI. CONCLUSIONS

Interfacial hole layers on Si, created by 2D-depositions containing B and/or Al, were studied using ring-shaped test structures and straightforward I-V measurements. By using differential measurements, the R_{sh} of the hole layer and the electron injection into the layer from the n-substrate were determined. Unlike R_{sh}, the latter is independent of N_{sub}, thus giving a more reliable method of comparing different deposition conditions. For example, if the nominal sheet resistance is 35 kΩ/sq and $N_{sub} = 10^{15}$ cm⁻³, then variations in N_{sub} from 10^{14} cm⁻³ to 10^{15} cm⁻³ cause a spread in R_{sh} from 32 kΩ/sq to 47 kΩ/sq. The presented test structures give a low turnaround time, non-destructive way of optimizing the deposition conditions with respect to electron blocking power.

ACKNOWLEDGMENT

The authors would like to thank A.A.I. Aarnik for cleanroom support and S.M. Smits for measurement support.

REFERENCES

[1] L. Qi, and L.K. Nanver, "Conductance along the interface formed by 400°C pure boron deposition on silicon", IEEE Electron Device Letters, Vol. 36, no. 2, 2015, pp.102-104.

[2] L. Qi, and L.K. Nanver, "Sheet resistance measurement for process monitoring of 400°C PureB depo. on Si," ICMTS 2015, pp. 169-174.

[3] S. Enderling, M.H. Dicks, J.T.M. Stevenson, A. Ross, S. Smith and A.J. Walton, "Suspended Greek cross test structures for measuring the sheet resistance of non-standard cleanroom materials," ICMTS 2005, pp. 1-5.

[4] T. Harada; K. Ito; T. Shibata; Y. Mita, "A bulk knife-edged as-deposition self-patterning structure for Greek-cross and organic thin film transistors," ICMTS 2006, pp.149-154.

[5] L.K. Nanver, L. Qi, V. Mohammadi, K. Mok, W. de Boer, N. Golshani, A. Sammak, T. Scholtes, A. Gottwald, U. Kroth, F. Scholze, "Robust UV/VUV/EUV PureB photodiode detector tech. with high CMOS compatibility," J. Sel. Topics Quantum Electron., 20, 6, 2014, pp. 1-11.

[6] L.Shi, S. Nihtianov, X. Sha, L.K. Nanver, A. Gottwald, F. Scholze, "Electrical and optical performance investigation of Si-based ultrashallow-junction VUV/EUV photodiodes", IEEE Transactions on Instrumentation and Measurement, Vol. 61, No. 5, 2012, pp. 1268-1277.

[7] A. Sakic, G. van Veen, K. Kooijman, P. Vogelsang, T. L. M. Scholtes, W. B. de Boer, J. Derakhshandeh, W. H. A. Wien, S. Milosavljevic and L. K. Nanver, "High-efficiency silicon photodiode detector for sub-keV electron microscopy," IEEE TED, vol. 59, no. 10, 2012, pp. 2707-2714.

[8] M.A. Juntunen, J. Heinonen, et al., "N-type induced junction black Si photo-diode for UV detection," Proc. of SPIE. 1024901, 2017, pp.1-7.

[9] Y. Ji, Y. Zhai, H. Yang, J. Liu, W. Shao, J. Xu, W. Li, and K. Chen "Improved device performances based on Si quantum dot/Si nanowire hetero-structures by inserting an Al_2O_3 thin layer," Nanoscale, 2017, 9, pp. 16038–16045.

[10] Sentaurus Device User Guide, Synopsys, Mountain View, CA, USA, 2016.

[11] D.S. Jeon and D.E. Burk, "MOSFET electron inversion layer mobilities - a physically based semi-empirical model for a wide temperature range," IEEE TED, vol. 36, no. 8, pp. 1456-1463, 1989.

[12] S.N. Bystrova, S.M. Smits, J.H. Klootwijk, R.A. M. Wolters, A.Y. Kovalgin, L.K. Nanver, and Jurriaan Schmitz, "Dealing with Leakage Current in TLM and CTLM Structures with Vertical Junction Isolation," ICMTS 2017, pp. 1-6.

[13] S. B. Evseev, L. K. Nanver, S. Milosavljević, "Ring-gate MOSFET Test Structures for Measuring Surface-Charge-Layer Sheet Resistance on High-Resistivity-Silicon Substrates," ICMTS 2006, pp.3-8.

Test structures for evaluating Al₂O₃ dielectrics for graphene field effect transistors on flexible substrates

Xinxin Yang, Marlene Bonmann, Andrei Vorobiev, Kjell Jeppson, and Jan Stake

Teraherz and Millimeter Wave Laboratory, Department of Microtechnology and Nanoscience
Chalmers University of Technology
Gothenburg, Sweden
xinxiny@chalmers.se

Abstract — **We have developed a test structure for evaluating the quality of Al₂O₃ gate dielectrics grown on graphene for graphene field effect transistors on flexible substrates. The test structure consists of a metal/dielectric/ graphene stack on a PET substrate and requires only one lithography step for the patterning of the topside metal electrodes. Results from measurements of leakage current, capacitance and loss tangent are presented.**

Keywords—test structures, graphene, field effect transistors, leakage current, capacitance, loss tangents, hysteresis

I. INTRODUCTION

Graphene is a two-dimensional material consisting of a single layer of carbon atoms arranged in a hexagonal lattice. Based on its unique electrical properties in combination with its chemical stability and mechanical flexibility, graphene has emerged as a promising material for flexible electronics in a wide range of applications. One such example is the flexible graphene terahertz detector previously reported in [1]. An important component in this context is the graphene-based field effect transistor (GFET), the properties of which when based on rigid SiO₂/Si substrates have developed rapidly in recent years. However, fabrication of high-performance GFETs on flexible substrates like polyethylene terephthalate (PET) substrates is a different story not yet well documented.

One crucial step in making high performance GFETs on flexible substrates is finding reproducible methods for fabricating dielectric films of high quality. In this paper, the design and fabrication of a test structure for evaluating the electrical properties of the gate dielectric will be described, and results from its use for evaluating the quality of flexible GFET Al₂O₃ dielectrics will be presented.

II. TEST STRUCTURE DESIGN AND FABRICATION

For evaluating the quality of the gate dielectric, a simple test structure was designed according to the principle shown in Fig. 1. The test structure is a metal-Al₂O₃-graphene capacitor placed on a flexible PET substrate. This type of test structure has been previously used with good results on solid substrates [2-3], but not before for evaluating the dielectric properties of

This work was financially supported in part by the EU Graphene Flagship, by the Swedish Foundation of Strategic Research (SSF), and by the Knut and Alice Wallenberg Foundation (KAW).

aluminum oxide (Al₂O₃) when the bottom electrode is a thin layer of graphene on a flexible PET substrate.

Also shown in Fig. 1 is the equivalent electrical circuit of the test structure. Since the dielectric was not patterned, there is no external contact to the graphene bottom plate. Instead, its dc potential is set through resistive voltage division. Since $R_P \gg R_S + R_O$, most of the applied dc voltage occurs across the circular capacitor with capacitance C under test. Therefore, both the leakage current and the capacitance of the circular capacitor can be measured as a function of voltage. Similarly, during capacitance measurements the ac signal occurs across the circular capacitor under test since $C_O \gg C$ because of the large area of the outer surrounding capacitor compared to the area of the inner capacitor. The only requirement for evaluating the dissipation losses is that the sheet resistance of the bottom graphene electrode must be known and uniform.

A. Test structure design

Test structures with three different diameters of the circular top electrode, 30, 40 and 50 μm, were designed for the purpose of evaluating the quality of the dielectric. The diameter of the outer circle defining the metal ground was 220 μm. The top Ti/Au electrode consists of a 100 nm layer of gold on top of a 5 nm adhesion layer of titanium, and was patterned by use of standard lift-off technology. With only the top metal layer being patterned, only one photolithographic process step is needed. As described above, probing during measurements is made from the topside of the test structure, with the potential of the bottom electrode being determined by resistive and capacitive voltage division. In the micro photo of the test structure shown in Fig. 2, the dark area surrounding the center top electrode is where the top metal was removed. For comparison, two types of structures are shown in Fig. 2. At the top of the photograph some complete test structures are shown with graphene as the bottom electrode, while at the bottom some open test structure are shown without the graphene electrode. The ripples below the outer electrode on the top structures are associated with the delamination of graphene from the PET substrate due to strain caused by the large area of the surrounding metal electrode. However, no ripples are to be seen below the center top electrode. Previous investigations indicate that there is usually no delamination of graphene under metal electrodes smaller than 100×100 μm².

978-1-5386-5072-1/18 $31.00 © 2018 IEEE

2018 INTERNATIONAL CONFERENCE ON MICROELECTRONIC TEST STRUCTURES, MARCH 19-22, 2018, AUSTIN, TEXAS, USA

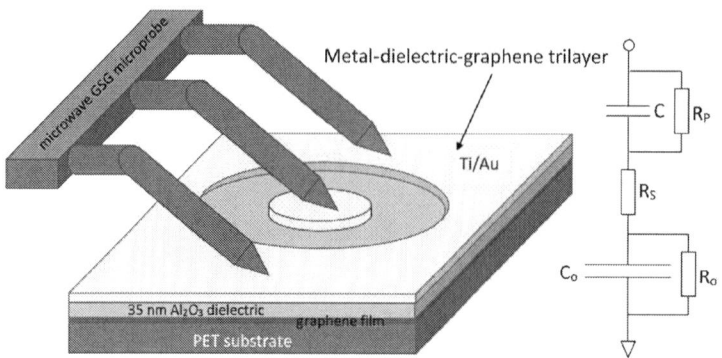

Fig. 1. Metal-dielectric-graphene trilayer test structure and equivalent circuit.

B. Test structure fabrication

Initially, the graphene for the bottom electrode was grown by chemical vapor deposition (CVD) on copper foil. The graphene was separated from the copper foil using electrochemical bubbling and transferred to the PET substrate using a supporting polymer PMMA frame that was dissolved after transfer. However, due to low yield, measurement results to be reported in this paper are from CVD graphene films on PET substrates acquired from Graphenea. In this case, the graphene film was separated from the copper foil and transferred to the PET substrate using a wet process indicating that the copper foil was removed by wet etching. When separated from the copper foil, the graphene film was placed on the target PET substrate to which it adheres by van der Waals forces. One of the main problems of any transfer method, and the disadvantage of not being able to deposit the graphene film directly on the target substrate, is that the graphene after transfer often shows some tears, wrinkles, and holes [4]. The dielectric film was formed by natural oxidation of a 3 nm thick aluminum film followed by e-beam evaporation of a 30 nm layer of Al_2O_3. The final thickness of the dielectric film was around 35 nm.

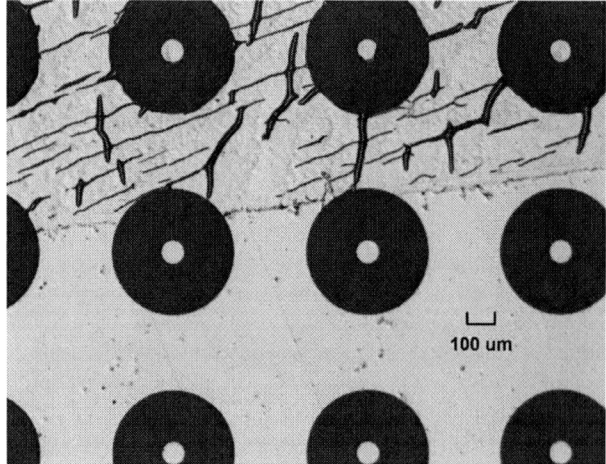

Fig. 2. Photo of the test structures. The test structures at the top are placed on the graphene film, while the structures below are placed directly on the PET substrate. Graphene wrinkles are clearly visible below the large outer metal contact area.

III. EXPERIMENTS

In this section the evaluation of the test structures concerning leakage and capacitance will be evaluated and discussed. The leakage current was measured versus the applied voltage using a Keithley 2604B SourceMeter. During measurements the voltage was increased in steps of $20 - 50$ mV with 100 ms delay times. The capacitance of the test structures was measured using a HP 4285A LCR meter and a semiconductor parameter analyzer at 1 MHz (Agilent B1500A).

A. Leakage measurements

Test structures with different diameters of the inner circular electrode were first characterized with respect to leakage current. Because of the large ratio between the area of the outer ground electrode and the area of the inner circular electrode, the resistance R_P between the inner electrode and the graphene film is much larger than resistance R_O between the graphene film and the outer electrode. Therefore, most of the applied dc voltage appears across the dielectric between the inner electrode and the graphene film. As previously reported in [5], leakage currents were found to be less than 1 $\mu A/cm^2$ for most non-defect capacitors in a voltage range from -2 to 2 V. In the voltage range from -5 to 5V the leakage current density was found to be less than 100 $\mu A/cm^2$ - a level considered negligible compared to drain and photo currents typical for terahertz detectors based on GFETs [6].

Furthermore, a pronounced hysteresis effect was observed in some samples except for low voltages where the leakage current is determined by thermal noise. The hysteresis effect is clearly visible in the graph shown in Fig. 3 where experimental data from [5] has been replotted on a semi-logarithmic scale. GFET hysteresis effects are well known, and are generally believed to be due to the charging and discharging of dielectric interface states [7-10].

Another observation based on the experimental data in Fig. 3 is the exponential dependence of the leakage current on the applied voltage. The most probable mechanisms controlling the nonlinear dc current through the dielectric film are Schottky emission, Poole-Frenkel emission, and space-charge limited currents (SCLC) - all showing a linear dependence on a semi-logarithmic scale as discussed in [11]. In this case we believe that Poole-Frenkel emission will dominate due to charge hopping between oxygen vacancies in the dielectric.

978-1-5386-5072-1/18 $31.00 © 2018 IEEE

2018 INTERNATIONAL CONFERENCE ON MICROELECTRONIC TEST STRUCTURES, MARCH 19-22, 2018, AUSTIN, TEXAS, USA

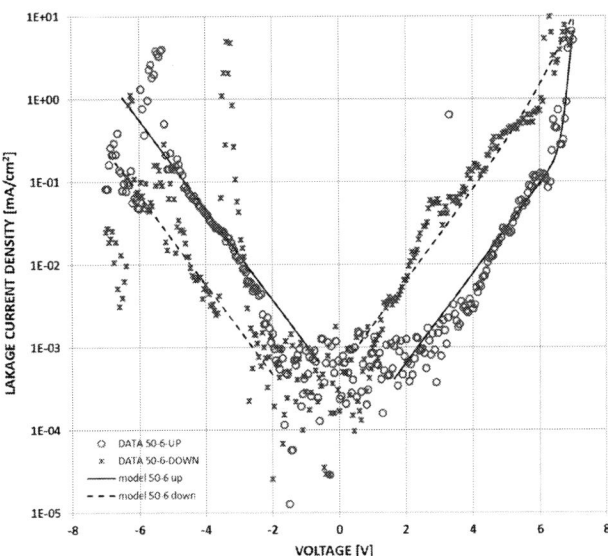

Fig. 3. Leakage current density vs. voltage for sample 50-6 with a top electrode diameter of 50 μm. Experimental data (symbols) and exponential models (lines). Characteristics obtained for increasing voltages are shown with circles (model: solid lines), while those obtained for the downwards voltage sweep are shown with crosses (model: dashed lines). Data replotted from [5].

However, not all samples exhibited such hysteresis effects while sweeping the voltage upwards and downwards as shown in Fig. 4. Here two samples, 30-5 and 30-6 with inner electrode diameters of 30 μm, show similar I/V characteristics without hysteresis. This is also true for sample 50-5 with a diameter of 50 μm. This sample shows very low leakage, less than 100 μA/cm^2, for voltages up to 7 V. Measurements were then repeated with different sweep rates. Some results from

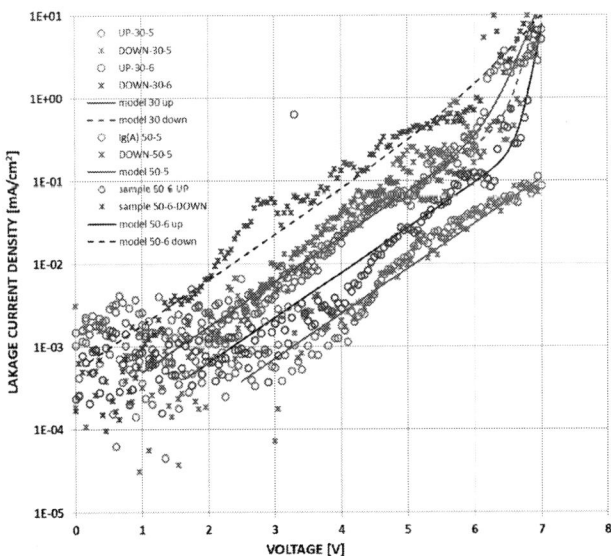

Fig. 4. Leakage current density vs. voltage. Experimental data (symbols) and exponential models (lines). While a considerable hysteresis effect is observed for sample 50-6, no hysteresis is observed for the two samples 30-5 and 30-6 with a diameter of 30 μm, or for sample 50-5 with a diameter of 50 μm.

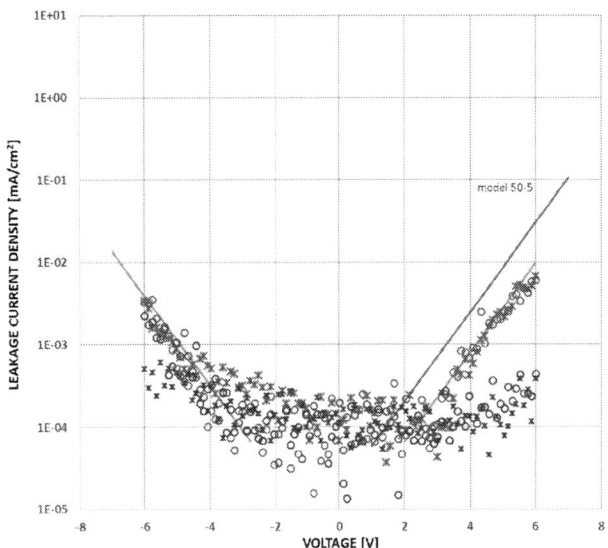

Fig. 5. Leakage current density vs. voltage. Experimental data (symbols) and exponential models (lines).

measurements with two different sweep rates are shown above in Fig. 5 for a sample with a top electrode with a diameter of 30 μm. The initial measurements done with a delay time of 100 ms showed the same type of leakage currents exponentially dependent on the voltage as already discussed. No hysteresis effects were observed. However, by increasing the delay time to 1 s, the leakage current was reduced about one order of magnitude for positive voltages around 6 V, while not much difference was observed for negative voltages.

B. Capacitance measurements

Capacitance measurements were first performed at zero bias for determining the relative permittivity of the dielectric. These measurements were done at 1 MHz on test structures with three different diameters (30, 40 and 50 μm) using an HP 4285A LCR meter. An average capacitance per unit area of slightly less than 200 nF/cm^2 was obtained, indicating a relative permittivity of 7.6 assuming an insulator thickness of 35 nm. This value is slightly below the value for bulk Al_2O_3 which is typically 9 [12].

A set of capacitance versus voltage (CV) measurements was then performed on test structures with a top electrode diameter of 30 μm using an Agilent B1500A semiconductor parameter analyzer. As shown in Fig. 6, the applied bias was swept forward from -6 V to 6 V and backwards from 6 V to -6 V with four different sweep rates (delay times) to track the hysteresis effects. The CV curves obtained for the four different forward sweep rates coincide with each other within ±0.5% as indicated by the error bars. This means that the Dirac point stays almost constant independent of the sweep rate, indicating that there is no accumulation of holes in the charge traps. However, for the four backward sweeps a significant hysteresis effect was observed, and the slower the sweep rate, the larger the hysteresis. This means that injected charges shift the Dirac point to values that are more negative.

978-1-5386-5072-1/18 $31.00 © 2018 IEEE

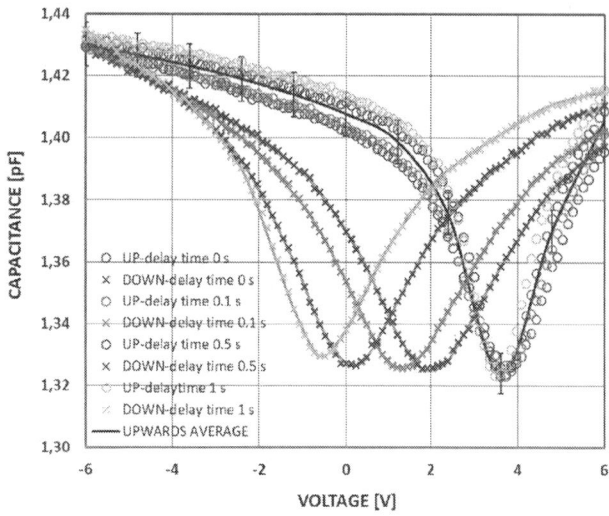

Fig. 6. Capacitance vs. voltage for four different voltage sweep rates. A voltage step of 0.12 V was used during measurements at 1 MHz.

C. Measurements of dielectric losses

The carrier mobility in graphene FETs has previously been shown to be strongly correlated to the loss tangent of the dielectric; the smaller the loss tangent, the larger the charge carrier mobility [13]. Therefore, the factors determining the losses of the test structure was investigated. Measurements of the total loss tangent are shown in Fig. 7 together with theoretically calculated series and parallel loss tangents using the equivalent electrical circuit model from Fig. 1. In the model, a capacitance per unit area of 200 nF/cm² was used. An area dependent series resistance on the order of 1000 Ω was estimated from measurements forcing the probes in direct contact with the graphene as described in [5] for structures 30 μm in diameter. While the loss tangents of both the parallel resistance R_P and the PET substrate were found negligible, the loss tangent of the series resistance was found to dominate. The differences between experimental data and losses derived from the circuit model are due to dielectric losses caused by defects in the Al_2O_3 dielectric. Obviously, there is room for improvement of the dielectric. For comparison, the loss tangent of bulk alumina is as low as 10^{-5} [14].

IV. CONCLUSIONS

A test structure, originally proposed and evaluated for investigating dielectric properties of capacitors on solid-state substrates, has been shown to be useful for evaluating the properties of the dielectric of GFETs on flexible polymer substrates. The test structure has topside contacts only, and its fabrication does not require any patterning of the dielectric. Low level leakage was observed for most non-defect samples at low voltages. As the voltage was increased beyond ±2 V, currents exponentially dependent on the applied voltage were observed and believed to be due to Poole-Frenkel emission. Confusing hysteresis effects were observed during leakage current and capacitance measurements in some samples under test - probably due to the charging and discharging of interface traps located at the many interfaces of the test structure.

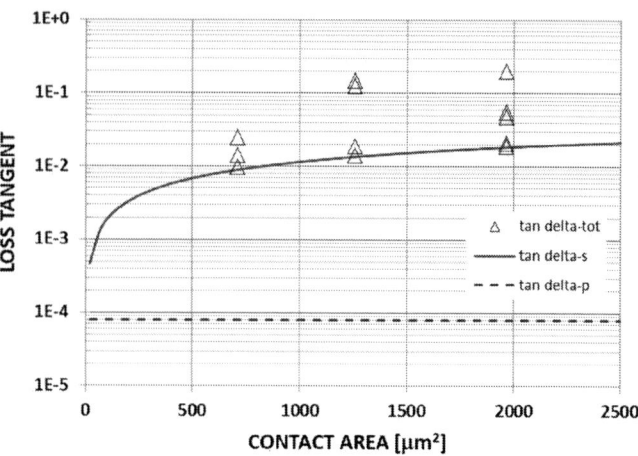

Fig. 7. The measured total loss tangent (symbols) together with the calculated series dissipation factor (solid line) and parallel loss tangents (dashed line) derived from the equivalent circuit in Fig. 1 vs. circular contact area.

REFERENCES

[1] X. Yang, A. Vorobiev, A. Generalov, M. A. Andersson, and J. Stake, "A flexible graphene terahertz detector", Appl. Phys. Lett. 111, 021102, 2017, doi: 10.1063/1.4993434

[2] Z. Ma, A. J. Becker, P. Polakos, H. Huggins, et al., "RF Measurement Technique for Characterizing Thin Dielectric Films", IEEE Trans. on Electron Devices, Vol. 45, No. 8, 1998.

[3] A. Vorobiev, and S. Gevorgian, "Tunable thin film bulk acoustic wave resonators with improved Q-factor", Appl. Phys. Lett. 96, 212904, 2010, doi:10.1063/1.3441413

[4] J. Sun et al., "Electrochemical Bubbling Transfer of Graphene Using a Polymer Support with Encapsulated Air Gap as Permeation Stopping Layer", J. of Nanomaterials, 2016, 7024246, doi: 10.1155/2016/7024246

[5] X. Yang, M. Bonmann, A. Vorobiev, J. Stake, "Characterization of Al_2O_3 Gate Dielectric for Graphene Electronics on Flexible Substrates", Global Symposium on Millimeter Waves & ESA Workshop on Millimetre-Wave Technology and Applications, 2016.

[6] L. Vicarelli, M. Vitiello, D. Coquillat, et al. "Graphene field-effect transistors as room-temperature terahertz detectors," Nature materials, 11, 10, pp. 865–871, 2012, doi:10.1038/nmat3417

[7] H. Wang, Y. Wu, C. Cong, J. Shang, and T. Yu, "Hysteresis of electronic transport in graphene transistors," ACS nano, vol. 4, no. 12, pp. 7221–7228, 2010, doi: 10.1021/nn101950n.

[8] G. Kalon, Y. J. Shin, V. G. Truong, A. Kalitsov, and H. Yang, "The role of charge traps in inducing hysteresis", Appl. Phys. Lett., 99, 083109, 2011, doi: 10.1063/1.3626854.

[9] M. Winters, E. Ö. Sveinbjörnsson, and N. Rorsman, "Hysteresis modeling in graphene field effect transistors", Journal of Applied Physics 117, 074501, 2015, doi: 10.1063/1.4913209.

[10] M. Bonmann, A. Vorobiev, J. Stake, and O. Engström, "Effect of oxide traps on channel transport characteristics in graphene field effect transistors", J. of Vacuum Science & Technology B, Nanotechnology and Microelectronics, 35, 2017, doi: 10.1116/1.4973904

[11] A. Vorobiev, P. Rundqvist, and K. Khamchane, "Microwave loss mechanisms in Ba0.25Sr0.75TiO3Ba0.25Sr0.75TiO3 thin film varactors", Journal of Applied Physics 96, (2004), doi: 10.1063/1.1789631

[12] R. Singh and R. Ulrich, "High and low dielectric constant materials,"Electrochemical Society Interface, 8, 2, pp. 26–30, 1999.

[13] S. Bidmeshkipour, A. Vorobiev, M. Andersson, A. Kompany, and J. Stake, "Effect of ferroelectric substrate on carrier mobility in graphene field-effect transistors", Appl. Phys. Lett., 107, 173106, 2015, doi: 10.1063/1.4934696

[14] X. Aupi et al., "Microwave dielectric loss in oxides: Theory and experiment", J. of Applied Physics, 95, 2004, doi: 10.1063/1.1605810

2018 INTERNATIONAL CONFERENCE ON MICROELECTRONIC TEST STRUCTURES, MARCH 19-22, 2018, AUSTIN, TEXAS, USA

Design of Ultraflexible Organic Differential Amplifier Circuits for Wearable Sensor Technologies

Masaya Kondo[1,2,3], Takafumi Uemura[1], Mihoko Akiyama[1], Naoko Namba[1], Masahiro Sugiyama[1,2,3], Yuki Noda[1], Teppei Araki[1,2], Shusuke Yoshimoto[1], and Tsuyoshi Sekitani[1,2]

[1] The Institute of Scientific and Industrial Research, 8-1 Mihogaoka, Ibaraki, Osaka 567-0047, Japan

[2] Graduate School of Engineering, Osaka University, 2-1 Yamada-Oka, Suita, Osaka 565-0871, Japan

[3] Advanced Photonics and Biosensing Open Innovation Laboratory, National Institute of Advanced Industrial Science and Technology (AIST)

2-1 Yamada-Oka, Suita, Photonics Center P3 Bldg. 2-1, Osaka University, Osaka 565-0871, Japan

Telephone: +81(0)668798400, Fax: +81(0)668798404

Email: kondo68@sanken.osaka-u.ac.jp

Abstract—We have designed and evaluated ultraflexible organic differential amplifier circuits for wearable sensor technologies. Transistor modeling for both p-type and n-type organic thin-film transistors was prepared for circuit simulations. The developed organic amplifier shows high gain of 60 dB and operates with 3 V: it can be applied to imperceptible sensor circuits for biomedical applications.

Keywords—organic thin-film transistors, low-voltage operation, differential amplifier, transistor modeling and assessment

I. Introduction

Flexible sensors to detect strain, temperature, pressure, magnetic fields, and biomedical phenomena have been developed recently, realizing a new smart sensor society [1–5]. Flexible sensor technology consists not only of signal detection elements, but also of flexible analog and digital circuits, which are fundamentally important for processing the detected signals. Among the many components of signal processing circuits, differential amplifier circuits are the most fundamental building blocks for signal amplification and simultaneous noise reduction. Organic thin-film transistors (OTFTs) present numerous benefits related to flexibility, light weight, and compatibility with low-cost fabrication processes on ultrathin flexible substrates. They are a suitable technology for realizing flexible differential amplifiers [6–7]. Nevertheless, conventional differential organic amplifiers have been difficult to drive using a supply voltage available from mobile batteries such as lithium ion batteries (3.7 V) [8–13]. Furthermore, realization of high-gain amplification remains challenging for OTFT based amplifier because no optimum simulation is available.

This report describes the fabrication of a low-voltage operating flexible organic differential amplifier with ultrahigh gain. To control the amplifier characteristics, we introduced a circuit simulation method for OTFT circuits based on an inorganic TFT simulation model. Moreover, we discuss the disagreement of simulation results with measured

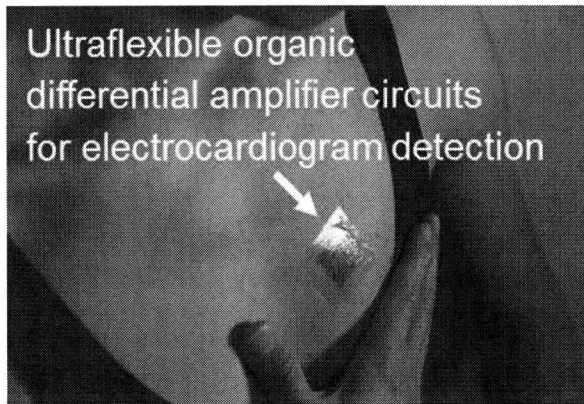

Fig. 1. Conceptual image used for electrocardiogram monitoring with an ultraflexible imperceptible sensor sheet on the human body.

characteristics of practical devices, focusing on input offset voltages. Finally, we demonstrate that consideration of threshold voltage variation and bias stress effects is important to realize practicable organic differential amplifiers that can be consistent with circuit simulation.

II. Fabrication Process

A. Ultraflexible organic circuits

Fig. 1 portrays a conceptual image of electrocardiographic (ECG) monitoring using an ultraflexible imperceptible sensor sheet with complementary-type organic circuits on a 1-µm-thick polymer substrate. The OTFT structure is presented in Fig. 2. For device fabrication, 100-nm-thick Al was thermally evaporated as gate electrodes on a 1-µm-thick polymer substrate supported by thin glass. The glass was coated by a fluorinated polymer as a sacrifice layer in advance. Next, the Al surface was anodized to form an AlOx layer, and was treated with a solution-processed self-assembled monolayer (SAM) of n-tetradecylphosphonic acid. The anodizing process

978-1-5386-5072-1/18 $31.00 © 2018 IEEE

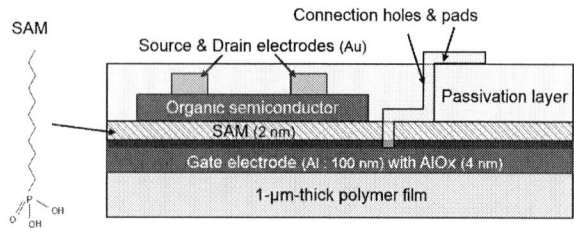

Fig. 2. Cross-sectional view of an ultraflexible organic thin-film transistor.

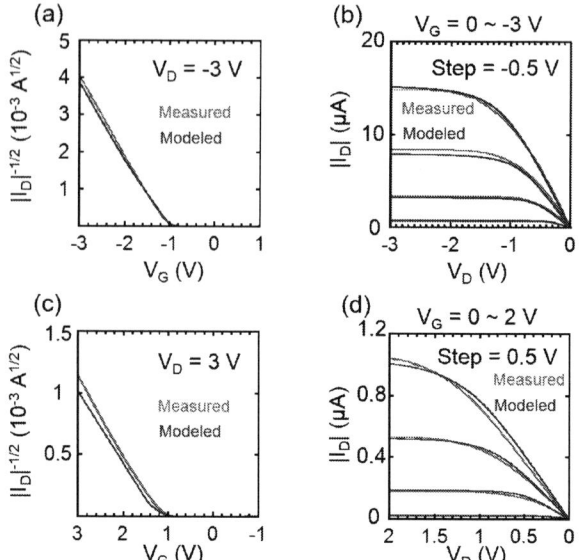

Fig. 3. Characteristics of p-channel and n-channel OTFTs: (a) transfer characteristic of an p-channel, (b) output characteristic of an p-channel, (c) transfer characteristic of an n-channel, and (d) output characteristic of an n-channel.

provides insulating high-k AlOx with low gate-leakage current of pA order. The total thickness and average capacitance of the AlOx-SAM hybrid gate insulator are, respectively, approximately 6 nm and 555 nF/cm^2. The capacitance value of 555 nF/cm^2 is higher than those of conventional flexible polymer insulators used in OTFTs. It is noteworthy that the large capacitance of the gate dielectric enables to accumulate large number of charge carriers in transistor channel between the gate dielectric and organic semiconductor at the low supply voltage of 3 V. Air-stable organic semiconductors (p-channel, DNTT; n-channel, XNRD60S01) and Au source/drain electrodes with 38 μm channel length were thermally evaporated on the hybrid insulator to complete complementary-type organic circuits. After the organic circuit fabrication, 500-nm-thick parylene was formed as an encapsulating film to suppress degradation associated with the atmosphere. Laser ablation technique was used (T-Centric Laser marker MD-T1000W; Keyence Co.) to make circuit interconnection holes for respective layers in devices. Finally, the devices were delaminated from the support glass. Even when the devices were delaminated, the electrical performance never changed. All electrodes and organic semiconductors were deposited through stencil shadow masks. Transistor modeling and circuits simulation

B. Electric characteristics of OTFTs

Fig. 3 depicts electrical characteristics of both p-type and n-type OTFTs. The channel width of each OTFT is 500 μm. All measurements were taken in a dark room in air. We used Cascade probe station and semiconductor parameter analyzer B1500A for measurements. The transistor operates in a saturation regime at no more than 3 V in p-type and n-type OTFTs. The threshold voltages were -1.07 V for p-channel and 1.20 V for n-channel. The OTFTs show good linearity for small V_{DS} and plateaus in saturation regime, suggesting ohmic contact between organic semiconductors and Au. Because of these good characteristics, the hole and electron mobility were extracted respectively from Figs. 3a and 3c as 1.27 and 0.24 cm^2/Vs, based on gradual channel approximation.

C. Transistor modeling

For simulations for OTFTs circuits, basic MOS-type FET models based on silicon technology have long been applied, although the conventional models lack compatibility with the multiplicity of process and structure in OTFT technologies. However, the proposed OTFTs showed analogous characteristics to those of amorphous silicon TFTs. Fundamentally, OTFTs and amorphous silicon TFTs are the same accumulation type transistors, and the mobility is similar in the scope of this research. Therefore, we used a standard SPICE model for transistor modeling: Level 61 PRI amorphous-silicon TFT model. To adjust the models, we firstly consider static behavior. Model parameter extraction was conducted through fitting transfer and output curves between the models and practical devices. Optimized results are shown in Fig. 3, which includes simulated curves (blue) overlaid on practical characteristics (red). In p-type OTFTs, the difference of threshold voltages between the model and measured characteristic is 30 mV. In n-type OTFTs, the maximal difference of saturation current between the model and measured characteristic is 270 nA. The models were implemented in a SPICE simulator to simulate an organic differential amplifier.

D. Circuits simulation and design

In the circuit design, we adopted a commonly used differential amplifier design with an active n-type current mirror load to fabricate high-gain organic differential amplifiers, as portrayed in Fig. 4a. For application of an organic differential amplifier to a wearable sensor, low power consumption is important to realize wireless long-term monitoring. Based on this practical point, we fixed the circuit supply voltage to 3 V and the total current to 300 nA from the supply voltage terminal to ground. As a result, the power consumption was estimated for 0.9 μW. This supply voltage and power loss are the lowest values ever reported for an organic differential amplifier [9-10,13]. We specifically examined the voltage-signal gain optimization in the low-frequency region of 1-100 Hz in the circuit simulation. This frequency region must be borne in mind to design imperceptible bio-potential sensors such as ECG sensors. Following these important limitation, the channel sizes of

2018 INTERNATIONAL CONFERENCE ON MICROELECTRONIC TEST STRUCTURES, MARCH 19-22, 2018, AUSTIN, TEXAS, USA

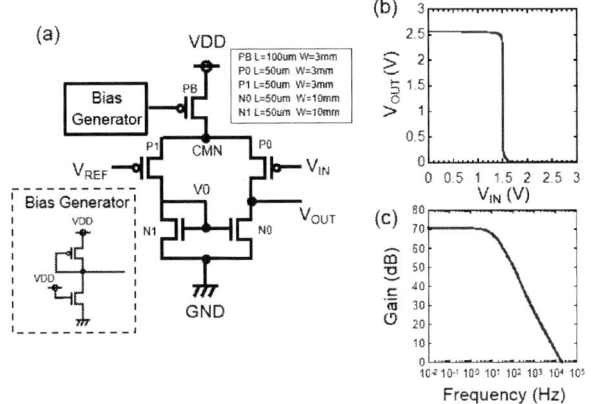

Fig. 4. Simulated organic differential amplifier: (a) circuit schematics, (b) transfer-voltage characteristics, and (c) gain-frequency characteristics.

OTFTs built in the amplifier were optimized as described in the inset of Fig. 4a. The channel width of n-type current mirror load is larger than those of p-type differential input pairs because of the superior characteristics of p-type OTFTs.

Figs. 4b and 4c respectively show simulated voltage-transfer and gain-frequency characteristics of the proposed amplifier with optimum parameters. The static transition voltage corresponds to $V_{REF} = 1/2 \times VDD$ (1.5 V), which means the input offset voltage is inevitably zero in the ideal simulation. And the maximal gain is the amplitude of 70.59 dB. This single-stage organic differential amplifier shows the highest amplifier gain among reported organic amplifier including two-stage design due to the following two reasons. One is that the large transconductance (g_m) of p-type OTFTs in differential pairs because of larger gate capacitance than that described in earlier reports [13]. The other is that n-type OTFTs belonging to a current mirror load exhibit better properties, such as mobility and current saturation (plateau) behavior, than other reported OTFTs using different materials.

E. Circuits layout

Fig. 5 depicts the test circuit layouts of organic differential amplifiers. For evaluation of variation in the circuit characteristics, we arranged eight single-stage organic differential amplifiers and eight p-type and five n-type OTFTs

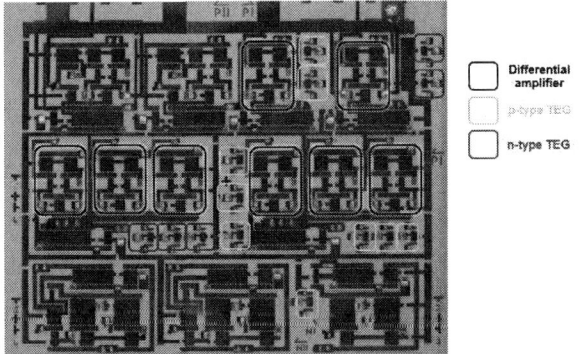

Fig. 5. Test circuit layout for orgnaic differential amplifiers.

as a test element group (TEG) in the same substrate. The channel length and width in the transistor TEG are 38 μm and 600 μm, respectively. The total device area is 38×36 mm^2 and the size of a differential amplifier is 6×7 mm^2

III. ORGANIC DIFFERENTIAL AMPLIFIER MEASUREMENT AND EVALUATION

A. Characteristics of organic differential amplifiers

Fig. 6a presents a micrograph of the organic differential amplifier. And the DC characteristics of the organic differential amplifier are depicted in Fig. 6b. The supply current was generated by a bias circuits in the amplifier. The typical voltage transfer characteristics (V_{IN} to V_{OUT}) obtained at $V_{REF} = 1.5$ V and $VDD = 3$ V show the voltage trip at 1.3 V. This result means that the offset voltage between input two terminals (V_{REF} and V_{IN}) is 200 mV in the practical device. In fact, this offset voltage is very large for practical applications aimed at detection of weak signals, such as mV level ECG signals.

In Fig. 6c, the gain-frequency characteristic is depicted and maximum gain of 60 dB at 0.02 Hz was obtained. In this measurements, we input fixed small signals to V_{IN} for 2 mV$_{p-p}$ AC sine wave with 1.5 V DC offset voltage. And V_{REF} was timely adjusted to the right amplification point for compensating the offset voltages. While measuring the gain characteristics, the amplification point changed as time advances, so that it was difficult to adjust V_{REF} to the ideal amplification point. This might be one of the possible reason that the measured gain of 60 dB is lower than the simulation value of 70.59 dB. However, the amplifier holds gain over unity up to 100 Hz and this amplification gain of the AC signal is the highest value of any reported organic differential amplifier. This frequency response is insufficient for high-

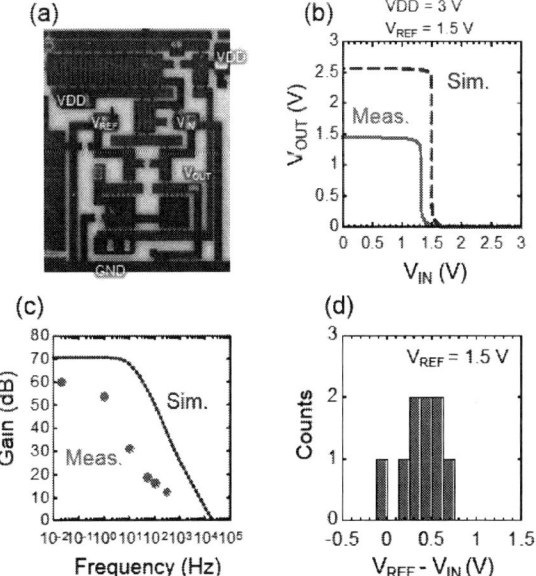

Fig. 6. Characteristics of an organic differential amplifier: (a) optical micrograph of an amplifier, (b) static characteristic, (c) gain-frequency characteristic, and (d) variation of offset voltage.

frequency application, but it is applicable for detecting low-frequency signals such as biological signals of the human body (*e.g.* ECG).

In these measurements, DC and AC characteristics were never consistent with the simulation results. Also, it is found that the both of large offset voltage and operational instability is the problem for the practical organic differential amplifier. Therefore, to investigate the characteristic dispersion, we measured several amplifiers. Figure 6d depicts the dispersion of offset voltages of differential amplifiers. The maximal and minimal offset variations are in the range from 651 mV to -123 mV. In the following sections, the reason for the large offset voltages and operational instability is carefully discussed.

B. Variation in DC characteristics of OTFTs

In the differential amplifier, the offset voltage is related to the threshold voltage of p-type OTFTs in differential input pairs and the saturation current of n-type OTFTs in the current mirror in general. Although optimized transistor modeling reasonably reproduces typical transistor characteristics as shown in Fig. 3, the variation of DC characteristics is not considered adequately. Therefore, we measured the TEG of p-type and n-type OTFTs in the same substrate. Fig. 7 presents the variation of transfer characteristics in the TEG. The maximal difference of threshold voltage among p-type TEG and saturation current among n-type TEG in the same V_G (V) are 170 mV and 400 nA. The current deviation is converted to be the offset voltage of 40 mV. As a result, total offset voltage provoked by characteristics variation is estimated to be 210 mV.

The variation in DC characteristics of OTFTs are caused by following several reasons. One possible reason is the variation of channel length, however the maximum deviation of channel length in the pair transistors was less than 3%. Other reason is the deviation of gate capacitance. Among the several substrate, the deviation of gate capacitance was less than 10%. The remaining reason might be related to the non-uniformity of the interface between semiconductor and insulator. It is well known that the condition of solution-processed SAM affects the morphology of polycrystalline organic semiconductor film. These device variability would be partially resolved by the improvement of fabrication processes.

These reasons for static variation causes at most 210 mV for offset voltage. It is still not enough to explain the

experimental large offset voltage and operational instability. In Fig. 6d, there is a notable feature of offset voltages. If the variation of OTFTs occurs randomly, the offset voltages are distributed around zero. However, the median value of offset voltages was not zero. This characteristic indicate that unbalance of pair transistors is caused by the evaluation process. Therefore, different biasing effect will be discussed in the following section.

C. Bias stress effect for operational instability

In this section, we discuss the bias stress effect of OTFTs to understand the operational instability. It is well known that voltage bias affects the threshold voltage and saturation current of OTFTs. We conducted bias stress tests using TEG transistors. First, we applied voltage of ±3 V to the gate terminal and drain electrode for 3 hours on the n-type and p-type OTFT. Before and after application of the bias stress, the transfer characteristics were measured and compared as shown in Fig. 8. That figure illustrates that the shifts of threshold voltage are approximately 1 V for p-type OTFTs and 0.5 V for n-type OTFTs, indicating that the operational instability of transistors is due to the bias stress effect. Also, these results mean that different stress condition at differential pairs is likely to produce large offset voltage in amplifier measurements.

For more detailed analysis, we conducted more practical bias stress tests imitating a bias condition while measuring DC (voltage-transfer) characteristics of differential amplifiers. Although the bias stress effect is a significant factor to consider operational instability of circuits, practical working conditions of devices have not been presumed in previous reports [14–17]. Therefore, we conducted bias stress tests using practical bias conditions. For these measurements, we applied presumed bias conditions to differential pair transistors and measured the drain current as shown Fig. 9. In Figs. 9b-9c, we assumed two different bias conditions for the differential pair of p-type OTFTs. It should be mentioned that almost the same bias condition during DC measurements will be applied to n-type OTFTs in the current mirror load. In the bias stress test, we applied DC offset voltage of 2.5 V, 1.5 V, 0 V, and 1.5 V for CMN, VOUT, VIN, and V0, respectively. These values were estimated by the practical amplifier operation and the start point (VIN=0 V) of DC measurement conditions. As shown in Fig. 9d, the current starts to decrease after stress application, and the current changes depending on bias conditions. Again, this is an obvious reason for the operational instability of

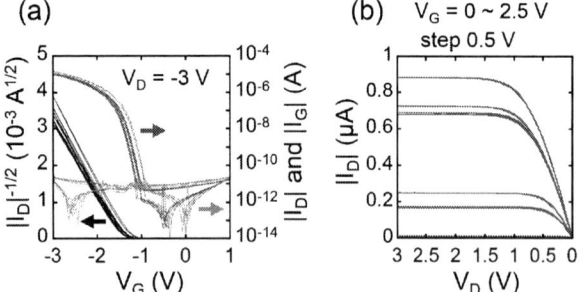

Fig. 7. Variation characteristics of OTFTs: (a) transfer characteristic of p-type OTFTs and (b) output characteristics of n-type OTFTs.

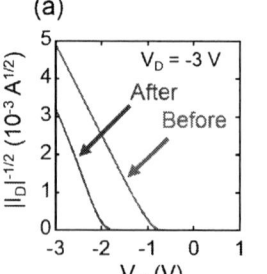
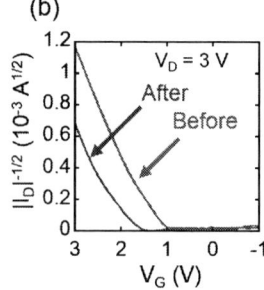

Fig. 8. OTFTs characteristics before and after bias stress for 3 hr : (a) transfer characteristic of p-type OTFTs and (b) transfer characteristics of n-type OTFTs.

978-1-5386-5072-1/18 $31.00 © 2018 IEEE

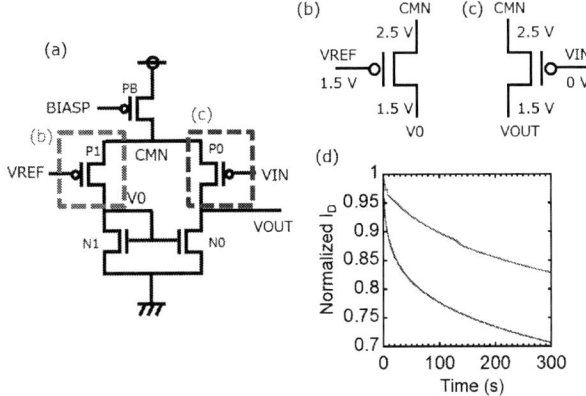

Fig. 9. Bias stress test in practical situations: schematics of (a) an organic differential amplifier and (b)–(c) imitation models, and graphs of (d) current shift by bias stress for (b) and (c).

differential amplifier. In addition, this result indicates that the balance of differential pairs changes from symmetric to asymmetric balance at the start of voltage-transfer measurement. This asymmetric characteristic causes large offset voltage of differential amplifiers. For more detail, as shown in Fig. 8, stress effect induces enhancement shift that is a negative shift of threshold voltage in p-type OTFTs. The different bias condition creates more than 10% difference in the current characteristics, which is consistent with the shift of offset voltages as shown in Fig. 6d and estimated to be ~200 mV offset voltage. Similarly, bias stress decreases the performance of current mirror, as a result, measured AC characteristics is also lower than the simulated results.

IV. CONCLUSION AND FUTURE PROSPECTS

For this work, we fabricated complimentary-type organic differential amplifiers based on transistor modeling and circuit simulations. We have successfully fabricated low-voltage operating organic differential amplifiers with the highest signal gain ever reported at the specific operation point. For transistor modeling, we used a standard TFT model in SPICE, a Level 61 PRI amorphous-silicon TFT model. Although the transistor model fitting was reasonably reproduces typical experimental characteristics, the divergence between simulated and practical amplifier characteristics were not negligible. We carefully analyzed the origin of the divergence, as a result, it was found that not only of the variation of TEG-OTFTs but also of bias stress effect strongly influenced the large offset voltages and operational instability.

Fundamentally, further improvements of OTFTs characteristics are required for practical utilization of the organic flexible differential amplifier. Firstly, bias stress effect should be suppressed. However, improved technologies and new materials to suppress bias stress have been reported and demonstrated [18-20]. More recently, we have also successfully developed very stable OTFTs by introducing new molecules, therefore the development of the differential amplifier with improved stability would be promising. In addition, further process optimization would reduce the device variability. As a result, the simulation method and results

reported in this paper will approach to the actual measurement characteristics.

ACKNOWLEDGMENTS

This study was partially supported by the Advanced Photonics and Biosensing Open Innovation Laboratory, National Institute of Advanced Industrial Science and Technology (AIST). We thank BASF Japan Co. Ltd and Nippon Kayaku Co. Ltd for the generous supply of DNTT, and XNRD60S01. We also thank Asahi Glass Co. Ltd. for the generous supply of fluorinated polymer, and Daisan Kasei Co. Ltd. for high-purity parylene (diX-SR).

REFERENCES

[1] D. J. Lipomi, M. Vosgueritchian, B. C.-K. Tee, S. L. Hellstrom, J. A. Lee, C. H. Fox, and Z. Bao, "Skin-like pressure and strain sensors based on transparent elastic films of carbon nanotubes," Nat. Nanotechnol., vol. 6, no. 12, pp. 788–792, Dec. 2011.

[2] T. Yokota, Y. Inoue, Y. Terakawa, J. Reeder, M. Kaltenbrunner, T. Ware, K. Yang, K. Mabuchi, T. Murakawa, M. Sekino, W. Voit, T. Sekitani, and T. Someya, "Ultraflexible, large-area, physiological temperature sensors for multipoint measurements.," Proc. Natl. Acad. Sci. U. S. A., vol. 112, no. 47, pp. 14533–8, Nov. 2015.

[3] S. Lee, A. Reuveny, J. Reeder, S. Lee, H. Jin, Q. Liu, T. Yokota, T. Sekitani, T. Isoyama, Y. Abe, Z. Suo, and T. Someya, "A transparent bending-insensitive pressure sensor," Nat. Nanotechnol., vol. 11, no. 5, pp. 472–478, Jan. 2016.

[4] M. Melzer, M. Kaltenbrunner, D. Makarov, D. Karnaushenko, D. Karnaushenko, T. Sekitani, T. Someya, and O. G. Schmidt, "Imperceptible magnetoelectronics," Nat. Commun., vol. 6, p. 6080, Jan. 2015.

[5] T. Sekitani, T. Yokota, K. Kuribara, M. Kaltenbrunner, T. Fukushima, Y. Inoue, M. Sekino, T. Isoyama, Y. Abe, H. Onodera, and T. Someya, "Ultraflexible organic amplifier with biocompatible gel electrodes," Nat. Commun., vol. 7, p. 11425, Apr. 2016.

[6] M. Kondo, T. Uemura, T. Matsumoto, T. Araki, S. Yoshimoto, and T. Sekitani, "Ultraflexible and ultrathin polymeric gate insulator for 2 V organic transistor circuits," Appl. Phys. Express, vol. 9, no. 6, p. 61602, Jun. 2016.

[7] Y. Takeda, K. Hayasaka, R. Shiwaku, K. Yokosawa, T. Shiba, M. Mamada, D. Kumaki, K. Fukuda, and S. Tokito, "Fabrication of Ultra-Thin Printed Organic TFT CMOS Logic Circuits Optimized for Low-Voltage Wearable Sensor Applications," Sci. Rep., vol. 6, no. 1, p. 25714, Sep. 2016.

[8] H. Marien, M. S. J. Steyaert, E. van Veenendaal, and P. Heremans, "A Fully Integrated ΔΣ ADC in Organic Thin-Film Transistor Technology on Flexible Plastic Foil," IEEE J. Solid-State Circuits, vol. 46, no. 1, pp. 276-284, 2011.

[9] M. Guerin, A. Daami, S. Jacob, E. Bergeret, E. Bènevent, P. Pannier, and R. Coppard, "High-gain fully printed organic complementary circuits on flexible plastic foils," IEEE Trans. Electron Devices, vol. 58, no. 10, pp. 3587-3593, 2011.

[10] G. Maiellaro, E. Ragonese, A. Castorina, S. Jacob, M. Benwadih, R. Coppard, E. Cantatore, and G. Palmisano, "High-Gain Operational Transconductance Amplifiers in a Printed Complementary Organic TFT Technology on Flexible Foil," IEEE Trans. CIRCUITS Syst. Regul. Pap., vol. 60, no. 12 pp. 3117-3125, 2013.

[11] J. Chang, X. Zhang, T. Ge, and J. Zhou, "Fully printed electronics on flexible substrates: High gain amplifiers and DAC," Org. Electron. Physics, Mater. Appl., vol. 15, no. 3, pp. 701-710, 2014.

[12] V. Vaidya, D. M. Wilson, X. Zhang, and B. Kippelen, "An organic complementary differential amplifier for flexible AMOLED applications," in Proceedings of 2010 IEEE International Symposium on Circuits and Systems, 2010, pp. 3260–3263.

[13] V. Pecunia, M. Nikolka, A. Sou, I. Nasrallah, A. Y. Amin, I. McCulloch, and H. Sirringhaus, "Trap Healing for High-Performance Low-Voltage Polymer Transistors and Solution-Based Analog Amplifiers on Foil," Adv. Mater., vol. 29, no. 23, 2017.

[14] J. B. Chang and V. Subramanian, "Effect of active layer thickness on bias stress effect in pentacene thin-film transistors," Cit. Appl. Phys. Lett. Appl. Phys., vol. 88, no. 92, pp. 233513–6028, 2006.

[15] U. Zschieschang, R. T. Weitz, K. Kern, and H. Klauk, "Bias stress effect in low-voltage organic thin-film transistors," Appl. Phys. A Mater. Sci. Process., vol. 95, no. 1, pp. 139–145, 2009.

[16] K. Fukuda, T. Suzuki, D. Kumaki, and S. Tokito, "Reverse DC bias stress shifts in organic thin-film transistors with gate dielectrics using parylene-C," phys. status solidi, vol. 209, no. 10, pp. 2073–2077, Oct. 2012.

[17] S. Bisoyi, U. Zschieschang, M. J. Kang, K. Takimiya, H. Klauk, and S. P. Tiwari, "Bias-stress stability of low-voltage p-channel and n-channel organic thin-film transistors on flexible plastic substrates," Org. Electron., vol. 15, no. 11, pp. 3173–3182, Nov. 2014.

[18] K. Fukuda, T. Suzuki, T. Kobayashi, D. Kumaki, and S. Tokito, "Suppression of threshold voltage shifts in organic thin-film transistors with bilayer gate dielectrics," Phys. Status Solidi Appl. Mater. Sci., vol. 210, no. 5, pp. 839–844, 2013.

[19] T. sekitani, S Iba, Y. Kato, Y. Noguchi, T. Someya, and T. Sakurai, "Suppression of DC bias stress-induced degradation of organic field-effect transistors using postannealing effects," Appl. Phys. Lett. Appl. Phys., vol. 87, no. 7, p. 073505, 2005.

[20] K. Fukuda, T. Suzuki, D. Kumaki, and S. Tokito, "Reverse DC bias stress shifts in organic thin-film transistors with gate dielectrics using parylene-C," Phys. status solidi, vol. 209, no. 10, pp. 2073–2077, Oct. 2012.

Session 5

Mismatch and Variability

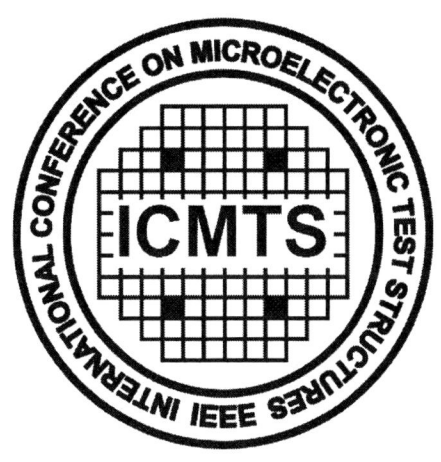

March 21, 2018

09:00-10:20

Co-Chairs:

Satoshi HABU, Keysight Technologies, Japan

Hi-Deok LEE, Chungnam National University, Taiwan

978-1-5386-5072-1/18 $31.00 © 2018 IEEE

A test structure to reveal short-range correlation effects of mismatch fluctuations in backend metal fringe capacitors

Hans Tuinhout, Adrie Zegers-van Duijnhoven and Ihor Brunets

NXP Semiconductors

High Tech Campus 46, 5656 AE Eindhoven, The Netherlands

Email: hans.tuinhout@nxp.com

Abstract—This paper presents a set of test structures that revealed a thus far unknown (or at least unreported) CMP-related short-range correlated mismatch fluctuation effect on the matching of backend metal fringe capacitors. It is shown that an apparent degradation of mismatch standard deviations at medium-range distances is in fact due to an improvement of matching for devices placed at very small distances.

Keywords— matching, backend metal fringe capacitors, distance effects, chemical-mechanical polishing

I. INTRODUCTION

Backend metal fringe capacitors (FringeCaps) are made of stacked interdigitated metal combs (Fig. 1, 2). In contemporary 6 to 10 metal-layer IC-technologies, FringeCaps are fabricated through copper damascene processing utilizing chemical-mechanical polishing (CMP). FringeCaps can have a relatively high capacitance per unit area, low voltage- and temperature dependencies and presumably have good matching. FringeCaps are therefore attractive components for realizing high-precision signal processing circuits, such as Analog-to-Digital Converters (ADCs), often based on switched binary multiples of unit capacitors. The size of the unit capacitor determines the speed/power/accuracy trade-off for the convertor. The linearity and the achievable resolution of the ADC depend heavily on the matching of these capacitors.

The matching performance of back-end metal FringeCaps is sensitive to many layout variables. Good matching is indeed achievable for large capacitors, see for instance [1], but only when the two (larger) capacitors of a matched pair are created through dense (common-centroid) checkerboards of many small unit cells. This is confirmed by guidance as given in design manuals of silicon wafer foundries. However, using checkerboarded (matched) common-centroid capacitors is not always feasible for high precision mixed-signal circuit realizations. Therefore, it is necessary to characterize the matching also for capacitor pairs with different areas, different aspect ratios and different placements. To our surprise we found that when the unit cell-size is increased substantially, the mismatch fluctuation standard deviations of FringeCaps degrades substantially to equivalent Pelgrom factors larger than 2%√(fF), even when using our proven common-centroid DUT-1-2-1-2 test structures [2]. Moreover, the resolution of a capacitive ADC is determined by the total number of unit cells (easily going up to several hundred!). It is therefore important to be able not only to model the matching properties of FringeCaps

of "regular" equal sized matched capacitor pairs at minimum distances, but also of the random variations of binary ratios. The resulting arrays of 2^6 to 2^{10} capacitors in high resolution ADC's inevitably span sizable areas, implying that all matched capacitor cells cannot be placed at minimum distance from each other even when they can be placed common-centroid. Large distances (say 100 µm or larger) are generally known to give rise to significant mismatch deterioration due to deterministic parametric wafers gradients. Common-centroid layout solutions are usually sufficient to mitigate effects of parametric gradients on matching [2]. However, this did not to work for backend metal FringeCap arrays in our studies.

To unravel the intricacies of FringeCap matching, we did a number of experiments to find the optimum layouts for matched FringeCaps. Part of this was a set of conventional DUT-1-2-1-2 matched FringeCaps test structure that revealed a -thus far-unknown (or at least unreported) distance-related matching mechanism. These structures and the -at first sight- astonishing results are discussed in this paper.

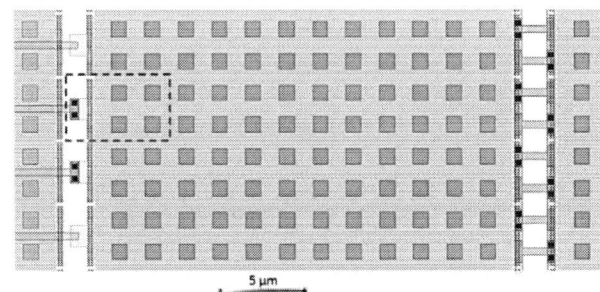

Figure 1. Capture of gds of FringeCap in the test structure. For clarity, only the M5 layer of the combs, two VIA layers and the M6 for the connections to the DUTs are drawn. Larger squares are the intentionally placed dummy tiles for meeting M6 density rules. The dashed region is enlarged and depicted in Fig.2.

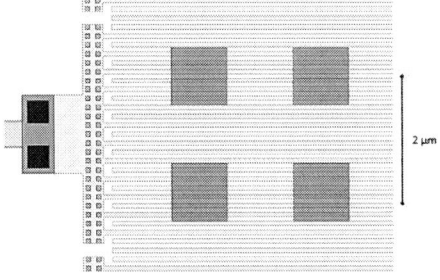

Figure 2. Capture of gds of detail of FringeCap (dashed rectangle of Fig 1.).

2018 INTERNATIONAL CONFERENCE ON MICROELECTRONIC TEST STRUCTURES, MARCH 19-22, 2018, AUSTIN, TEXAS, USA

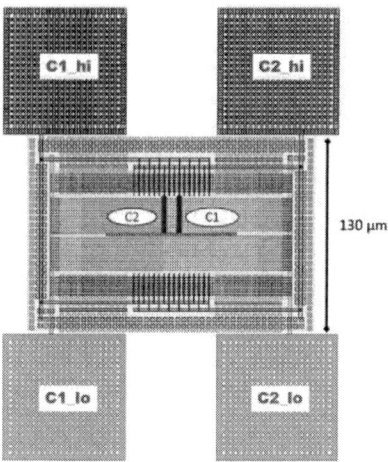

Figure 3. Capture of gds of FringeCap matched pair for DUT-1-2-1-2 mismatch characterization. Connection highlights depict which capacitor is connected to which pad. The upper left (blue) pad (labelled C1_hi) is connected to the right capacitor (C1), while the upper right (red) pad (C2_hi) is connected to the left capacitor (C2). Lo connections to the capacitors are not highlighted.

II. TEST STRUCTURES

A. DUT-1-2-1-2 matched capacitor pairs

The basic idea of the test structures used in this work is depicted in Fig. 3. A standard 2x2 probe-pad module contains two rows of 16 equally laid-out capacitors of which, in this case, only two capacitors are actually connected (C1 and C2). This layout configuration mimics the type of implementation as used in switched capacitor arrays for ADC applications. For the DUT-1-2-1-2 characterization, the capacitors are connected to the pads through so-called "outside quadruple rails", a modified version from what was discussed in [2]. The connection frame is fixed for the 32 capacitors and by placing the VIA5's (see Fig.1 and 2), the desired DUT capacitors are selected. Note that the two capacitors in Fig. 3 are (intentionally) not placed at minimum distance, whereas those in Fig.1 are.

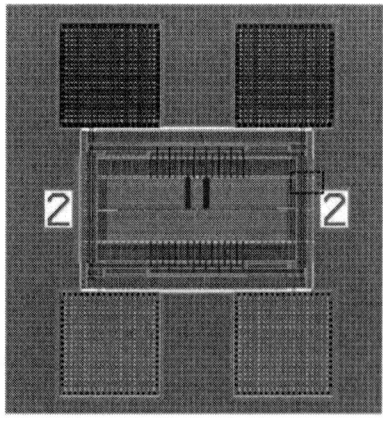

Figure 4. The matched pair is placed in a sea of metal with dummy capacitors in an area of 200 x 130 µm² and scripted dummy tiles in the remainder of the module to avoid density gradients. The dashed rectangle region is enlarged and depicted in Fig.5.

B. DUTs in a "sea of metal"

One of the main concerns for extreme high-precision mismatch characterization of backend metal FringeCaps is assuring exact equality of the layout environment in the vicinity of the supposedly identical components. To exclude effects of layout density gradients possibly affecting CMP and etching homogeneity, the DUTs are placed in so called "seas of metal", in this case formed by 200x130 µm² fields of DUTs and dummy capacitors with identical lines and spaces (see Fig.1). In Fig. 4, the (chip finishing) scripted dummy tiles are also plotted, demonstrating that the layer density is maintained within the required density design rules over the entire 4-pad module (and even beyond). Obviously, the module ID must be kept free of dummy tiles. Fig. 5 depicts an enlargement of the region in the module (far away from the DUTs), where the dummy DUTs region ends and the scripted dummy tiling starts.

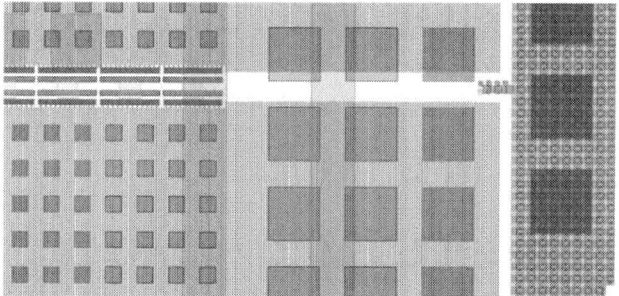

Figure 5. Capture of gds of detail of module (dashed rectangle of Fig 4.) showing the density regularity of the dummy FringeCaps and the transition to the scripted dummy tile region.

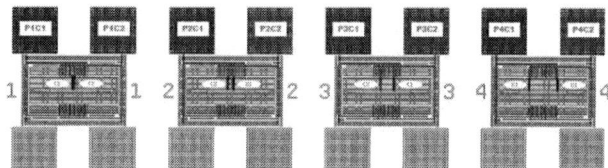

Figure 6. Four DUT-1-2-1-2 matched capacitor pairs placed at varying distances within their "sea of metal" region. A connection highlight tool, is used to indicate which pad is connected to which capacitor. Blue is used for the C1 capacitors, red for the C2's. Pad labels indicate which pad is connected to which DUT.

C. The four short-distance matched pairs

Fig. 6 shows the four DUT-1-2-1-2 matched capacitors pairs used for this study. All pairs are placed in exactly the same "sea of metal" environment. As described above, the pairs are formed through placing the VIA5's on the desired hi and lo taps of two of the FringeCaps. Fig. 7 shows a detail of the two connected capacitors of Pair 2. The distance between the two connected capacitors is approximately 9.8 µm for this pair. The distance between the two capacitors in a module varies from about 3 µm for P1 to 49 µm for P4. This is in line with short-range distance effects as encountered in typical SAR-ADC capacitor arrays.

978-1-5386-5072-1/18 $31.00 © 2018 IEEE   88

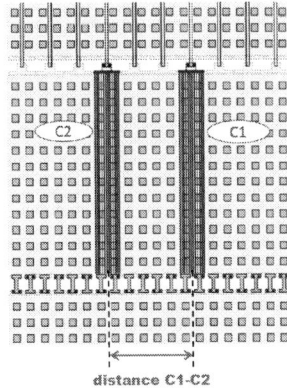

Figure 7. Capture of gds of detail of pair 2. The distance between C1 and C2 is approximately 9.8 µm for this pair

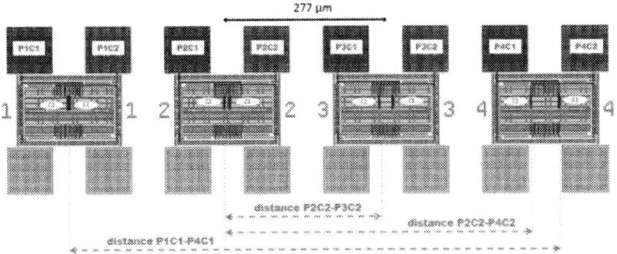

Figure 8. Capture of gds of the four pairs. The module to module distance is approximately 277 µm (on silicon). The arrows indicate three of the possible combinations, covering distances of about 280, 560 and 840 µm respectively.

D. The long-distance matched pairs

Fig. 8 shows how the four "short-distance" DUT-1-2-1-2 matched capacitors modules can be utilized to characterize the impact of much larger distances. As the connected capacitors in each module have exactly the same ("sea of metal") environments and are measured within minutes of each other, it is possible to treat the C1's (or the C2's) of two neighboring modules on each chip again as a pair. The idea is that the devices of these pairs are so far away from each other, that parametric wafer gradients affect the effective mismatch standard deviations. One of the purposes of the study described in this paper was (also) to assess whether the commonly expected parametric wafer gradient impacts could be observed for these FringeCaps. With an on silicon distance of 277 µm between the modules, and a typical parametric gradient contribution of 1 to 2 ppm/µm, its impact should be distinguishable.

III. MEASUREMENTS

The FringeCaps were designed and fabricated in a standard 40 nm CMOS foundry technology. Each of the M1-to-M5 fringe capacitors is approximately 3×25 µm^2 large (Fig. 1). The combs have the minimum allowed finger-widths and -spacings of 70 nm (Fig. 2). The capacitors were targeted at a capacitance value of 280 fF. Note that at this capacitance level, the expected mismatch fluctuation standard deviation is of the order of 0.1 %, implying the need for careful high-precision measurements.

The devices were measured with the capacitance meter of a Keithley 4200-SCS (@ 1 MHz, 99 mV AC, quiet integration,

99 samples) as described in [1]. The (random) measurements system noise contribution to the mismatch observation with the capacitance meter in the above settings is about 30 to 40 aF (after taking the median of the 99 samples). This corresponds to about 0.01 % of mismatch standard deviation for the targeted capacitance value.

All results discussed in this paper come from a 300 mm wafer, probed using an atto-guarded S300 Cascade-Microtech semi-auto wafer prober. Each capacitor is probed twice using two DCM-coaxial needles which are connected to the Hi and Lo terminals of the 4200-CVU capacitance meter. The DUT-1-2-1-2 measurement sequence is achieved through subsite stepping forth-back-forth with the prober. The population size consists of 138 pairs distributed over the 300 mm wafer (Fig. 9).

Figure 9. A 3-D wafer map of the (P4C1) capacitance variation of the measured 138 dies on a 300 mm wafer.

IV. RESULTS

A. Wafer maps and sanity checks

Fig 9. shows an example of the capacitor variation for C1 of pair 4 (P4C1) across the wafer. The median capacitance is approximately 277 fF. The (non-random) spread across the wafer corresponds to a standard deviation of 1.90 %. All measured C1's and C2's from pairs 1 to 4 (obviously) show the same characteristic pattern of a hill surrounded by a valley.

Fig. 10 shows the 3-D wafer map of the relative capacitance mismatch $\Delta C/C = 200 \times (C1 - C2) / (C1 + C2)$ of the pair number 4. Note the (expected) random pattern across the wafer, a prerequisite for proper interpretation of mismatch fluctuations.

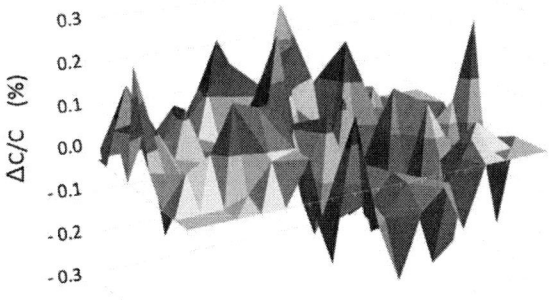

Figure 10. 3-D wafer map of the (P4C1-P4C2) capacitance mismatch variation across the wafer.

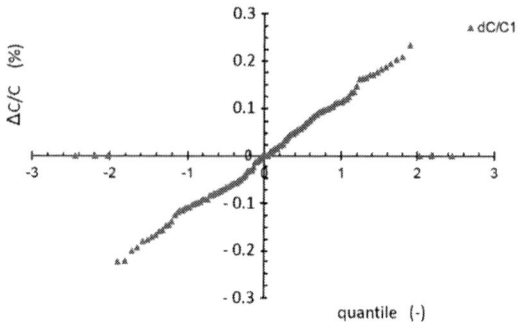

Figure 11. Cumulative probability plot (Q-Q plot) of the capacitance mismatch as measured for pair 4. A straight line indicates that a Normal distribution is appropriate for this population and the dy/dx of the line represents the standard deviation.

Fig. 11 shows the normal-scaled cumulative probability plot of the capacitor mismatch of the 138 sample population of pair number 4. Note the almost perfect Normal (Gaussian) distribution. The robustly estimated standard deviation of the mismatch is 0.11 % for this population.

B. Short range mismatch degradation

The key result of this study is depicted in Fig 12. When the spacing between the devices of a pair increases from the minimum of 3.3 µm for pair 1, to 49 µm for pair 4, the capacitor mismatch standard deviation increases from about 0.033 %, which is quite good, to about 0.107 %, which is substantially higher than originally hoped for. Such an increase is much more than what can normally be expected from parametric gradients. If we assume a conservative 2%/cm (or 2ppm/µm) gradient contribution, a distance of 50 µm between the two DUTs of a pair can add no more than 0.01 % to the mismatch standard deviation. The observed increase in Fig. 12 is clearly substantially (and statistically significant) higher.

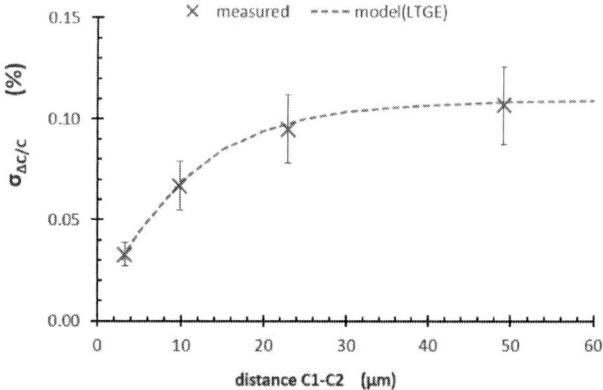

Figure 12. Capacitor mismatch standard deviations of pairs 1 to 4 as a function of distance between the capacitors of the pair. Error bars represent the (3σ) statistical uncertainty associated with a Gaussian distributed population of 138 samples. The blue dashed line is a line to guide the eye (LTGE), discussed in section V.

C. Long-range mismatch degradation

Using the approach as described in section II-D the expected (inevitable) impact of very large distances on the matching of the FringeCaps was determined. The result is depicted in Fig. 13. As expected, the mismatch increases for distances of the order of 300 to 900 µm, but it is clear that these parametric gradients start contributing significantly at distances well above a hundred microns. Also note that the parametric-gradient-induced-increase for large distances has a very different characteristic length compared to the drop that is seen below 100 µm, suggesting that we are looking at the effects of two very different physical mechanisms. A plausible explanation for this drop below 100 µm is discussed below in section V.

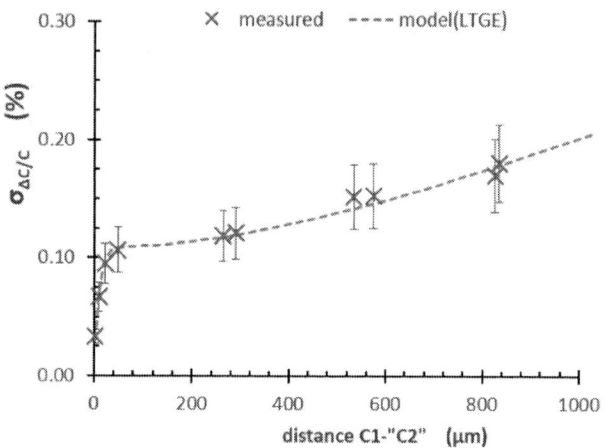

Figure 13. Capacitor mismatch standard deviations of pairs 1 to 4 and of mixed module pairs as a function of distance between the capacitors of the pair. Error bars represent (3σ) statistical uncertainty associated with a Gaussian distributed population of 138 samples. The blue dashed line is a line to guide the eye, discussed in section V.

V. DISCUSSION

While a result as shown in Fig. 12 was actually in line with what we were looking for in this study, it still came as a (pleasant) surprise. The mechanisms causing unexpectedly large mismatch fluctuations in some of our previous FringeCap mismatch test structures were not understood, but some of the seemingly contradicting results led to designing the new structures as discussed in this paper.

The result as depicted in Fig. 12 was initially interpreted as very good matching (0.033 %) at short distances with an additional fluctuation mechanism becoming dominant for larger distances. However, the practical question then became: which (random) physical mechanism can be held responsible for a perfectly random capacitor mismatch variability at larger distances (Fig. 10 and 11), while not affecting DUTs placed at shorter distances?

Past experiences with similar unexpected results were that one should always think first of test structure artefacts related to the layout environment of the DUTs (in the arrays). This is why the test structures were so painstakingly placed in their "seas of metal", assuring that local gradients of lithography, etching, and CMP of the copper damascene technology could have absolutely no influence on the observed mismatch fluctuations. Each capacitor in the four DUT arrays is truly identical at the M1 to M5 levels that determine the capacitance values and their matching.

So, in the end it was finally realized that what was originally seen as a matching **degradation** at larger distances due to an additional -not understood- fluctuation mechanism, should in fact interpreted as a matching **improvement** for pairs that are very close together, due to local correlation of the mismatch cause. This may sound as a silly "chicken or egg" argument, but this in fact should be seen as a paradigm change in the thinking about (FringeCap) matching. The theoretical statistical explanation now becomes as follows: Assume there is a random mismatch fluctuation mechanism associated with physical disturbances that are significantly larger than the distance between the pairs. Then it makes sense that parametric changes caused by such disturbances become correlated, and hence yielding smaller mismatch standard deviations when the devices are small and placed close together.

Our hypothesis is that the physical cause of these correlated fluctuations must be found in the CMP processing that is applied for creating the FringeCaps dense copper combs. CMP can be seen as an elegant, but rather coarse processing step involving a large spongy polishing pad and a slurry with chemicals and abrasive particles. To maintain process control/uniformity, the polishing pad is regularly cleaned/roughened with a conditioner disc with diamonds of the order of 10 μm diameter. The reconditioning will create or open-up (clotted) trenches in the polishing pad that are essential for dispersing the slurry and removing polished material.

It is well-known [3,4] that CMP of copper causes copper line thickness variations due to dishing of wide lines (Fig. 14 a.) and erosion of narrow oxide pillars (Fig 14 b.). Our physical model is that this erosion shows a variability pattern (Fig. 15) with a range (correlation distance) of the order of 10 μm. This explains why small capacitors at small distances can be in the same valley or hill and hence vary less (or rather vary together), whereas at larger distances they vary uncorrelated.

One could argue that the sort of variability as depicted in Fig. 15 would have easily been spotted on SEM or TEM cross-sections, but this is in fact not so likely. Fig. 15 is obviously not drawn to scale. The real magnitude of local thickness variations required to explain a 0.1 % mismatch standard deviation (Fig. 12) amounts to no more than a fraction of a nanometer on the 100 to 200 nm thick copper lines used in these FringeCaps. Such thickness variations will not be spotted, and certainly not be a major concern for fabs. Remember that the typical (and accepted) overall wafer spread on the capacitor values is a few percent (Fig. 9) corresponding to a +/- 3 to 5 nm thickness control for each metal layer over the wafer. However, when the matching magnifying glass is used, with larger devices, appropriate test structures and ditto measurement methods, such seemingly minor effects do become significant.

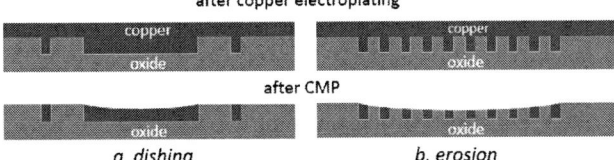

Figure 14. Schematic representation of differences between dishing (a.) and erosion (b.), well know artefacts of CMP on vertical dimensions of copper lines in contemporary copper damascene back-end technologies.

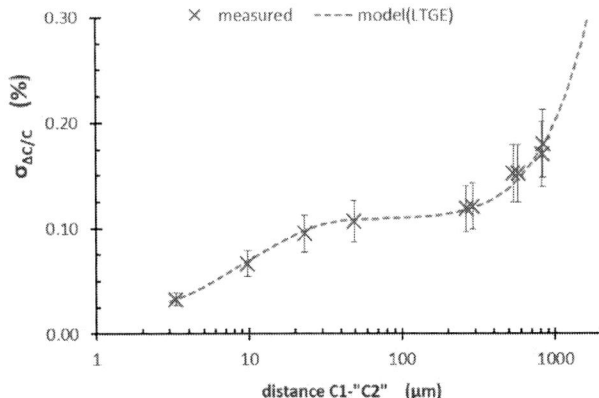

Figure 15. Schematic representation of variability of erosion due to inhomogeneities of the polishing pad. Vertical copper thickness variability is **not** drawn to scale.

With the two new insights obtained in this study, namely the occurrence of short-range correlated fluctuations in the region below 100 μm and the emergence of a long-range parametric gradient contribution above (say) 200 μm, a heuristic equation can be derived that was in fact added to Figures 12, 13 and 16 as the line to guide the eye (LTGE). This line is based on the square-root of the quadratic sum of the variances of three variability elements. With d being the effective distance between the two capacitors in μm and C the median of the capacitor (277 fF in this case), the LTGE as plotted in Fig 12, 13 and 16 is given by

$$LTGE(d) = A_{eff}(d) / \sqrt{(C)}$$

$$= \sqrt{(A^2_{micro} + A^2_{EV}(d) + A^2_{grad}(d))} / \sqrt{(C)}$$

The three A-factors are defined as follows:

- A_{micro} represents the conventional intrinsic random mismatch standard deviation area scaling Pelgrom-factor, here 0.2 %√fF.

- The factor A_{EV} represents the CMP erosion variability that reduces when the distance (d) between the two capacitors of a pair becomes small. For the LTGE (model) was chosen for a negative exponential correlation equation with a characteristic distance D_0.

  $$A_{EV}(d) = A_{CMP} \times (1-\exp(-d/D_0)) ,$$

  with $A_{CMP} = 1.8$ %√fF and $D_0 = 10$ μm.

- The parametric gradient contributor A_{grad} that scales linear with distance d.

  $$A_{grad}(d) = \sqrt{(C)} \times 1E\text{-}4 \times P_{grad} \times d$$

  with $P_{grad} = 1.7$ %/cm.

Figure 16. Same data as in Fig. 13 but plotted using a Log-scale for the X-axis.

Although the LTGE line in the figures 12, 13 and 16 is based on a sensible combination of realistic physical effects, it should be noted that this is not intended as a physical model based on first principles. It fits well for the DUT layout in this experiment, but it is not claimed that the same parameters can be used for any other geometric realizations of backend metal FringeCaps. However, the results obtained through this study do provide valuable insights and therefore guidelines on what can, at least qualitatively, be expected for other layouts and has helped to understand some of the originally surprising results of other FringeCap mismatch studies not further discussed in this paper.

The most important findings of this study are that the intrinsic microscopic mismatch fluctuations are in fact really small. Note that an A_{micro} of 0.2 %√fF results in a mismatch standard deviation of no more than 0.012 % for a 277 fF capacitor. On the other hand, the A_{CMP} of 1.8 %√fF describing the CMP erosion variability component is in fact much larger than hoped for by many mixed-signal system architects. This work does explain why very low mismatch standard deviations for large devices can (indeed) only be achieved through very tight common-centroid checkerboarding. And finally, the mismatch estimators calculated from populations of pairs formed by taking devices placed in adjacent modules quite convincingly confirm the expectation that long-range distance effects associated with parametric wafer gradients act at a much larger distance scale than the tens of microns observed in Fig. 12 and are hence usually negligible for standard matched pairs. However, this factor must be taken into account for large arrays of DUTs when the components cannot be divided into small unit cells and placed in a common-centroid checkerboard layout.

VI. CONCLUSION

A new set of test structure provides new insights in the matching properties of backend metal FringeCaps. The matched pairs are placed in "seas of metal" to assure absolute environmental identicality of each device, no matter where the devices are placed in their DUT array. This averts unpredictable artefacts on the mismatch from the damascene backend processing. High-precision capacitance measurements are used to correctly asses mismatch standard deviation levels down to 0.03 %. Although not proven, a reasonable hypothesis is that the matching improvement of pairs at short to medium distances is caused by a previously unknown -at least unreported- impact of CMP erosion variability. This results in correlated, and hence smaller, random mismatch fluctuations for closely spaced FringeCap matched pairs. This interpretation is a paradigm change in the way of thinking about matching and random variability effects.

ACKNOWLEDGMENT

Robert Jones, Kurt Juncker, Erik Olieman, Alphons Litjes, Ibrahim Candan, Ashish Panpalia, Nicole Wils and Céline Detcheverry of NXP are kindly acknowledged for their fruitful discussions.

Thiago Hanna Both of Universidade Federal do Rio Grande do Sul in Porto Alegre, Brazil is particularly acknowledged for planting the statistical correlation seed during his internship at NXP, without realizing which magnificent tree would grow out of this seed…

REFERENCES

[1] H. Tuinhout, F. van Rossem, N. Wils, "High precision on-wafer backend capacitor mismatch measurements using a benchtop semiconductor characterization system", ICMTS-2009, Proceedings 2009 IEEE Conference on Microelectronic Test Structures, pp. 3-8, 2009.

[2] H. Tuinhout, N. Wils, "A cross-coupled common centroid test structures layout method for high precision MIM capacitor mismatch measurements", ICMTS-2014, Proceedings 2014 IEEE Conference on Microelectronic Test Structures, pp. 243-248, 2014.

[3] R. Chang, Y. Cao, C.J. Spanos, "Modelling electrical effects of metal dishing due to CMP for on-chip interconnect optimization", IEEE Transactions on Electron Devices. Vol. 51 Issue 10, pp. 1577-1583, October 2004.

[4] A. Toffoli, S. Maitrejean , J. Duport de Pontcharra, F. de Crecy, D. Bouchu, L. Arnaud, F. Boulanger, "Test structure for characterizing metal thickness in damascene CMP technology", ICMTS-2008, Proceedings 2008 IEEE Conference on Microelectronic Test Structures, pp. 210-213, 2008.

2018 INTERNATIONAL CONFERENCE ON MICROELECTRONIC TEST STRUCTURES, MARCH 19-22, 2018, AUSTIN, TEXAS, USA

Monte Carlo Analysis by Direct Measurement using V_{th}-shiftable SRAM Cell TEG

Shogo Yamaguchi, Daisuke Nishikata, Hitoshi Imi and Kazuyuki Nakamura
Center for Microelectronic Systems, Kyushu Institute of Technology
680-4 Kawazu, Iizuka, Fukuoka, JAPAN 820-8502
Telephone: +81-948-29-7584 Fax: +81-948-29-7586
Email: shogo_yamaguchi@cms.kyutech.ac.jp

Abstract— The measurement system in which the Monte Carlo analysis of SRAM operation can be performed in actual measurement using V_{th}-shiftable SRAM cell TEG (VTST) was developed. The dynamic V_{th}-shift circuit (DVSC) using electrolytic capacitors and mechanical relays for setting individual V_{th}-shift voltages for six MOSFETs in a memory cell enables to share a programmable external voltage source. The measured results of the Monte Carlo analysis for SRAM function test and the static noise margin evaluation were agreed well with the simulated results. The proposed method can compactly cope with the recently proposed SRAM with a larger number of transistors.

Keywords—SRAM, Monte Carlo analysis, SPICE, static noise margin, ratioless SRAM

I. INTRODUCTION

For SRAM using state-of-the-art processes of deep sub-micron generation, it is becoming difficult to secure operating margins such as static noise margin (SNM) due to the lowering of power supply voltage and the increase in the device characteristic variation [1]. To overcome this problem, we had proposed the 12-transistor ratio less SRAM (RL-SRAM) design [2,3]. In order to demonstrate the immunity of the RL-SRAM operation for the device characteristic variability by the measurement, we had developed MOSAIC SRAM cell TEGs with intentionally added device variability within the gate width [4]. We had also developed the V_{th}-shiftable SRAM cell TEGs (VTSTs) to demonstrate the immunity for the device variability in the threshold voltage (V_{th}) of MOSFETs [5]. Figure 1 shows simple models for V_{th}-shifts in the MOSFETs in 6-transistor SRAM (6T-SRAM) cell. Each MOSFET may have the inherent V_{th}-shift caused by the device characteristic variability. Figure 2 shows the VTST for 6T-SRAM in which the all gate terminals of MOSFETs are extracted to the chip boundary and tapped with the analog I/O (AIO) buffers. In our previous research, six external voltage sources (EVSs) were manipulated to apply the fixed V_{th}-shift to each MOSFET to evaluate the fail condition maps (FCMs) and critical-ΔV_{th}s [5].

In this paper, we construct the advanced measurement system to evaluate the statistical operation margin for SRAM in measurement, as well as the Monte Carlo analysis in SPICE simulation.

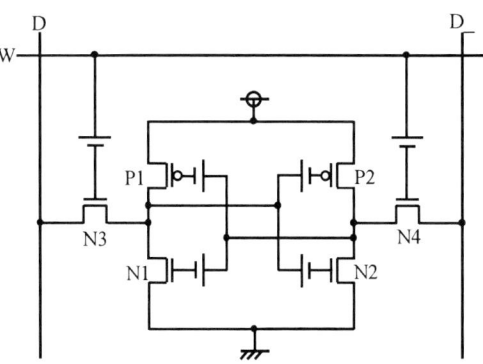

Fig. 1. Model of V_{th}-shifts in 6T-SRAM.

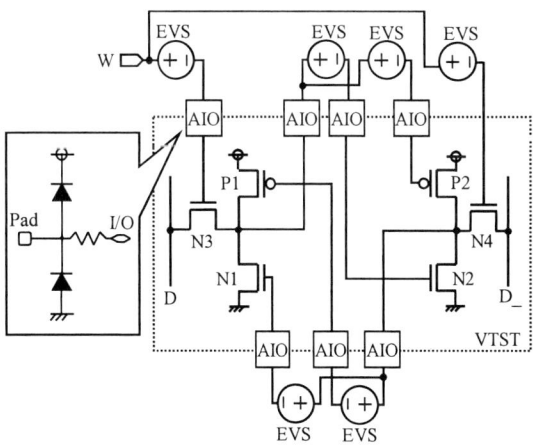

Fig. 2. VTST for 6T-SRAM.

II. DYNAMIC V_{TH}-SHIFT CIRCUIT

The Monte Carlo analysis performed in the SPICE is to repeat simulations by randomly modifying design values to be

978-1-5386-5072-1/18 $31.00 © 2018 IEEE   93

on the Gaussian distribution. In our previous research as shown in Fig. 3 (a), the V_{th}-shift was set manually by using multiple EVSs, however, in order to perform the Monte Carlo analysis in measurement, it is necessary to automate the setting of the individual V_{th}-shift values. Therefore, as shown in Fig. 3 (b), we employ only one programmable EVS (P-EVS) that generates V_{th}-shift values that are successively charged in target capacitors connected to the gates of the respective MOSFETs. In this method, EVSs as many as the number of transistors are not required, and further, it is possible to avoid influences such as characteristic variations among the EVSs. Figure 4 shows the time dependence of the discharge characteristics of some capacitance components. Since mechanical relays are used in our measurement system to switch connections, the retention time of sub-second order is required. For that purpose, an aluminum electrolytic capacitor of 1000-μF or more should be employed.

Figure 5 shows a dynamic V_{th}-shift circuit (DVSC) for setting individual Vth-shift voltages for each MOSFET. It consists of an electrolytic capacitor (Cvs) and two double-pole double-throw (DPDT) mechanical relays (MR1 and MR2). The input voltage of the DVSC is supplied by the common P-EVS with the values of the Gaussian distribution. Then it applies to the Cvs in DVSC via MR1 to store the respective Vth-shift by the charge select (CS) signal (Fig. 5 (a)). Since the aluminum electrolytic capacitor is the polarized electronic component, the second DPDT relay (MR2) controlled by the polarity select (PS) signal is further required to support the negative V_{th}-shifts (Fig. 5 (b)).

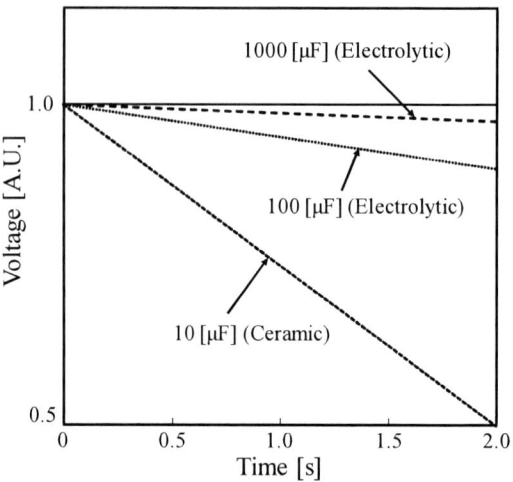

Fig. 4.   Discharge characteristics for capacitor components.

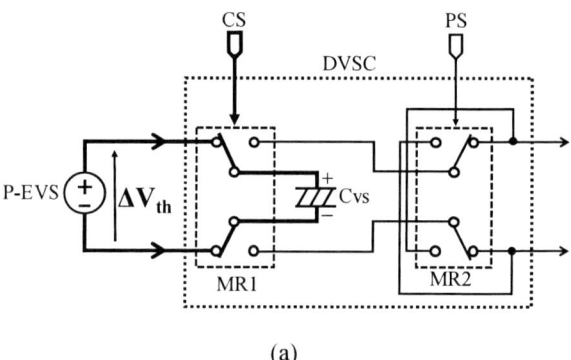

(a)

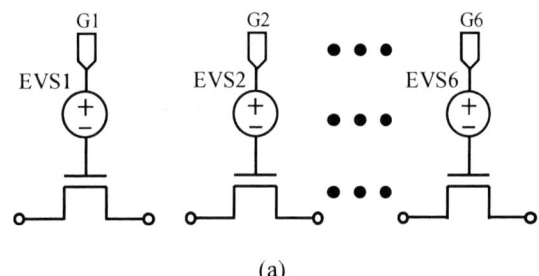

(a)

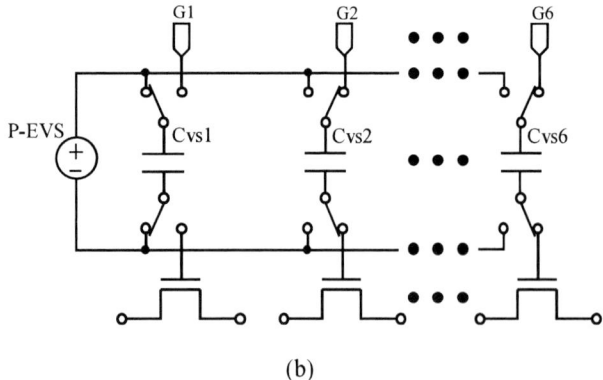

(b)

Fig. 3. V_{th}-shift settings for MOSFETs. (a) with Multiple EVSs. (b) with Programmable EVS (P-EVS) and capacitors.

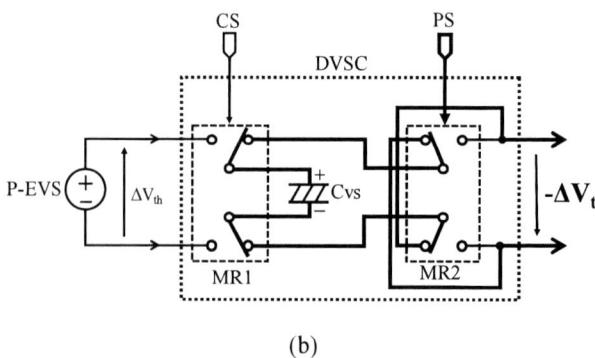

(b)

Fig. 5.   Dynamic V_{th}-shift circuit (DVSC). (a) Precharge for Cvs. (b) Polarity selection.

2018 INTERNATIONAL CONFERENCE ON MICROELECTRONIC TEST STRUCTURES, MARCH 19-22, 2018, AUSTIN, TEXAS, USA

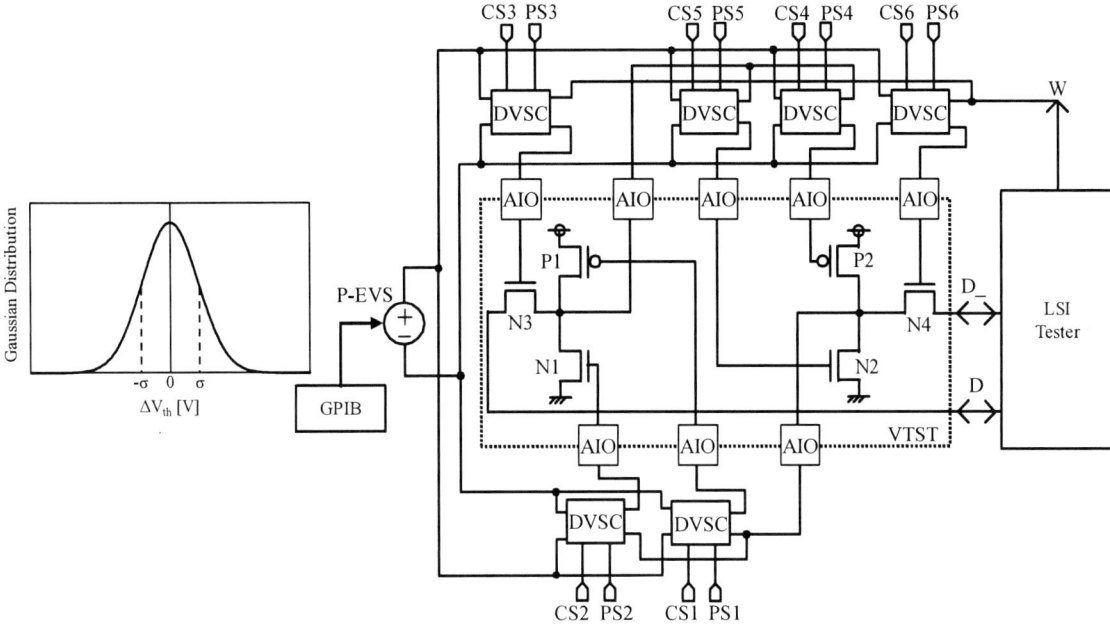

Fig. 6. Configuration for performing Monte Carlo analysis for SRAM function test.

III. SRAM FUNCTION TEST

Figure 6 shows a configuration diagram for performing Monte Carlo analysis for the SRAM function test using the VTST. Six DVSCs are connected to the gate of six transistors in the VTST. By sequentially selecting DVSCs, the Vth-shift value of the Gaussian distribution is set for each MOSFET. Then the function test is performed by LSI tester. By repeating this procedure, Monte Carlo analysis is performed by actual measurement. Figure 7 shows the results of the Monte Carlo analysis in which the function test was repeated 2000 times. The measurement and the simulation results agree well with in both Vdd = 1.8 V and Vdd = 0.6 V.

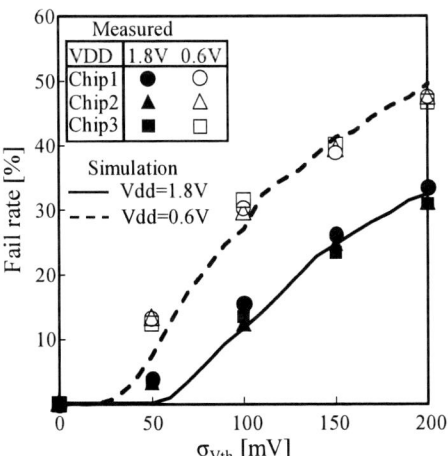

Fig. 7. Result of Monte Carlo analysis of SRAM function test (Frequency = 100 Hz, N = 2000).

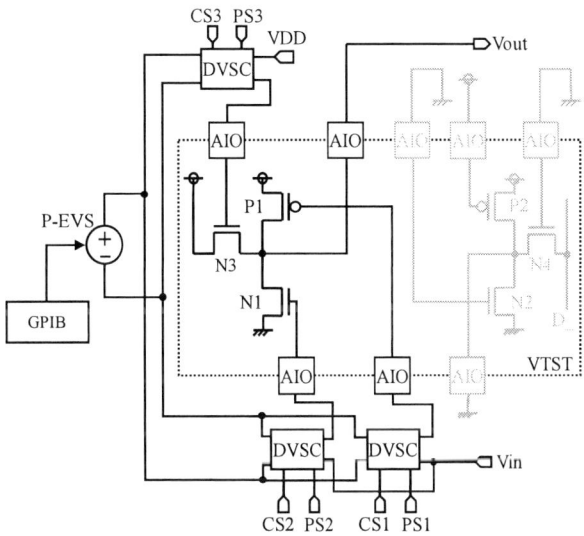

Fig. 8. Configuration for SNM Measurement.

IV. STATIC NOISE MARGIN ANALYSIS

In the design of the SRAM memory cell, SNM evaluation (so-called Butterfly Curve) is commonly used to confirm its operation margin. Our measurement system is also applicable to the SNM evaluation. Using three DVSCs as shown in Fig. 8, the DC input / output characteristics of an inverter consisting the SRAM cell can be repeatedly measured. Figures 9 (a) and (b) show the butterfly curves obtained by the simulation and measurement, respectively. Figure 10 summarizes the relationships between the V_{th}-shift for one standard deviation (1σ) and the SNM. In both results, the simulation result and the

978-1-5386-5072-1/18 $31.00 © 2018 IEEE

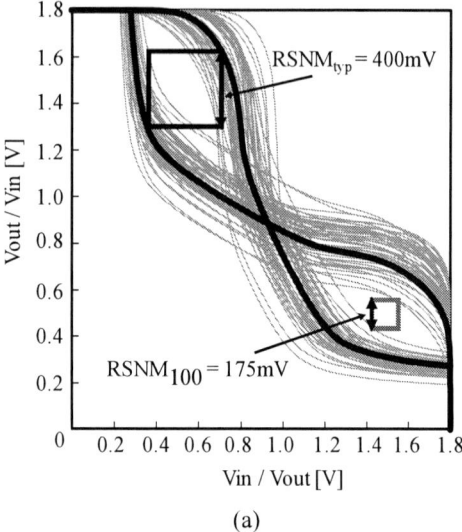

(a)

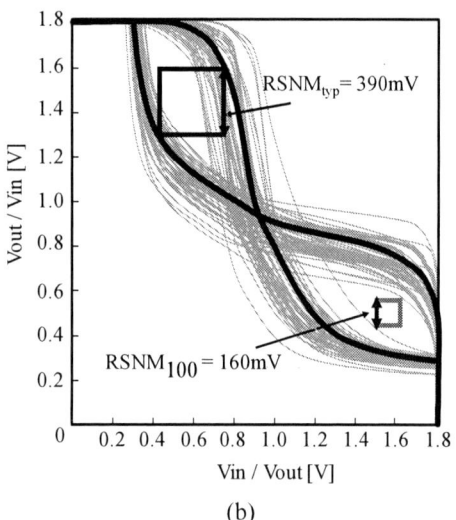

(b)

Fig. 9. Butterfly Curves for SRAM cell (Vdd = 1.8 V, σ_{Vth} = 0.1 V, N = 100). (a) Simulated. (b) Measured.

measurement result very well agree with each other. It was confirmed that the Monte Carlo analysis similar to the simulation could be performed by actual measurement.

V. CONCLUSIONS

We have successfully achieved the Monte Carlo analysis for SRAM in actual measurement using VTST, P-EVS and DVSCs. Since our proposed method shares the P-EVS, it can

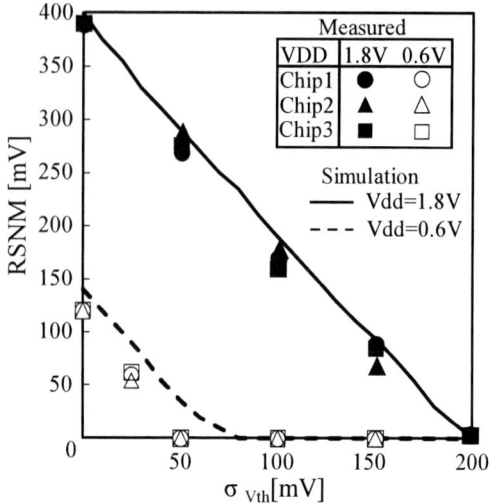

Fig. 10. Result of Monte Carlo analysis of SNM.

compactly cope with SRAM having a larger number of transistors which are proposed in recent years.

ACKNOWLEDGMENT

This work was supported by JSPS KAKENHI Grant Number JP15K06021 and VLSI Design and Education Center (VDEC), the University of Tokyo, in collaboration with Cadence Design Systems, Inc., Mentor Graphics, Inc., the Rohm Corporation and the Toppan Printing Corporation.

REFERENCES

[1] E. Seevinck , F. J. List, and J. Lohstroh "Static-noise margin analysis of MOS SRAMcells", IEEE J. Solid-State Circuits, vol. SC-22, no. 5, October 1987.

[2] T. Saito, H. Okamura, H. Yamamoto, and K. Nakamura, "Ratio-less 10Transistor Cell and Static Column Retention Loop Structure for Fully Digital SRAM Design", IEEE International Memory Workshop (IMW), pp.167-170, May 2012.

[3] T. Kondo, H. Yamamoto, H. Imi, H. Okamura, and K. Nakamura, "A Measurement of Ratio-less 12-transistor SRAM cell Operation at Ultra-low Supply-voltage", International Conference on Solid State Devices and Materials (SSDM), pp.82-83, Sep. 2014.

[4] H. Okamura, T. Saito, H. Goto, M. Yamamoto, and K. Nakamura, "Mosaic SRAM Cell TEGs with Intentionally-added Device Variability for Confirming the Ratio-less SRAM Operation", IEEE International Conference on Microelectronic Test Structures, pp.212-215, Mar. 2013.

[5] S. Yamaguchi, H. Imi, S. Tokumaru, T. Kondo, H. Yamamoto, and K. Nakamura, "V_{th}-shiftable SRAM Cell TEGs for Direct Measurement for the Immunity of the Threshold Voltage Variability", IEEE International Conference on Microelectronic Test Structures, pp.50-52, Mar. 2017.

2018 INTERNATIONAL CONFERENCE ON MICROELECTRONIC TEST STRUCTURES, MARCH 19-22, 2018, AUSTIN, TEXAS, USA

Process Variation Estimation using A Combination of Ring Oscillator Delay and FlipFlop Retention Characteristics

Takuma Konno†, Shinichi Nishizawa‡ and Kazuhito Ito‡
†Faculty of Engineering, Saitama University
‡Graduate School of Science and Engineering, Saitama Univesity
255, Shimo-Okubo, Sakura-Ku, Saitama 338-8570, JAPAN
TEL: +81-48-858-3468
E-Mail: nishizawa@mail.saitama-u.ac.jp, kazuhito@ees.saitama-u.ac.jp

Abstract—We propose an extraction method of process variation utilizing D-Flip-Flop (DFF) data retention characteristics and Ring Oscillator(RO) oscillation delay. Extracted process variation is modeled as PMOS and NMOS threshold voltage variations. Retention characteristics of the DFF circuit has different sensitivity to threshold voltage variation from the RO circuit. A DFF circuit is newly introduced as a complementary test structure of the conventional RO circuit for process variation extraction. By combining the RO circuit and the DFF circuits, we can accurately estimate the shift of global process variation. The test structure is implemented into silicon chip and the amount of global variation shift is extracted from measured data.

I. INTRODUCTION

The progressive scaling of CMOS transistor fabrication technology achieves both faster device operation speed and lower energy consumption, and it results in the continuous improvement of VLSI circuit performance. Meanwhile, transistor performance variation significantly increases and it becomes a serious issue for VLSI design. In order to overcome this issue from a circuit design perspective, process variation compensation schemes have been proposed [1] [2]. However, these schemes require to measure the amount of process variation shift from each chip, therefore an accurate and simple process monitor circuit is required.

Several test structures have been proposed for the device characterization. A common method is to directly measure the transistors currents using device matrix arrays [3] [4] or operational amplifier [5]. These measurements provide the desired measured values, however the measurement is analog thus they require long test time, large circuit area, and expensive measurement equipment inside and/or outside of the chip.

Ring oscillator (RO) circuit is another solution for device performance characterization. Since RO circuit converts device performances to the oscillation frequency, easy and high speed device characterization are possible. Reference [6] uses RO circuits with specially designed inverters such as inverter with pass-transistors, pass transistor loads to extract transistor parameter variations (threshold voltages, gate length, gate capacitances) and metal wire variations. However, since the oscillation frequency is determined by both rising delay and falling delay of each logic stage, it is difficult to separate the variations of PMOS transistors and NMOS transistors

using the RO circuits. In reference [7], RO circuits with different sensitivities to transistor parameter variations (PMOS threshold voltage, NMOS threshold voltage, gate length) and proposed to estimate these variations solving the simultaneous equation. However, since this technique uses RO circuit, it is difficult to separate the variations of PMOS transistor and NMOS transistor from measurement results.

To overcome this problem, a latch circuit inside the D-FlipFlop (DFF) is newly introduced as a complementary circuit of the RO circuit for process variation estimation. A latch circuit is a memory element for a digital circuit, and the retention characteristics of the latch circuit strongly depends on the PMOS transistor variation and NMOS transistor variation [8].

In this paper, we propose to use the retention characteristics of a latch circuit for device parameter extraction. Both the DFF circuit and RO circuit are used for device parameter extraction, and the measurement result is converted to die-to-die global threshold voltage variation. Scannable storage elements (scan latchs and scan flipflops) are widely used in digital VLSI circuits for testing purpose. If these scannable storage elements can be used for device parameter extraction, retention characteristics of latch circuits can be obtained without implementing extra latch circuits

The rest of this paper is organized as follows. Section II briefly describes the retention failure in latch circuit. Section III describes proposed variation estimation method. Section IV describes measurement result. Section V concludes this paper.

II. RETENTION FAILURE IN LATCH CIRCUIT

We briefly explain the mechanism of retention failure in latch circuit. Fig. 1 is a structure of a Transmission Gate FlipFlop (TGFF) circuit. A TGFF consists of two latch circuits, data transfer logics, and clock drivers. The latch circuit inside the DFF consists of an inverter of the datapath and a tri-state inverter as a keeper. When the tri-state inverter works as an inverter, the latch circuit works to hold the data. When the supply voltage is high, the on-resistance of the transistor is sufficiently low, and the off-resistance of the transistors are sufficiently high. Since the output voltage of an inverter is determined by the on-resistance and the off-resistance of transistors thus its output voltage is close to its rail-to-rail

978-1-5386-5072-1/18 $31.00 © 2018 IEEE   97

2018 INTERNATIONAL CONFERENCE ON MICROELECTRONIC TEST STRUCTURES, MARCH 19-22, 2018, AUSTIN, TEXAS, USA

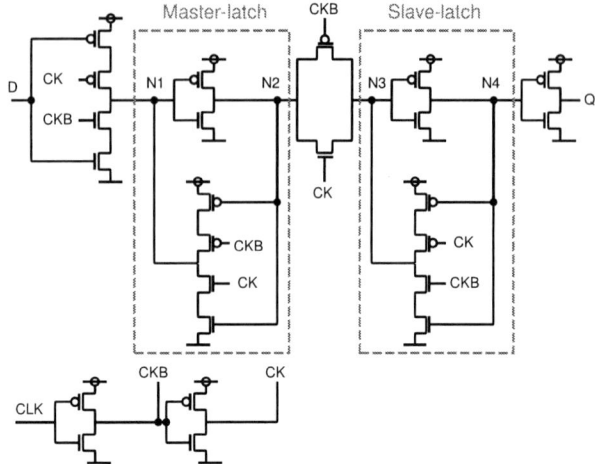

Fig. 1. Schematic of a Transmission Gate FlipFlop circuit, composed of two latch circuits.

Fig. 2. Retention failure of the latch circuits with different process variations.

voltage. In this case, the latch circuit successfully holds the data.

When supply voltage is lowered and the transistors are operated in a subthreshold region, both the on-resistance and the off-resistance are determined by subthreshold characteristics of the transistors.

In this operation condition, the output voltage of the inverter disengage from rail-to-rail voltage since the on-resistance and the off-resistance are close each other. The subthreshold current varies depending on the threshold voltage variation, and the threshold voltage variation strongly affects the output voltage. When the output voltage of the inverter circuit exceeds from the logical threshold value of the other inverter circuit, the stored data is flipped and the data retention fails. This retention failure voltage of each latch circuit strongly depend on threshold voltage variation. Figure 2 shows an example of the retention failure of the latch circuit having different threshold voltage variations. As the amount of the threshold voltage variation increases, the retention voltage increases since these latch circuits have larger imbalance in the PMOS transistor performance and the NMOS transistor performance. This characteristics is determined by the balance of the PMOS transistor performances and the NMOS transistor performances [10] [11]. Hence, we can extract the amount of the threshold voltage variation from the FlipFlop circuit characteristics.

III. VARIATION ESTIMATION METHOD

We extract the die-to-die global variation as the PMOS threshold voltage shift ($\Delta V_{th,p}$) and the NMOS threshold voltage shift ($\Delta V_{th,n}$), respectively. [1] In this paper, we estimate the amount of PMOS threshold voltage shifts and NMOS threshold voltage shifts from the measured RO frequency

[1]In the following discussion, we assume source voltage is zero for both the PMOS transistor and the NMOS transistor. Thus, the NMOS threshold voltage is a positive value and the PMOS threshold voltage is a negative value.

and the retention characteristics of DFF. First, a simultaneous equation is build which expresses the relationships of a RO frequency, retention voltage of a DFF, and threshold voltage of transistors. Next, the amount of PMOS and NMOS threshold voltage shifts are calculated solving by the simultaneous equation substituting the measured RO frequency and the retention voltage of DFF. We use following simultaneous equation for the variation extraction,

$$F_{meas.} - F_{sim.} = K_{F,1}\Delta V_{th,p}\Delta V_{th,n} + K_{F,2}\Delta V_{th,p} + K_{F,3}\Delta V_{th,n} \quad (1)$$

$$V_{DR,meas.} - V_{DR,sim.} = K_{V,1}\Delta V_{th,p}\Delta V_{th,n} + K_{V,2}\Delta V_{th,p} + K_{V,3}\Delta V_{th,n}, \quad (2)$$

where $F_{meas.}$ and $F_{sim.}$ are the oscillation frequencies of RO from silicon measurement and simulation, $V_{DR,meas.}$ and $V_{DR,sim.}$ are the retention voltages of DFF from silicon measurement and simulation, $K_{F,1}$, $K_{F,2}$, $K_{F,3}$, $K_{V,1}$ $K_{V,2}$ and $K_{V,3}$ are the sensitivity coefficients of the PMOS and NMOS transistors, respectively. Sensitivity coefficients are calculated using transistor level simulation. Using these two equations, two unknown parameters $\Delta V_{th,p}$ and $\Delta V_{th,n}$ are estimated. Figure 3 shows a fitting result of equations (1),(2) using circuit simulation. The results show that the two equations a show good fitting result. It is also found that the two circuits have different sensitivities to the PMOS threshold voltage shift and the NMOS threshold voltage shift.

In order to evaluate the proposed method, the estimation accuracy is evaluated by simulation. For evaluation, we use a commercial 65 nm FDSOI process. In the simulation, we assume that the same threshold voltage variations are applied to both the RO and DFF. Estimate the amount of threshold voltage variations by the proposed method, and evaluate the difference of estimated value from golden data (applied values of threshold voltage variation in simulation). Figure 4 shows applied threshold voltage variations and estimation results based on the proposed method. Note that the range of the threshold voltage variation is approximately half of the process

978-1-5386-5072-1/18 $31.00 © 2018 IEEE   98

2018 INTERNATIONAL CONFERENCE ON MICROELECTRONIC TEST STRUCTURES, MARCH 19-22, 2018, AUSTIN, TEXAS, USA

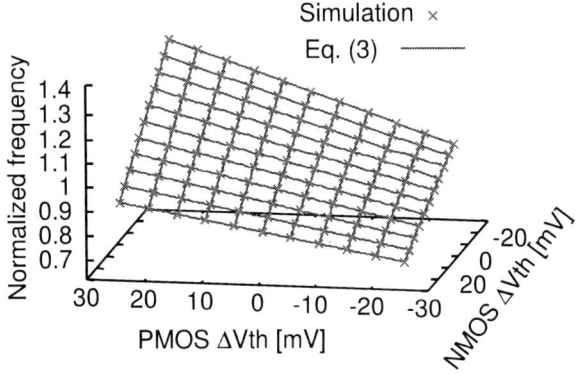

(1) RO oscillation frequency.

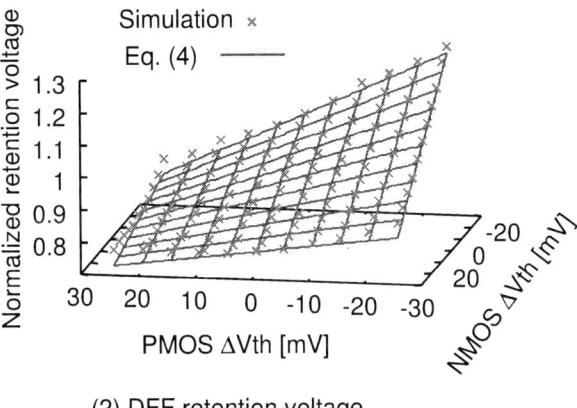

(2) DFF retention voltage.

Fig. 3. Fitting result of equations (1) and (2) for the RO oscillation frequency and the DFF retention voltage.

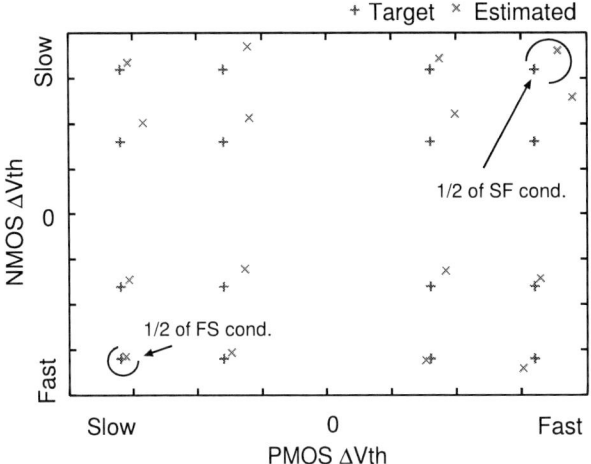

Fig. 4. Evaluation of the estimation accuracy of the threshold voltages.

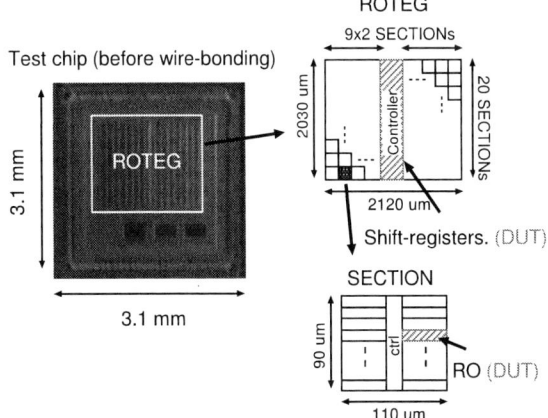

Fig. 5. Microphotograph and structure of the test chip.

corner. The result shows that the estimated threshold variation voltage is close to the value applied in the simulation. When the PMOS threshold voltages and NMOS threshold voltages deviate far from a typical condition, estimation error increases since we use same sensitivity coefficients for wide process condition space. The maximum estimation error is calculated as 3.02 mV.

IV. EXPERIMENTAL RESULTS

A. Test Chip Structure

Figure 5 shows the microphotograph of a test chip fabricated in 65 nm FDSOI process. The test chip contains a large macro named ROTEG. A ROTEG has basic macro named SECTION. SECTIONs are arranged into a set of 20-by-19 arrays, so ROTEG has 380 SECTIONs in total. A SECTION contains RO circuits with different logic types and stages. In this experiment, we use a 19 stage RO composed of standard inverter cells. The oscillation signal is divided by 128 and counted outside of the test chip.

Figure 6 shows a block diagram of the test chip. SECTIONs are controlled by a control circuit located at the center of the test chip. There are three inputs, serial input configuration,

serial clock, and macro enable. The configuration data is stored into a shift-register circuit. When the enable signal is low, all of the ROs are disabled. Therefore, any data can be input to the shift register without activating ROs. We use DFFs which compose the shift register circuit as device under test for the retention characteristics evaluation.

We measure DFFs retention characteristics in the following steps;

1) Shift in values 1 or 0 to the shift register at a nominal voltage.
2) Set the clock signal high for the master-latch test, or low for the slave-latch test.
3) Decrease supply voltage from a nominal voltage to a target low voltage.
4) Return to the nominal voltage, and shift out the stored values from DFFs.
5) Update the target low voltage and iterate this operation (return to 1)).

2018 INTERNATIONAL CONFERENCE ON MICROELECTRONIC TEST STRUCTURES, MARCH 19-22, 2018, AUSTIN, TEXAS, USA

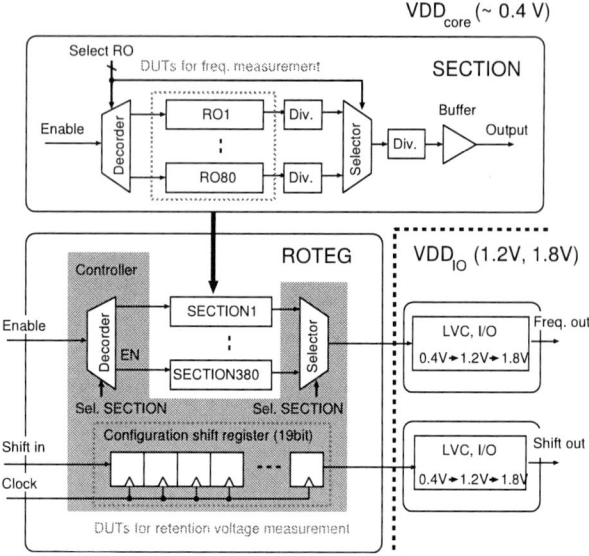

Fig. 6. Block diagram of the test structure.

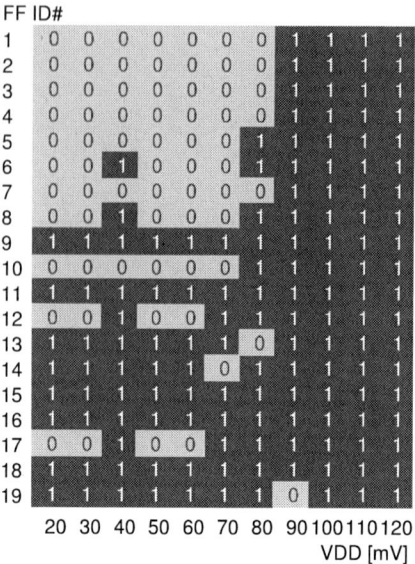

(1) Case store "1" for initial value of master-latch.

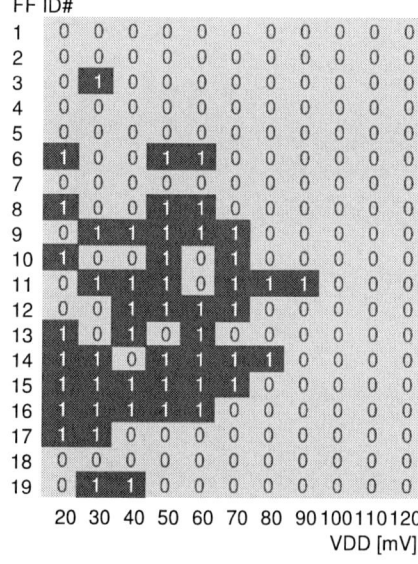

(2) Case store "0" for initial value of master-latch.

Fig. 7. Measured result of retention characteristics of DFFs (chip #1). (1) Case master-latch holds "1". (2) Case master-latch holds "0".

We use Keithley 2280S precision DC supply as a power source for RO measurement and DFF retention characteristics evaluation. Since the measurement result varied depending on the environment, we measure three times for each supply voltage and take a majority vote for robust data analysis.

B. Measurement Results

Figure 7 shows the measurement result of DFFs master-latch retention characteristics. We change the supply voltage from 120 mV to 20 mV at 10 mV intervals. As the supply voltage decreases, some latches failed its data retention. For example, as the supply voltage becomes lower, FlipFlop ID#1 which holds "1" flips "0" at 80 mV Measured result shows 19 latches have different retention characteristics, depending on transistor performances variation. Some latches failed its data retention in higher supply voltage but succeeded its data retention in lower voltages. This is because environmental noise affects to the retention measurement, even if the majority is taken. In this paper, we define the retention failure voltage as the highest supply voltage when the latch circuit failed its data retention. 380 samples of 19 stage RO with standard inverter is used and obtain these oscillation frequencies at 0.4 V operation. To estimate global process variation, we use average value of RO oscillation frequency and data retention voltage of the measured chips.

Table I shows the measured results of a retention voltage of DFF and an RO oscillation frequency from two chips. Also, table I includes simulation result at TT condition. Measured results show some difference from a simulation result and these differences are converted into threshold voltage shift of PMOS transistor and NMOS transistor.

Table II shows the estimation results of a PMOS threshold voltage variation and a NMOS threshold voltage variation, which are calculated from the RO oscillation frequency and

the DFF retention characteristics. The result shows that two measured chips have similar threshold voltage variations.

V. CONCLUSION

We propose an extraction method of process variation utilizing retention characteristics of DFFs. Retention characteristic of DFF circuit is newly introduced as complement test structure for standard RO circuit, and it enables accurate estimation of global process condition shift. Test structures

TABLE I
SIMULATED AND MEASURED RESULTS OF RETENTION VOLTAGE OF DFFs(MASTER-LATCH) AND RO CIRCUITS.

	Retention voltage[V] (Hold "1")	RO freq. [MHz]
Chip 1	0.835	167
Chip 2	0.929	157
Simulation (Typical)	0.880	173

TABLE II
ESTIMATION RESULTS OF THRESHOLD VOLTAGE VARIATIONS (FROM TWO CHIPS).

Chip	$\Delta V_{TH,p}$	$\Delta V_{TH,n}$
Chip 1	-1.27%	0.85%
Chip 2	-3.57%	3.01%

are implemented into silicon chips and result shows we can estimate global variation shift from measured data.

ACKNOWLEDGMENTS

This work has been partly supported by JSPS KAKENHI JP17K12657. This work is also partly supported by VLSI Design and Education Center (VDEC), the University of Tokyo in collaboration with Synopsys, Inc., Cadence Design Systems, Inc., and Menter Graphics, Inc.

REFERENCES

[1] S. M. Martin, K. Flautner, T. Mudge, and D. Blaauw, "Combined dynamic voltage scaling and adaptive body biasing for lower power microprocessors under dynamic workloads," in *International Conference on Computer-Aided Design*, 2002, pp. 721–725.

[2] J. Tschanz and S. Narendra, "Effectiveness of Adaptive Supply Voltage and Body Vias for Reducing Impact of Parameter Variations in Low Power and High Performance Microprocessors," *IEEE Journal of Solid-State Circuits*, vol. 38, no. May, pp. 826–829, 2003.

[3] K. Agarwal, S. Nassif, F. Liu, J. Hayes, and K. Nowka, "Rapid Characterization of Threshold Voltage Fluctuation in MOS Devices," in *International Conference on Microelectronic Test Structures*, 2007, pp. 74–77.

[4] W. Zhao, F. Liu, and K. Agarwal, "Rigorous Extraction of Process Variations for 65-nm CMOS Design," *IEEE Transactions on Semiconductor Manufacturing*, vol. 22, no. 1, pp. 196–203, 2009.

[5] B. L. Ji, D. J. Pearson, I. Lauer, F. Stellari, D. J. Frank, L. Chang, and M. B. Ketchen, "Operational Amplifier Based Test Structure For Transistor Threshold Voltage Variation," in *International Conference on Microelectronic Test Structures*, 2008, pp. 3–7.

[6] M. Ketchen, M. Bhushan, and D. Pearson, "High speed test structures for in-line process monitoring and model calibration," in *International Conference on Microelectronic Test Structures*, vol. 18, April, 2005, pp. 33–38.

[7] I. A. Mahfuzul, A. Tsuchiya, K. Kobayashi, H. Onodera, I. A. Mahfzul, A. Tsuchiya, K. Kobayashi, and H. Onodera, "Variation-sensitive Monitor Circuits for Estimation of Die-to-Die Process Variation," in *International Conference on Microelectronic Test Structures*, no. 1, 2011, pp. 153 – 157.

[8] A. Datta, M. Abu-Rahma, S. Dasnurkar, H. Rasouli, S. Tamjidi, M. Cai, S. Sengupta, P. Chidambaram, R. Thirumala, N. Kulkarni, P. Seeram, P. Bhadri, P. Patel, S. S. Yoon, and E. Terzioglu, "Analysis, Modeling and Silicon Correlation of Low-voltage Flop Data Retention in 28nm Process Technology," in *International Symposium on Quality Electronic Design*, 2013, pp. 580–584.

[9] N. H. E. Weste and D. M. Harris, *CMOS VLSI Design (4th edition)*, Addison Wesley, 2010.

[10] H. Fuketa, S. ichi O'uchi, and T. Matsukawa, "A Closed-Form Expression for Minimum Operating Voltage of CMOS D Flip-Flop," *IEEE Transactions on Very Large Scale Integration (VLSI) Systems*, vol. 25, no. 7, pp. 2007–2016, 2017.

[11] T. Kamakari, J. Shiomi, T. Ishihara, and H. Onodera, "Analytical Stability Modeling for CMOS Latches in Low Voltage Operation," *IEICE Transactions on Fundamentals of Electronics, Communications and Computer Sciences*, vol. E99.A, no. 12, pp. 2463–2472, 2016.

NPN Mismatch Dependence on Layout

Cory Compton

Macom, 4000 MacArthur Blvd., Newport Beach, CA 92660

Abstract— **Mismatch structures are normally designed to look at pairs of identical devices with near ideal layouts. In this paper we look into the effects of orientation and NPN density on the mismatch results of NPNs in two 0.18um SiGe BiCMOS process. The mismatch structures were added to scribeline PCM modules, which allowed us to look at the results from multiple mask sets.**

Keywords—NPN, BJT, bipolar, matching, mismatch, test structure

I. INTRODUCTION

It is well known that analog IC design is heavily dependent on the mismatch of the various components used in the circuit. The only way to have some certainty of the circuit performing as expected is to follow the known best practices in both the schematic and layout aspects of the design. While it is expected that analog circuit designers will have some familiarity with the basic requirements to achieve good matching, it is still fairly common to see circuit issues due to poor layout. For the more advanced CMOS processes, some of the layout dependence on mismatch can to first order be seen in simulation due to the extraction and modeling of many layout dependent parameters, but there are still likely many layout effects that are not modeled. In the case of NPNs in BiCMOS processes, there are generally not any NPN specific layout dependent parameters that can alert circuit designers to layout issues.

At this point it is common for SPICE models to provide at least some Monte Carlo mismatch simulation support for the primary devices in a given process, but the models are usually based on relatively ideal test structure layouts, and the data is usually obtained from a limited number of lots and wafers from a single testchip maskset. In comparison to CMOS devices, there have been relatively few papers written about layout dependence of NPNs, and those are usually based on test structures implemented in relatively large process or modeling testchips [1,2]. In production, foundries often only place a small set of relatively ideal test structures in the scribeline Process Control Monitor (PCM) modules in order to monitor mismatch (assuming they have added any at all). Ideally mismatch is determined by variations inside of devices that are intrinsic to the process and are independent of things like device layout, maskset, or process variation. However, in practice it is well known that just about anything can have some influence on the mismatch of devices by affecting the means or standard deviations, and this influence can vary across a wafer, wafer-to-wafer, lot-to-lot, or maskset-to-

maskset [3,4]. The goal of this work is to investigate the effect of some common layout issues on NPN mismatch by placing matching structures in customer scribeline PCM modules, such that they can be added to multiple masksets, and provide a more comprehensive view of the mismatch.

II. TEST STRUCTURE DESIGN

In order to design effective test structures, one needs to know the goals for the structures, the design and layout constraints dictated by the process and foundry, and the capabilities of the measurement system that will be used to make the measurements. Once these are known, the device configurations to be tested, and the test structure topologies used to measure them can be defined [5].

A. PCM Module Definition

These NPN matching structures were designed in the same 0.18μm SiGe BiCMOS foundry process as used in [6] (process A), and a subsequent generation of this process with higher performance NPNs (process B). The PCM module pads are 70μm x 70μm with a 100μm pitch. In order to support larger structures than can fit in the narrow 30μm spaces between the pads, there are four regions where two pad pitches have been skipped in order to create 230μm wide gaps. Representative schematic and layout examples for a module dedicated to NPN mismatch are shown in Fig. 1 and Fig. 2 respectively.

B. Measurement System Definition

Our test structures are primarily intended to be measured on a custom test system based on an Agilent (now Keysight) 4156C, Keithley 707 switch matrix with 7174A low leakage switch matrix cards, and a Cascade Summit 12K semi-automatic prober with thermal chuck. One advantage of this system over some of the production parametric testers is that the 4156C has Kelvin SMUs that can be used to circumvent both probe and interconnect resistance. While production testers may actually use Kevin SMUs internally in order to circumvent switch matrix resistance, the Kelvin functionality is often not made available outside of the tester for use by the customer. However, the 4156C SMUs are not truly Kelvin, as the force and sense terminals are internally connected through a network that allows the SMUs to be easily used in non-Kelvin mode by only using the force terminals. The down side of this approach is that the sense terminal no longer has an ideal high impedance, so the resistance between the force and sense needs to be fairly low in order to guarantee accurate results (Agilent specifies no more than 10 Ohms for the force and sense terminals for currents up to 100mA).

2018 INTERNATIONAL CONFERENCE ON MICROELECTRONIC TEST STRUCTURES, MARCH 19-22, 2018, AUSTIN, TEXAS, USA

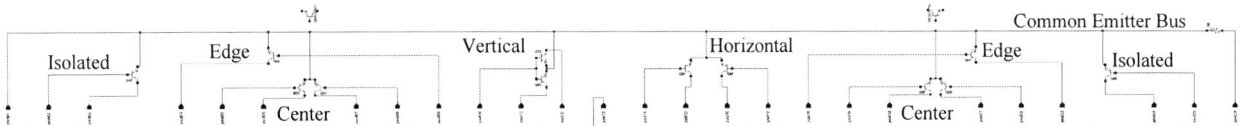

Fig. 1.   Schematic of a full PCM module dedicated to NPN mismatch.

Fig. 2.   Layout of a full PCM module dedicated to NPN mismatch.

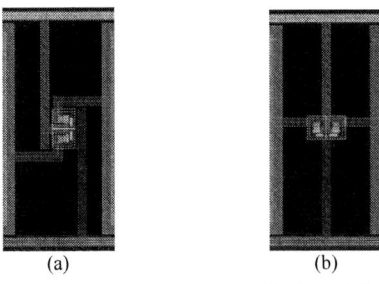

(a)   (b)

Fig. 3.   NPN matched pairs in vertical (a) and horizontal (b) orientations.

Fig. 4.   NPN array with two center and one corner device padded out.

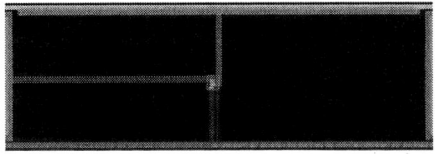

Fig. 5.   Isolated NPN to look at worst case isolated to dense mismatch.

C. Test Structure Definition

For this paper we are primarily looking at the effects of orientation and NPN density on mismatch as a function of NPN emitter length (0.76μm, ~2μm, and ~10μm). In order to look at orientation, we have simply included two sets of nearly ideal matched pairs with the emitters orientated in vertical or horizontal orientations (Fig. 3). Note that due to the relatively narrow spaces between the pads, we were not able to place the 10.16μm long emitter devices horizontally. In order to look at NPN density effects, we have placed fairly large NPN arrays with the two center NPNs and one corner NPN padded out (Fig. 4). In addition, we have placed single isolated NPNs adjacent to the arrays in order to represent the worst case difference in NPN density (Fig. 5). For the 2μm array, the isolated device is 1.5μm long in order to look at how much additional mismatch is created when the sizes are not matched. Due to the constraints dictated by the pad configuration, we have placed the array structures in the two inner 230μm wide gaps, and the isolated NPNs are placed in the outer 230μm wide gaps. The foundry generated fill in our PCM modules

appears to be consistent across the different mask sets, so this should not add additional uncertainty to the results.

As the emitter connections are generally the most important in terms of setting the bias points of the transistors, we like to use Kelvin connections for the emitter terminals whenever possible. Since using two pads for each emitter terminal would be very expensive, we have instead chosen to connect all of the emitters together through a single shared Kelvin emitter bus. In order to minimize the bus resistance, we used all the routable metals for the bus (1-5), and for the 230μm gaps we made the bus wider and added metal 6. By connecting the two outermost pads of the NPN matching structures together, we are able to Kelvin bias all of the emitters at the cost of only two pads per module. This means we can implement Ic mismatch structures with 3 additional pads (2 collectors and 1 common base) per pair, or Ic and beta mismatch structures with 4 additional pads (2 collectors and 2 bases) per pair. The most important issue we see with the Kelvin emitter bus approach for these test structures is that simultaneous measurements using the array corner or isolated devices will be impacted by Vbe offsets due to IR drop in the bus, as the emitters will be connected to the bus at different points. The expectation is that this effect will be negligible for low current measurements and it may become significant for high current measurements.

III. MEASUREMENT METHOLOGY

A. Test Architecture

Since the focus for these measurements is on analog design, and since analog circuits are normally biased by current, our characterization measurements are made as a function of Vce and emitter current density rather than Vbe. The downside of this approach is that the Vce bias conditions need to found iteratively, as they depend on the measured Vbe, so a special test program (DeltaIc) needed to be developed in our test environment. The basic measurement configuration is shown in Fig. 6, and it should be noted that the test is base referenced (Vb = 0V). While the 0V base reference is not that important for the DeltaIc test (as the base and emitter terminals are connected to common SMUs), it is beneficial for the corresponding beta matching test (DeltaBeta), as setting the two independent base supplies to 0V should minimize the voltage offset between them. In addition to the Ic mismatch, it is important to also report the measured Vbe, as this is the only measure of whether the devices were biased properly. Note that this test would be considered a simultaneous measurement per [5], as both devices are biased at the same time, and the Ic measurements are made almost simultaneously (the 4156 can only measure one unit at a time). One advantage of simultaneous measurements is that they are relatively immune

978-1-5386-5072-1/18 $31.00 © 2018 IEEE   103

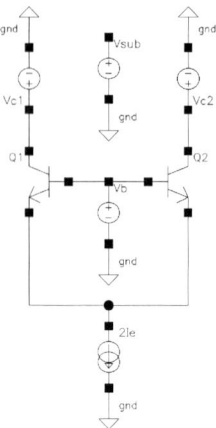

Fig. 6. Schematic of DeltaIc measurement configuration

to systematic shifts due to the relatively show temperature changes expected in the test environment. In addition, since the BJTs are current biased, temperature changes throughout the test sequence are less important, as they just cause relatively small changes in the measured Vbe, and the effect on Ic mismatch should be minimal, as the total current in the pair is fixed.

B. Test Parameters

For characterization purposes, we normally measure mismatch as a function of Vce and emitter current density (Je). Since the emitter widths are essentially fixed at the minimum value for the process, we find it easier to refer to Je in terms of current per emitter length (mA/μm) rather than the traditional units of area (mA/μm²). For these measurements we have swept Vce from 0.2V to 1.4V (the maximum voltage depends on the breakdown characteristics of the device). The maximum Je is intended to correspond to the bias conditions we expect to be used in high speed circuits (e.g. 1mA/μm for process A and 1.2mA/μm for process B), and we also measure at current densities one and two orders of magnitude below the maximum. The maximum Je roughly represents the high current region where parameters like Re may become dominant, the Je one decade lower represents the medium current region where the device should be fairly ideal (e.g. Re should have minimal impact), and the Je two decades lower represents the low current regions where non-ideal leakage currents may become significant.

IV. DATA ANALYSIS

For those not accustomed to matching measurements, it might seem that all one needs to do is measure some matched pairs at a fixed bias point and report the standard deviation of the mismatch. Unfortunately, it is rare that things are that simple, as there are many issues that may affect the results (e.g. device oscillation, defective devices, and tester limitations), such that the only thing that may be reported is the noise level of the tester used to make the measurements.

A. Oscillation Issues

It is common for high performance transistors to oscillate when biased in high gain regions when probed with low frequency or DC testers. The onset of oscillation can often be

seen in IV curves as a distinct jump in current or voltage. The problem usually occurs as a function of device performance and the total device size (e.g. combined FET width or emitter length), as large fast transistors normally have higher currents and transconductances. For NPNs, the onset of hard oscillation can often be seen as a large discontinuity in the Gummel curves. Fig. 7 shows the full wafer Gummel plot for a single device without obvious oscillation issues, and Fig. 8 shows the plot for a matched pair with hard oscillation starting at Vbe = 0.8V.

For the current based DeltaIC matching test methodology we are using, there are some obvious signs of oscillation issues. Tables I, II, and III give examples of relative 3σ mismatch, mean mismatch, and mean Vbe respectively for a couple of large device pairs. The tables on the left show the results for a pair that does not have obvious signs of oscillation, and the tables on the right show the results for a pair with obvious oscillation issues. One of the most concerning things about the 3σ mismatch results is that in isolation one could easily interpret the high current (Je = 1mA/μm) data in the right table as valid, since the values are not terribly far from the low and medium current ones, and it would be reasonable to see different mismatch at high currents. The significant variation across the different Vce voltages is the only thing that invites suspicion of the 3σ mismatch data on its own. However, the mean mismatch data indicates obvious issues with the data, as the mean mismatch increases from relatively insignificant values to very large and unstable offsets. A further confirmation of oscillation issues can be seen in the mean Vbe data, as the Vbe should have increased for the high current density column rather than decreased (the low Vbe measurements are equivalent to the excessive currents seen in the voltage based Gummel curves). One advantage of the forced current density method is that the Vbe values should be relatively constant across all of the devices (for a constant Vbe test it would be wise to convert the collector currents into current densities in order to make misbehaving devices more obvious). While cases of hard oscillation can be relatively easy to spot, there are also cases of mild oscillation that can corrupt the measurement data without causing large shifts in the device operating points (e.g. more subtle shifts in mean offsets). For the test structures presented in this paper, we have found that all of the matched pairs with independent base connections are much more sensitive to oscillation issues than the identically sized pairs with common base connections.

B. Screening Methodology

In comparison to the analysis of typical test data where the primary result is the mean and the secondary result is the standard deviation, for matching measurements the primary result is the standard deviation and the secondary results are the mean and variation of the standard deviation (e.g. standard deviation of the standard deviations from multiple wafers).

For matching measurements our data analysis methodology potentially uses three levels of screening. The first level is based on traditional screening limits, and it is primarily intended to screen out device measurements that return error codes (e.g. open devices or Vce non-convergence). The second level of screening is based on R-squared values, and it

2018 INTERNATIONAL CONFERENCE ON MICROELECTRONIC TEST STRUCTURES, MARCH 19-22, 2018, AUSTIN, TEXAS, USA

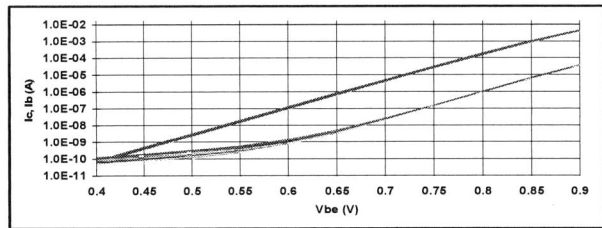

Fig. 7. Gummel plot without obvious oscillation issues.

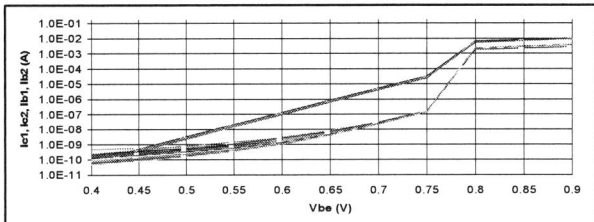

Fig. 8. Gummel plot with oscillation beginning at Vbe = 0.8V.

is intended for measurements that use linear least squares regressions, such as the floating gate capacitance mismatch technique [7] (for this technique good devices often have $1-R^2$ values of 1e-6 or better!). The third level of screening is based on a recursive 3σ filter approach similar to described in [8]. For the most part this technique works well, but no matter how you screen your data, it is very important to look at how many points have been screened out of each measurement. For typical datasets with tens to hundreds of samples, the 3 sigma screening should only remove at most a few points.

C. Repeatability Testing

For mismatch we are essentially measuring a noise signal, and the primary result is the standard deviation of the noise. Since the noise is determined by the measurement of many individual devices, it is not easy to quantify how much of the noise is due to the devices under test versus the tester itself. Many different methods have been used to try to quantify the noise level of the test. For instance, some focus on the very short term repeatability of the measurement for an individual device, but our method is to simply measure the complete test program across a full wafer twice. In this case we get two full sets of matching data for the wafer, including any effects from the test sequencing and probing. Normally the full matching statistics are computed for each set of data, and the means and standard deviations are compared. For those that are not accustomed to mismatch analysis, it may seem that if the mean and standard deviation are similar, then the test was repeatable. However, when measuring noisy signals such as mismatch, the fact that two sets of data have similar standard deviations may only mean that the tester noise was consistent over time. The way we determine how much of the mismatch is attributable to the device versus the tester is to subtract one data set from the other on a point by point basis. Then we can compute the same statistics on the difference data as were used for the original data. The mean of the difference data represents shifts in things like the measurement unit offsets, and the 3σ of the difference data can be interpreted as the test noise (due to the tester or the device itself [9]). Ideally the measurement noise should be at least one order of magnitude lower than that of the

device being measured. Tables IV and V show the 3σ mismatch repeatability and mean mismatch repeatability respectively for pairs without and with obvious oscillation issues. Note that the repeatability results for devices with oscillation are surprisingly not that bad on our tester.

V. MEASUREMENT RESULTS

As Macom is a fabless company with respect to silicon CMOS and BiCMOS processes, these PCM modules are only added to the scribes if there is room left over after all of the foundry PCMs have been placed. Since these PCMs are not part of the foundry's standard wafer acceptance testing, they are not normally measured in production by the foundry. At this point these structures are only being tested on engineering wafers that become available for circuit development and debugging, rather than on all of the wafers produced. Unfortunately, these wafers are often only available for a short time, so we try to test as many of the available structures as possible before we have to give the wafer up for assembly. In the interest of reducing test time, and also preventing tester issues from potentially ruining all the die on a wafer, we normally only test roughly half of the full reticle steppings on a wafer by using checkerboard die patterns.

A. Process A Results

The mismatch results for process A can be found in Table VI for the relative 3σ mismatch and Table VII for the mean mismatch. For process A we are presenting the checkerboard data for single wafers from 3 different masksets, which are denoted by Mask-Lot-Wafer (M-L-W) identifiers. In order to make the data set more manageable, we are only including the results for a single bias point in these tables (Vce = 1V and Je = 0.1mA/μm). The medium current density was chosen as it usually gives the cleanest data; normally without oscillation or significant emitter bias offset issues. Note that since wafer C-1-08 was still available, and it has some interesting results, we have included another set of data with all 73 measureable die on the wafer.

This mismatch data can loosely be analyzed by looking for three things. First, we should see the relative 3σ mismatch improve as the emitter length (Le) increases. Second, for a given emitter length, the vertical (Vert), horizontal (Horiz), and array center to array center (AC-AC) devices should ideally have identical mismatch, but in practice we would expect the AC-AC devices to have improved mismatch due to the large array of dummy transistors surrounding the center pair. Third, the array center to array edge (AC-AE) and array center to isolated (AC-Iso) devices should ideally have mismatch identical to the AC-AC devices, but given the differences in NPN density, we would expect to see the 3σ mismatch degrade and potentially see significant mean offsets for these cases.

The most surprising result for process A is the very large mean offset (~19%) for the 0.76μm AC-Iso structure on wafer C-1-08. Looking at the two datasets for this wafer, we can see that the mean offset for this pair does not change significantly when the measurement die count is increased from 30 to 73. Since the AC-Iso structures are the worst case in terms of Vbe offsets, we have included Table X with the full mean mismatch measurement results for both standard and reversed Kelvin

978-1-5386-5072-1/18 $31.00 © 2018 IEEE   105

force and sense terminal assignments. For these relatively small NPNs, the emitter currents are not large enough to cause significant mean offsets, as the results are almost identical for the low current density, slightly changed for the medium current density, and only offset by about 0.5% for the high current density. We would expect the actual offset to be close to the average of the two, so if the problem was more significant we could modify the test to measure in both configurations and report the average. Of course the Vbe offsets will add some additional noise and uncertainty to the measurement results, so this issue should be avoided for very high precision matching measurements. Note that the largest NPNs are expected to have severe offsets at high current densities due to this issue, but as the devices oscillate under these bias conditions the problem is moot.

Further analysis of the test results show that the mean offset of -19% is actually larger than the 3σ mismatch, such that the maximum mismatch value is still negative at -11.5%, and there is no clear pattern to the offset across the wafer. In addition, to eliminate any possible strange SMU-to-SMU mismatch effects, we computed the mismatch from individual Gummel curve measurements for each device and obtained a very similar mean offset. Note that the individual device measurements also greatly alleviate the Kelvin emitter bus Vbe offset issue, as the feedback loop should compensate for the different bus resistance to each NPN. Thus, we conclude that the data is likely valid, and there is something unusual going on with this device on this wafer. Unfortunately, we only have wafers from a single lot with this mask set, so we cannot see if there is any lot-to-lot variation. However, we did measure an additional somewhat leaky fast corner wafer from this lot, and it had a very similar mean offset for this device, so the problem may affect the whole lot.

B. Process B Results

The mismatch results for process B can be found in Table VIII for the relative 3σ mismatch and Table IX for the mean mismatch. For process B we have been able to measure multiple wafers for two lots with maskset A, so in the interest of space we are only presenting the mismatch for the combination of all of the checkerboard measurements on these wafers (A-1-All and A-2-All). The results generally show the expected trends of improved mismatch for the AC-AC devices relative to the vertical and horizontal pairs, and degraded mismatch for the AC-AE and AC-Iso devices. While not as extreme as the process A 0.76μm AC-Iso case, there are many cases of significant mean offsets.

VI. CONCLUSIONS AND FUTURE WORK

The significant variations in the mean and 3σ mismatch data for these two processes only reinforces the need for circuit designers to always follow best practices. As these test structures are far from ideal, on a subsequent tapeout for a different foundry, we revised the layout of the array edge devices to mitigate the Vbe offset issue by matching the emitter routing and connecting to a common point on the emitter bus. While the probecard we used for these measurements has reasonable DC leakage, we suspect that the lack of guarding inside of the probe card may be adding to the oscillation issues we are seeing for these matching test

structures and the ring oscillators in [6]. In order to try to alleviate the oscillation issues, and potentially improve capacitance measurements, we have ordered a higher performance (fully guarded) probe card for the next process.

TABLE I. RELATIVE 3σ MISMATCH EXAMPLES

Without Obvious Oscillation (%)

Vce (V)	Je (mA/μm)		
	0.01	0.1	1
0.2	1.00 [a]	0.95	0.96
0.4	0.97	0.94	0.97
0.6	0.97	0.94	0.98
0.8	0.97	0.94	0.99
1.0	0.97	0.95	1.00
1.2	0.97	0.95	1.02
1.4	0.97	0.95	1.03

With Obvious Oscillation (**Bold**) (%)

Vce (V)	Je (mA/μm)		
	0.01	0.1	1
0.2	1.00 [a]	0.93	**1.02**
0.4	0.94	0.91	**1.58**
0.6	0.94	0.91	**1.73**
0.8	0.94	0.91	**1.71**
1.0	0.94	0.91	**1.15**
1.2	0.94	0.91	**0.84**
1.4	0.95	0.92	**0.65**

[a.] The 3σ mismatch results have been normalized to the first bias point

TABLE II. MEAN MISMATCH EXAMPLES

Without Obvious Oscillation (%)

Vce (V)	Je (mA/μm)		
	0.01	0.1	1
0.2	-0.32	-0.30	-0.17
0.4	-0.34	-0.31	-0.18
0.6	-0.34	-0.31	-0.18
0.8	-0.35	-0.31	-0.18
1.0	-0.35	-0.32	-0.18
1.2	-0.35	-0.32	-0.19
1.4	-0.35	-0.32	-0.19

With Obvious Oscillation (**Bold**) (%)

Vce (V)	Je (mA/μm)		
	0.01	0.1	1
0.2	0.08	0.13	**27.92**
0.4	0.15	0.16	**36.14**
0.6	0.15	0.16	**35.34**
0.8	0.15	0.16	**32.66**
1.0	0.15	0.16	**25.46**
1.2	0.15	0.16	**26.09**
1.4	0.15	0.16	**26.64**

TABLE III. MEAN VBE EXAMPLES

Without Obvious Oscillation (V)

Vce (V)	Je (mA/μm)		
	0.01	0.1	1
0.2	0.719	0.783	0.851
0.4	0.719	0.783	0.850
0.6	0.719	0.783	0.850
0.8	0.719	0.783	0.849
1.0	0.719	0.783	0.849
1.2	0.719	0.782	0.848
1.4	0.719	0.782	0.848

With Obvious Oscillation (**Bold**) (V)

Vce (V)	Je (mA/μm)		
	0.01	0.1	1
0.2	0.722	0.786	**0.816**
0.4	0.722	0.786	**0.727**
0.6	0.722	0.786	**0.661**
0.8	0.722	0.785	**0.650**
1.0	0.722	0.785	**0.638**
1.2	0.722	0.785	**0.654**
1.4	0.722	0.785	**0.646**

TABLE IV. 3σ MISMATCH REPEATABILITY EXAMPLES

Without Obvious Oscillation (%)

Vce (V)	Je (mA/μm)		
	0.01	0.1	1
0.2	0.10	0.09	0.08
0.4	0.09	0.09	0.08
0.6	0.10	0.09	0.08
0.8	0.09	0.09	0.08
1.0	0.09	0.09	0.08
1.2	0.10	0.09	0.09
1.4	0.09	0.09	0.09

With Obvious Oscillation (**Bold**) (%)

Vce (V)	Je (mA/μm)		
	0.01	0.1	1
0.2	0.06	0.03	**1.14**
0.4	0.04	0.03	**2.66**
0.6	0.04	0.03	**2.81**
0.8	0.04	0.03	**1.37**
1.0	0.04	0.03	**2.43**
1.2	0.04	0.03	**2.62**
1.4	0.04	0.03	**2.41**

TABLE V. MEAN MISMATCH REPEATABILITY EXAMPLES

Without Obvious Oscillation (%)

Vce (V)	Je (mA/μm)		
	0.01	0.1	1
0.2	-0.01	-0.01	-0.02
0.4	-0.01	-0.01	-0.01
0.6	-0.01	-0.01	-0.01
0.8	-0.01	-0.01	-0.01
1.0	-0.01	-0.01	-0.01
1.2	-0.01	-0.01	-0.01
1.4	-0.01	-0.01	-0.01

With Obvious Oscillation (**Bold**) (%)

Vce (V)	Je (mA/μm)		
	0.01	0.1	1
0.2	0.01	0.00	**0.13**
0.4	0.01	0.00	**0.37**
0.6	0.01	0.00	**0.45**
0.8	0.01	0.00	**0.17**
1.0	0.01	0.00	**0.32**
1.2	0.01	0.00	**0.33**
1.4	0.01	0.00	**0.31**

TABLE VI. PROCESS A RELATIVE 3σ MISMATCH (VCE = 1V, JE = 0.1MA/μM) (%)

M-L-W	Dice	Le = 0.76μm					Le = 2μm (Iso 1.5 μm)					Le = 10.16μm			
		Vert	Horiz	AC-AC	AC-AE	AC-Iso	Vert	Horiz	AC-AC	AC-AE	AC-Iso	Vert	AC-AC	AC-AE	AC-Iso
A-1-02	30	1.00[b]	1.14	1.25	1.22	1.33	0.71	0.70	0.61	0.59	0.88	0.46	0.47	0.59	0.69
B-1-18	26	0.63	1.10	1.12	1.05	0.85	0.55	0.67	0.65	0.85	1.26	0.48	0.41	0.43	0.48
C-1-08	30	0.62	1.64	1.24	1.80	1.28	0.52	0.50	0.47	0.68	1.03	0.28	0.24	0.46	0.77
C-1-08	**73**	0.71	1.64	1.36	1.82	1.10	0.57	0.57	0.61	0.67	1.06	0.30	0.31	0.44	0.78

[b] The 3σ data has be normalized to the A-1-02 vertical 0.76um results.

TABLE VII. PROCESS A MEAN MISMATCH (VCE = 1V, JE = 0.1MA/μM) (%)

M-L-W	Dice	Le = 0.76μm					Le = 2μm (Iso 1.5 μm)					Le = 10.16μm			
		Vert	Horiz	AC-AC	AC-AE	AC-Iso	Vert	Horiz	AC-AC	AC-AE	AC-Iso	Vert	AC-AC	AC-AE	AC-Iso
A-1-02	30	-0.57	0.63	-0.56	1.01	-2.15	-0.55	-0.04	-0.25	-4.43	28.62	-0.65	0.03	-4.13	-7.51
B-1-18	26	0.36	-1.61	-1.90	2.05	-4.45	0.35	-0.08	-0.30	-1.01	33.22	0.20	-0.18	-2.83	-4.02
C-1-08	30	1.65	-3.41	0.35	-4.20	-19.20	0.18	-0.57	0.29	-1.46	27.94	-0.38	0.53	-3.07	-1.46
C-1-08	**73**	0.65	-2.73	1.08	-2.48	-18.82	0.15	-0.82	0.14	-1.06	27.78	-0.32	0.16	-2.89	-1.23

TABLE VIII. PROCESS B RELATIVE 3σ MISMATCH (VCE = 1V, JE = 0.12MA/μM) (%)

M-L-W	Dice	Le = 0.76μm					Le = 2μm (Iso 1.5 μm)					Le = 10μm			
		Vert	Horiz	AC-AC	AC-AE	AC-Iso	Vert	Horiz	AC-AC	AC-AE	AC-Iso	Vert	AC-AC	AC-AE	AC-Iso
A-1-All	150	1.00[c]	1.02	0.93	1.26	0.98	0.66	0.72	0.56	0.78	1.13	0.36	0.31	0.57	0.56
A-2-All	120	0.97	0.99	0.89	1.32	1.34	0.83	0.73	0.63	0.82	1.15	0.35	0.31	0.65	0.63
A-3-05	30	0.88	1.06	0.74	1.31	1.42	0.56	0.61	0.53	0.89	1.40	0.35	0.32	0.57	0.35
B-1-15	26	0.59	0.67	0.99	1.08	1.17	0.75	0.62	0.75	0.90	0.96	0.45	0.41	0.83	0.61
C-1-15	72	0.92	1.04	0.92	1.08	1.30	0.66	0.74	0.62	0.72	1.26	0.37	0.36	0.44	0.48

[c] The 3σ data has be normalized to the A-1-All vertical 0.76um results.

TABLE IX. PROCESS B MEAN MISMATCH (VCE = 1V, JE = 0.12MA/μM) (%)

M-L-W	Dice	Le = 0.76μm					Le = 2μm (Iso 1.5 μm)					Le = 10μm			
		Vert	Horiz	AC-AC	AC-AE	AC-Iso	Vert	Horiz	AC-AC	AC-AE	AC-Iso	Vert	AC-AC	AC-AE	AC-Iso
A-1-All	150	0.04	-0.44	0.62	5.54	1.46	-0.57	-0.46	0.01	0.52	34.73	-0.62	-0.34	-0.72	-1.24
A-2-All	120	1.05	-1.22	0.77	4.32	7.05	-0.13	-0.51	0.26	-0.55	37.46	-0.66	-0.46	0.54	-2.59
A-3-05	30	-0.30	-0.48	-0.60	3.43	9.35	-1.25	-1.31	-0.48	0.98	43.04	-0.80	0.03	4.42	-1.01
B-1-15	26	-0.19	-1.69	0.44	-0.07	-1.53	-0.92	0.39	0.25	-3.03	34.38	0.67	-0.29	-5.79	-4.83
C-1-15	72	0.18	-0.13	-0.70	-0.96	-3.93	-1.43	-0.02	-0.82	-0.67	32.07	-0.34	0.20	-1.66	-3.42

REFERENCES

[1] H.P. Tuinhout and W.C.M. Peters, "Measurement of lithographical proximity effects on matching of bipolar transistors," 1998 IEEE International Conference on Microelectronic Test Structures (ICMTS), pp. 7–12.

[2] H.P. Tuinhout, A. Bretveld, and W.C.M. Peters, "Current mirror test structures for studying adjacent layout effects on systematic transistor mismatch," 2003 IEEE International Conference on Microelectronic Test Structures (ICMTS), pp. 221–226.

[3] R.W. Gregor, "On the relationship between topography and transistor matching in an analog CMOS technology," IEEE Transaction on Electron Devcies, vol. 39, no. 2, pp. 275-282, Feb 1992.

[4] P.G. Drennan, "Device mismatch in BiCMOS technologies," 2002 Bipolar/BiCMOS Circuits and Technology Meeting (BCTM), pp. 104-111.

[5] H.P. Tuinhout, "Design of matching test structures," 1994 IEEE International Conference on Microelectronic Test Structures (ICMTS), pp. 21–27.

[6] C. Compton, "NPN CML ring oscillators for model verification and process monitoring," 2015 International Conference on Microelectronic Test Structures (ICMTS), pp. 103-108.

[7] H.P. Tuinhout, H. Elzinga, J.T. Brugman, and F. Postma, "Accurate Capacitor matching meaurements using floating gate test structures,"

TABLE X. MEAN MISMATCH FOR PROCESS A WAFER C-1-08 WITH STANDARD AND REVERSED KELVIN EMITTER BUS CONNNECTIONS

Standard Kelvin Emitter (%)

Vce (V)	Je (mA/μm)		
	0.01	0.1	1
0.2	-19.01	-18.87	-18.11
0.4	-18.83	-18.79	-18.13
0.6	-18.84	-18.80	-18.20
0.8	-18.84	-18.81	-18.27
1.0	-18.84	-18.82	-18.34
1.2	-18.85	-18.83	-18.41
1.4	-18.84	-18.84	-18.48

Reversed Kelvin Emitter Bus (%)

Vce (V)	Je (mA/μm)		
	0.01	0.1	1
0.2	-19.00	-18.82	-17.64
0.4	-18.83	-18.73	-17.66
0.6	-18.83	-18.74	-17.73
0.8	-18.83	-18.75	-17.80
1.0	-18.84	-18.76	-17.87
1.2	-18.84	-18.77	-17.93
1.4	-18.84	-18.78	-18.00

1995 IEEE International Conference on Microelectronic Test Structures (ICMTS), pp. 133–137.

[8] A. Cathigno, S. Bordez, K. Rochereau, and G. Ghibaudo, "From MOSFET matching test structures to matching data utilization: not an ordinary task," 2007 IEEE International Conference on Microelectronic Test Structures (ICMTS), pp. 230–233.

[9] H.P. Tuinhout, J.H. Klootwijk, W.C. Goeke, and L.K. Stauffer, "Impact of transistor noise on high precision parameter matchign measurements," 1995 IEEE International Conference on Microelectronic Test Structures (ICMTS), pp. 133–137.

978-1-5386-5072-1/18 $31.00 © 2018 IEEE

Session 6

On-Chip Characterization

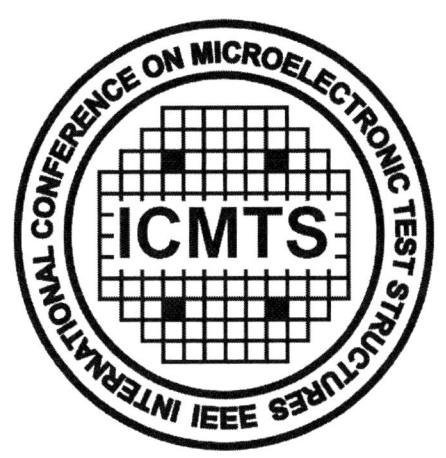

March 21, 2018

10:50-12:10

Co-Chairs:

Christopher HESS, PDF Solutions, USA

Greg YERIC, ARM, USA

978-1-5386-5072-1/18 $31.00 © 2018 IEEE

2018 INTERNATIONAL CONFERENCE ON MICROELECTRONIC TEST STRUCTURES, MARCH 19-22, 2018, AUSTIN, TEXAS, USA

On–Chip Reconfigurable Monitor Circuit for Process Variation and Temperature Estimation

Tadashi Kishimoto*, Tohru Ishihara* and Hidetoshi Onodera*
* Graduate School of Informatics, Kyoto University, Yoshida-honmachi, Sakyo-ku, Kyoto 606-8501, JAPAN
email: tkishimoto@vlsi.kuee.kyoto-u.ac.jp, {ishihara,onodera}@i.kyoto-u.ac.jp
Tel: +81-75-753-5948, Fax: +81-75-753-5343

Abstract—This paper proposes a monitor circuit that can estimate process variation and temperature by circuit reconfiguration. The circuit topology of the temperature monitoring is crafted such that the oscillation frequency is determined by the amount of leakage current which has an exponential dependency to temperature. The voltage dependence of this circuit is small in the configuration for temperature measurement, and the temperature dependence is small in the configuration for process variation estimation. A test chip fabricated in a 65 nm CMOS process demonstrates the temperature estimation capability with accuracy within -0.3 °C to 0.4 °C over a temperature range of 10 °C to 100 °C, as well as the ability for estimating process variations.

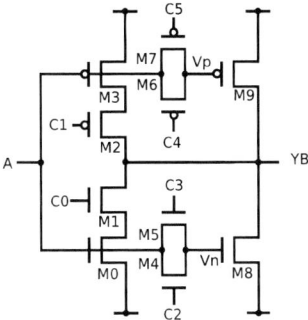

Fig. 1: An inverter delay cell in the reconfigurable RO proposed in Ref. [6].

I. INTRODUCTION

In the scaled technologies, the influence of parameter variations on LSI becomes larger. There are three major sources of variation: process (P), voltage (V) and temperature (T). They are commonly called PVT variations and are major factors restricting the performance of LSIs [1].

Process variations are primarily characterized by variations in threshold voltage and channel length. Process variation affects the amount of current flowing in the circuit, causing variations in the delay of the circuit. Threshold voltage variation affects the amount of leakage current which determines static power dissipation. Temperature affects power dissipation and reliability of LSI. For example, the performance of server processors is restricted by operating temperature [2]. It is therefore popular to perform dynamic temperature management in a recent processor [3]. Process variation and temperature vary depending on the location on a chip [4], [5], and therefore location-specific on-chip monitoring becomes a necessity.

Voltage fluctuation also affects the circuit operation, which is one of the important fluctuation factors. However, most of the voltage variation can be controlled by the power supply voltage. Therefore, this work focuses only on process and temperature variations.

A method is proposed that estimates the amount of process variation from the frequency of a reconfigurable ring oscillator (RO) [6]. The circuit takes advantage of the reconfiguration such that a single circuit can be configured to various topologies from which process variation can be characterized effectively. In this work, we propose a new topology–reconfigurable

RO that can estimate both process and temperature. The proposed monitor circuit can be configured to a circuit topology that has a high sensitivity to temperature with a very low sensitivity to supply voltage. It is also possible to suppress the temperature dependency of the frequency when the RO is configured for estimating process variation. We can thus estimate the deviation of process shift due to aging such as BTI [7] simultaneously with temperature estimation.

The rest of the paper is organized as follows. Section II explains topology reconfigurable ring oscillator and its configuration for temperature and process variation monitoring. Section III shows the measurement result of our proposed monitor circuit. Section IV shows the estimation result of temperature and process variation. Section V concludes the discuss.

II. RECONFIGURABLE RO

A Ring Oscillator (RO) can be easily implemented and its oscillation frequency can be easily measured. It is therefore one of commonly used circuits to evaluate parameter variations in P, V, and T [8]. In Ref. [6], a reconfigurable RO which can estimate multiple process variations by a single circuit has been proposed. A reconfigurable inverter cell in the RO is shown in Fig. 1.

In this paper, we propose an updated reconfigurable inverter cell that can estimate both process variation and temperature as shown in Fig. 2. For estimating temperature, we utilize the

978-1-5386-5072-1/18 $31.00 © 2018 IEEE   111

2018 INTERNATIONAL CONFERENCE ON MICROELECTRONIC TEST STRUCTURES, MARCH 19-22, 2018, AUSTIN, TEXAS, USA

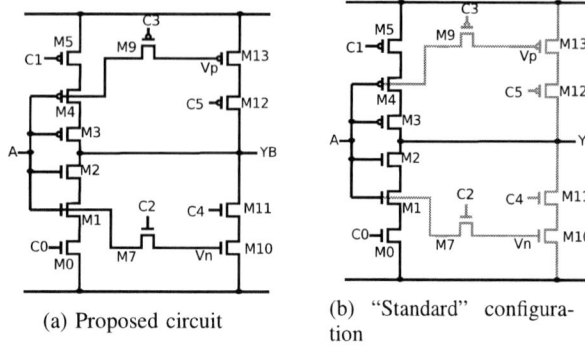

(a) Proposed circuit   (b) "Standard" configuration

Fig. 2: Proposed topology–reconfigurable inverter cell and configuration example.

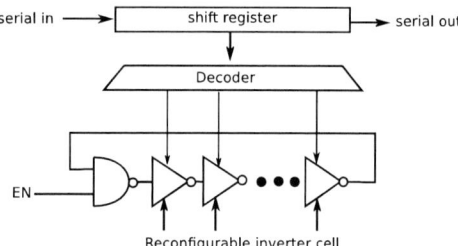

Fig. 3: Structure of reconfigurable RO.

TABLE I: Configurations of the reconfigurable inverter cell.

C0	C1	C2	C3	C4	C5	Mode
1	0	0	1	0	1	"Standard"
0	0	1	1	1	1	"N-sensitive"
1	1	0	0	0	0	"P-sensitive"
0	0	1	0	1	0	"N-leak"
1	1	1	0	1	0	"P-leak"

to these stacked transistors, we can better control leakage current. The circuit in Fig. 4 (a) is pulled down by leakage current and pulled up by strong inversion current ("N-leak configuration"). The circuit in Fig. 4 (b) is pulled up by leakage current and pulled down by strong inversion current ("P-leak configuration").

Here, we explain the principle of operation by taking "N-leak" configuration as an example. In the "N-leak" configuration, all control signals except C1 are set to turn off the transistors. Therefore, among the transistors to which the control signal is connected, only M5 is turned on. Let's consider the case where VSS is applied to input A. At this time, considering the first stage, M3 and M4 are turned on and M1 and M2 are turned off. Also, since the transistors M7 and M9 of the second stage are turned off, the second stage becomes inactive. Therefore, similarly to a normal stacked inverter, the output YB is pulled up by PMOS transistors M3, M4, M5 connected in series.

Next, consider the case where VDD is applied to input A. At this time, M3 and M4 of the first stage are turned off while M1 and M2 are turned on. In this situation, only M0 is OFF in the NMOS pull-down path. Regarding the PMOS pull-up path in the first stage, the two transistors M3 and M4 are OFF. The leakage current contributing to pull–up from the PMOS path in the first stage becomes smaller than the leakage current contribution to pull–down from the NMOS path in the first stage due to the stack effect. Regarding the pull-up path in the second stage, since M9 is OFF and VDD is applied to input A, the voltage of Vp is kept at VDD. Therefore, both of M12 and M13 are OFF, which produces reduced leakage current due to the stack effect. For the pull-down path in the second stage, a small voltage Vn appears at the gate of M10 similar to the original circuit of Fig. 1. The leakage current, however, is much smaller due to the stacking transistor M11, but larger than the pull-up leakage current by M12 and M13. Due to the leakage current flowing from the first and the second pull-down paths, the output YB is pulled down.

The configurations for estimating process variation is shown in Fig. 5. These configurations have high sensitivity to threshold voltage variation. Here, the principle of operation is briefly described. In this configuration, the first stage pulls up and pulls down output YB in the same way as a normal inverter. In the second stage, pull-up or pull-down signals are supplied via pass transistors M7 and M9. When the pass transistor M7 or M9 is ON, the pull-up or pull-down voltage supplied from the input A is caused to drop by the threshold voltage by pass transistor M7 or M9. Due to the influence of the pull-

characteristics of subthreshold leakage current which has an exponential relationship to temperature. Exploiting the characteristics of the subthreshold leakage current, we can derive a monitor circuit that has a strong sensitivity to temperature. Furthermore, the supply voltage sensitivity of the leakage current is lower than that of the strong inversion current. Therefore, it is possible to design a monitor circuit with lower supply voltage sensitivity utilizing the leakage current.

In estimating process variation, we can reduce sensitivity of temperature by utilizing mutual compensation on temperature dependence of mobility and threshold voltage. It is known that there is a bias voltage that suppresses the temperature dependence for a wide temperature range in MOSFETs [9]. By utilizing these characteristics, the frequency of the monitor circuit is less sensitive to temperature at a certain voltage.

Fig. 3 shows a block diagram of the topology–reconfigurable RO that consists of reconfigurable inverters and control logic for reconfiguration. Fig. 2 (a) shows the proposed reconfigurable inverter and Fig. 2 (b) shows the proposed "Standard" configuration which behaves like a stacked inverter. Figs. 4 and 5 explain each circuit configuration for estimating process variation and temperature. Configurations used for process variation and temperature estimation are listed in Table I.

The configurations for estimating temperature variation are shown in Fig. 4. In the proposed circuit topology, the second stage has cut–off transistors M11 and M12. The first stage also has a stacked structure to suppress leakage current of transistors other than the target transistors M0 and M5. Thanks

978-1-5386-5072-1/18 $31.00 © 2018 IEEE   112

2018 INTERNATIONAL CONFERENCE ON MICROELECTRONIC TEST STRUCTURES, MARCH 19-22, 2018, AUSTIN, TEXAS, USA

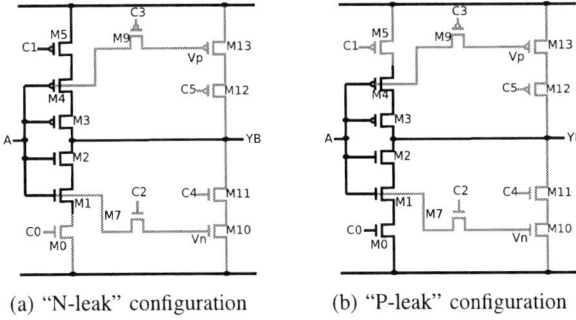

(a) "N-leak" configuration   (b) "P-leak" configuration

Fig. 4: Configurations for temperature estimation.

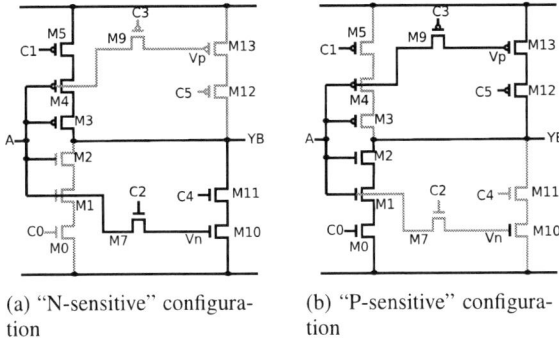

(a) "N-sensitive" configuration

(b) "P-sensitive" configuration

Fig. 5: Configurations for estimating process variation.

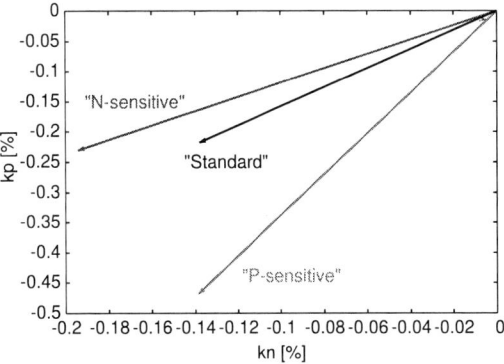

Fig. 6: Sensitivity to threshold voltage of the configuration for estimating process variation.

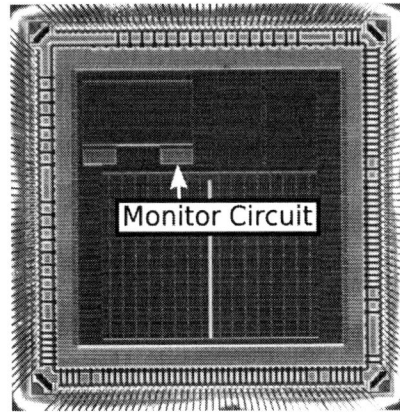

Fig. 7: Test chip structure fabricated in 65 nm CMOS process.

up or pull-down voltage dropped by the threshold voltage, the frequency of this configuration becomes sensitive to the threshold voltage fluctuation.

In this paper, let us assume we would like to monitor the threshold voltage shift of an NMOS transistor ΔV_{thn} and the threshold voltage shift of a PMOS transistor ΔV_{thp} due to aging effect. The sensitivity of the threshold voltage variation in each configuration is shown in Fig. 6, where k_{n} is the sensitivity of the frequency to ΔV_{thn} and k_{p} is the sensitivity of the frequency to ΔV_{thp}. Since the sensitivity to each threshold voltage is different in each configuration, estimation of the threshold voltage separately becomes easy.

III. MEASUREMENT RESULT

First, we explain our test chip. The proposed reconfigurable RO has been fabricated in a 65 nm CMOS technology. A micro-graph of the test chip is shown in Fig 7. In the following results, we use "N-leak" configuration with the gate width of NMOS transistor being 3 times larger than that of PMOS transistor such that stable operation is achieved under a wide range of PVT condition. "P-leak" configuration can also be applied if the gate width of NMOS and PMOS transistors are properly selected.

Fig. 8 shows the oscillation frequency of the "N-leak" configuration in Fig. 4 (a) as a function of the temperature. In a logarithmic scale, the oscillation frequency indicates almost linear dependence on the inverse of the temperature. Fig. 9 shows the oscillation frequency of "N-leak" and "Standard"

configurations as a function of supply voltage. The oscillation frequency of "N-leak" configuration keeps almost constant over a wide range of supply voltage from 0.4 V to 1.0 V. The frequency at 1.0 V supply is only 2.6% larger than the frequency at 0.4 V supply. On the other hand, the frequency of the "Standard" configuration exhibits a large voltage sensitivity. The frequency at 1.0 V supply is more than 1,000% larger than that at 0.4 V supply.

Fig. 10 shows the measured oscillation frequencies of the "N-sensitive" and "P-sensitive" configurations as well as "Standard" configuration as functions of the body-bias voltage of NMOS transistors. "N–sensitive" configuration has a higher sensitivity to the threshold voltage of NMOS transistors than "P-sensitive" and "Standard" configurations. The frequency of "N–sensitive" configuration at 0 V NMOS body bias is more than 500% larger than the frequency at −1.0 V NMOS body bias. On the other hand, the frequency of "P– sensitive" configuration is 24% larger than the frequency at −1.0 V NMOS body bias. Fig. 11 represents the temperature sensitivity of "N-sensitive" and "P-sensitive" configurations as well as "Standard" configuration at 1.0 V supply. These configurations have much less sensitive to temperature than

978-1-5386-5072-1/18 $31.00 © 2018 IEEE   113

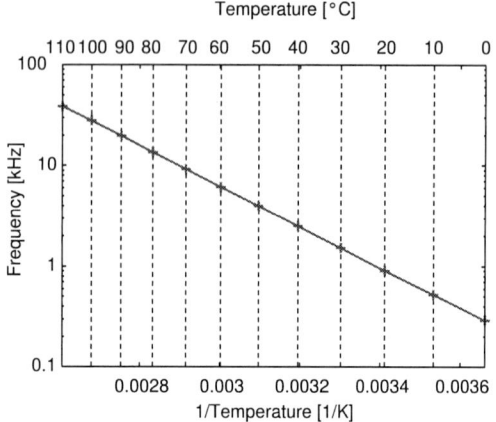

Fig. 8: Temperature dependence of the frequency of "N-leak" configuration at supply voltage 1.0 V.

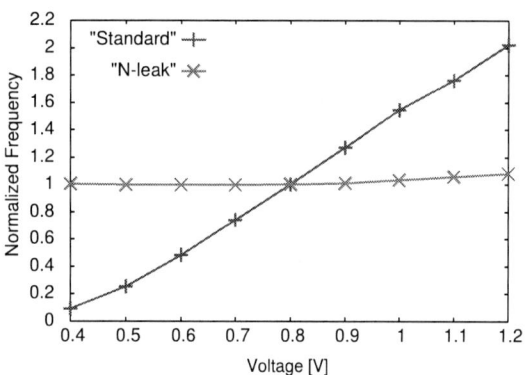

Fig. 9: Voltage dependence of the frequency of "N-leak" configuration and "Standard" configuration at room temperature.

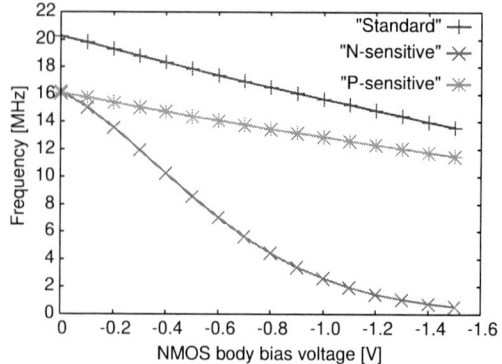

Fig. 10: Difference in frequency dependence with respect to NMOS body bias in each configuration.

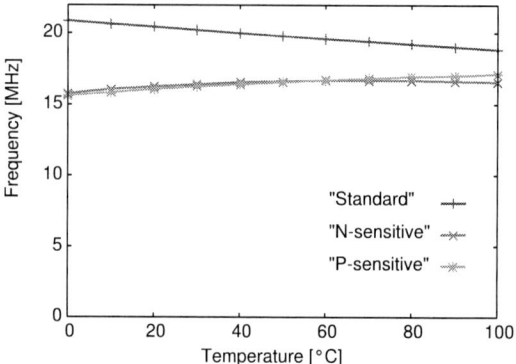

Fig. 11: Temperature dependence of frequency of the configuration for estimating process variation at supply voltage 1.0 V.

"N-leak" and "P-leak" configurations. For example, when the temperature changes from 10 °C to 100 °C, the frequency of "N–sensitive"configuration changes only by 3.2%. This characteristic depends on supply voltage. However, as long as the supply voltage does not change significantly, the temperature characteristic of the frequency does not change significantly. Fig. 12 shows the temperature sensitivity of "N-sensitive" and "P-sensitive" configurations at 1.2 V and 0.9 V supply. When the temperature changes from 10 °C to 100 °C, the frequency of "N–sensitive"configuration changes only by 8.2% at 1.2 V supply and 20.0% at 0.9 V supply.

IV. TEMPERATURE AND PROCESS VARIATION ESTIMATION

We can exploit the exponential sensitivity shown in Fig. 8 for temperature monitoring. We approximate the logarithm of the frequency as a linear function of the inverse of the temperature,

$$\ln(F) = a_T \cdot \frac{1}{T} + b_T, \tag{1}$$

where F is frequency, T is absolute temperature, a_T and b_T are temperature coefficients. If we apply a two-point

calibration, we can derive parameter values of a_T and b_T. Fig. 13 shows the amount of estimation error by the proposed monitor circuit after 2-point calibration at the supply voltage of 1.0 V. Calibration is performed at 20 °C and 80 °C. The amount of estimation error ranges from −0.3 °C to 0.4 °C over a temperature range of 10 °C to 100 °C. By obtaining the value of the parameter a_T in advance by fitting the model equation (1) to the characteristics of a specific chip or circuit simulation results, estimating temperature by calibration at one point becomes possible. Fig. 14 shows the amount of estimation error by the proposed monitor circuit after 1-point calibration at the supply voltage of 1.0 V. Calibration is performed at 50 °C. The amount of estimation error ranges from −1.2 °C to 0.9 °C over a temperature range of 10 °C to 100 °C.

A method to estimate process variation is proposed [6], [10]. This method can be also applied to this monitor circuit. In Ref [10], the following equation is used to estimate the process variation.

$$\Delta f = f_M - f_{Ref} = k_n \times \Delta V_{thn} + k_p \times \Delta V_{thp} + k_l \times \Delta L. \tag{2}$$

where f_M is the measured frequency and f_{Ref} is the reference frequency, and $\Delta V_{thn}, \Delta V_{thp}$ and ΔL are deviations from

2018 INTERNATIONAL CONFERENCE ON MICROELECTRONIC TEST STRUCTURES, MARCH 19-22, 2018, AUSTIN, TEXAS, USA

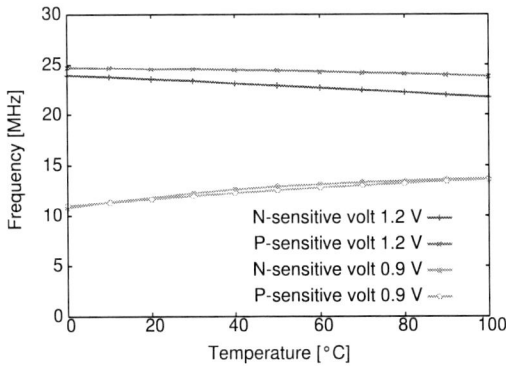

Fig. 12: Temperature dependence of frequency of the configuration for estimating process variation at 1.2 V and 0.9 V supply.

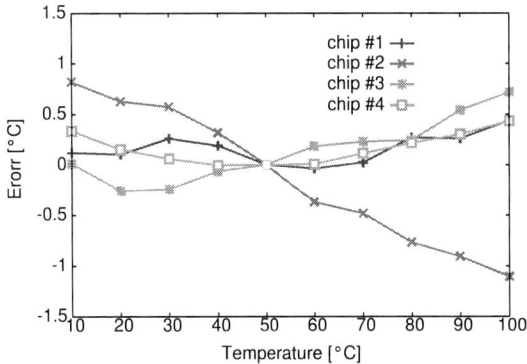

Fig. 14: Sensing error versus temperature after one-point calibration at the supply voltage of 1.0 V.

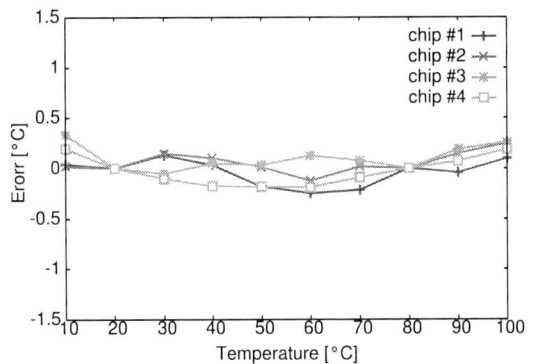

Fig. 13: Sensing error versus temperature after two-point calibration at the supply voltage of 1.0 V.

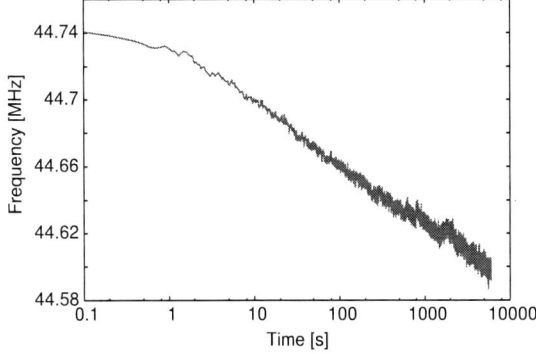

Fig. 15: Time change of frequency in "P-sensitive" configuration at 2.0 V supply and temperature 80 °C.

the reference values where the oscillation frequency becomes f_{Ref}. The value f_{Ref} can be obtained by circuit simulation using an RC extracted netlist from the layout. Parameters $k_{\mathrm{n}}, k_{\mathrm{p}}, k_{\mathrm{l}}$ are sensitivity coefficients. By solving simultaneous equations using three kinds of configurations with different sensitivities, the process variation can be estimated using Eq. (2). In reality, the frequency does not change linearly with respect to the threshold voltage and the channel length. Therefore, it is necessary to change the threshold voltage and channel length of reference frequency and iterate this procedure several times.

Table II lists the estimated variations of threshold voltages and channel length from the TT condition for four test chips.

Channel length L has a static variation and does not change over time. Therefore, with a fixed value of L, we can estimate the threshold voltage variation of PMOS transistors over time under aging condition. Fig. 15 shows a measured oscillation frequency of "P-sensitive" configuration over 600 seconds under elevated supply voltage of 2.0 V and temperature of 80 °C. Slight degradation of oscillation frequency can be observed. Assuming only the threshold voltage of PMOS transistors has increased equally in the circuit, the decrease in the oscillation frequency can be converted to the increase in the PMOS threshold voltage as listed in Table III.

Finally, we compare to some of the state-of-the-art temperature sensors, as listed in Table IV. Compared to the BJT monitor circuit [16], proposed monitor circuit has a lower

TABLE II: Estimation result of the threshold voltage and channel length shift.

measure chip	ΔVthn[mV]	ΔVthp[mV]	ΔL[nm]
chip #1	25	32	-6.6
chip #2	27	44	-10
chip #3	8.4	24	-5.6
chip #4	21	26	-5.6

TABLE III: Threshold voltage increase over time.

Time [s]	ΔV_{thp} [mV]
0	0
10	0.8
100	1.5
1000	2.1
6000	2.8

978-1-5386-5072-1/18 $31.00 © 2018 IEEE   115

TABLE IV: Performance comparison of the temperature sensor.

	This work	JSSC'16 [11]	VLSI'08 [12]	JSSC'15 [13]	JSSC'15 [14]	CAS'17 [15]	JSSC'13 [16]
Technology	65 nm	65 nm	65 nm	65 nm	65 nm	180 nm	160 nm
Type	MOSFET	MOSFET	MOSFET	MOSFET	MOSFET	MOSFET	BJT
Design	Cell-base	Cell-base	Custom	Custom	Cell-base	Custom	Custom
Area[mm^2]	0.000686[1]	0.0082[2]	0.0012[3]	0.000279[3]	0.0045[3]	0.45[4]	0.08[4]
V_{dd}	0.4–1.0	0.85–1.05	1.1	0.6–1.0	0.5–1.0	0.55–0.85	1.5–2
Range [$^\circ$C]	10–100	0–100	40–90	0–100	5–75	0–100	-55–125
Calibration Point	2-point	2-point	1-point	2-point	1-point	2-point	1-point
Error[$^\circ$C]	-0.3/0.4[5]	±0.9	<3.1	-2.4/1.5	-1.3/1.4	-1.33/0.62	±1.5(3σ)
Supply Sensitivity[$^\circ$C/mV]	0.000945	0.034	0.018	0.00008	N/A	0.00167	0.0005
Power[μW]	0.21	154	N/A	360	0.076	0.075	5.1

[1] Core ring only
[2] Thermal sensor, counter and correction logic
[3] Thermal sensor Only
[4] Front-end with ADC
[5] At supply voltage 1.0 V

accuracy and a narrower temperature range. However, the supply voltage requirement of the BJT monitor circuit is not compatible with an LSI in a scaled process where supply voltage is 1 V or less. Compared to other MOSFET based monitor circuits, the proposed monitor circuit exhibits lower voltage sensitivity or higher accuracy. Moreover, the proposed circuit has the widest supply voltage range with reduced design cost due to its compatibility with cell-base design. In addition, the proposed circuit can estimate not only temperature but also process variation.

V. CONCLUSIONS

We have proposed a topology–reconfigurable RO which can be used both estimating process variation and temperature. A test chip has been fabricated in a 65 nm CMOS process. The circuit can operate under a wide range of supply voltage from 0.4 V to 1.0 V. Measured oscillation frequency in "N-leak" configuration varies only 2.6 % when the supply voltage changes from 0.4 V to 1.0 V at room temperature. The estimation error in temperature sensing ranges from -0.3 $^\circ$C to 0.4 $^\circ$C over a temperature range of 10 $^\circ$C to 100 $^\circ$C. We have shown that the proposed circuit has ability to estimate threshold voltages and channel length simultaneously. It could be also used for sensing threshold voltage increase by aging effect.

ACKNOWLEDGMENT

This work has been partly supported by KAKENHI Grant-in-Aid for Scientific Research 16H01713. This work is also supported by VLSI Design and Education Center (VDEC), the University of Tokyo in collaboration with Cadence Design Systems, Inc. and Synopsys, Inc.

REFERENCES

[1] S. Borkar, T. Karnik, S. Narendra, J. Tschanz, A. Keshavarzi, and V. De, "Parameter Variations and Impact on Circuits and Microarchitecture," in *Proc of Design Automation Conference*, June 2003, pp. 338–342.

[2] C. R. Lefurgy, A. J. Drake, M. S. Floyd, M. S. Allen-Ware, B. Brock, J. A. Tierno, J. B. Carter, and R. W. Berry, "Active Guardband Management in Power7+ to Save Energy and Maintain Reliability," *IEEE Micro*, vol. 33, no. 4, pp. 35–45, jul 2013.

[3] D. Brooks and M. Martonosi, "Dynamic thermal management for high-performance microprocessors," *Proceedings HPCA Seventh International Symposium on High-Performance Computer Architecture*, pp. 171–182, 2001.

[4] M. Pedram and S. Nazarian, "Thermal Modeling, Analysis, and Management in VLSI Circuits: Principles and Methods," *Proceedings of the IEEE*, vol. 94, no. 8, pp. 1487–1501, 2006.

[5] K. A. Bowman, S. G. Duvall, and J. D. Meindl, "Impact of die-to-die and within-die parameter fluctuations on the maximum clock frequency distribution for Gigascale integration," *IEEE Journal of Solid-State Circuits*, vol. 37, no. 2, pp. 183–190, 2002.

[6] A. Islam, T. Ishihara, and H. Onodera, "Reconfigurable Delay Cell for Area-efficient Implementation of On-chip MOSFET Monitor Schemes," in *Proc.of Asian Solid-State Circuits Conference*, Nov 2013, pp. 125–128.

[7] D. K. Schroder and J. A. Babcock, "Negative bias temperature instability: Road to cross in deep submicron silicon semiconductor manufacturing," *Journal of Applied Physics*, vol. 94, no. 1, pp. 1–18, 2003.

[8] K. J. Kuhn, M. D. Giles, D. Becher, P. Kolar, A. Kornfeld, R. Kotlyar, S. T. Ma, A. Maheshwari, and S. Mudanai, "Process technology variation," *IEEE Transactions on Electron Devices*, vol. 58, no. 8, pp. 2197–2208, 2011.

[9] I. M. Filanovsky and A. Allam, "Mutual compensation of mobility and threshold voltage temperature effects with applications in cmos circuits," *IEEE Transactions on Circuits and Systems I: Fundamental Theory and Applications*, vol. 48, no. 7, pp. 876–884, Jul 2001.

[10] I. A. K. M. Mahfuzul, A. Tsuchiya, K. Kobayashi, and H. Onodera, "Variation-Sensitive Monitor Circuits for Estimation of Global Process Parameter Variation," *IEEE Transactions on Semiconductor Manufacturing*, vol. 25, no. 4, pp. 571–580, Nov 2012.

[11] T. Anand, K. A. Makinwa, and P. K. Hanumolu, "A VCO Based Highly Digital Temperature Sensor with 0.034 $^\circ$c/mV Supply Sensitivity," *IEEE Journal of Solid-State Circuits*, vol. 51, no. 11, pp. 2651–2663, 2016.

[12] E. Saneyoshi, K. Nose, M. Kajita, and M. Mizuno, "A 1.1V 35μm x 35μm thermal sensor with supply voltage sensitivity of 2C/10%-supply for thermal management on the SX-9 supercomputer," *IEEE Symposium on VLSI Circuits, Digest of Technical Papers*, pp. 138–139, 2008.

[13] T. Yang, S. Kim, P. R. Kinget, and M. Seok, "Compact and Supply-Voltage-Scalable Temperature Sensors for Dense On-Chip Thermal Monitoring," *IEEE Journal of Solid-State Circuits*, vol. 50, no. 11, pp. 2773–2785, 2015.

[14] A. Islam, J. Shiomi, T. Ishihara, and H. Onodera, "Wide-Supply-Range All-Digital Leakage Variation Sensor for On-Chip Process and Temperature Monitoring," *IEEE Journal of Solid-State Circuits*, vol. 50, no. 11, pp. 2475–2490, Nov 2015.

[15] X. Wang, P. H. P. Wang, Y. Cao, and P. P. Mercier, "A 0.6V 75nW All-CMOS Temperature Sensor with 1.67m$^\circ$ C/mV Supply Sensitivity," *IEEE Transactions on Circuits and Systems I: Regular Papers*, vol. 64, no. 9, pp. 2274–2283, 2017.

[16] K. Souri, Y. Chae, and K. Makinwa, "A CMOS temperature sensor with a voltage-calibrated inaccuracy of ± 0.15 $^\circ$ C (3σ) from -55 to 125 $^\circ$ C," *Digest of Technical Papers - IEEE International Solid-State Circuits Conference*, vol. 55, no. 1, pp. 208–209, 2012.

DFT-enabled Within-die AC Uniformity and Performance Monitor Structure for Advanced Process

Nui Chong, I-Ru Chen, Da Cheng, Amitava Majumdar, Ping-Chin Yeh and Jonathan Chang

Xilinx, Inc., San Jose, California 95124, USA

Email: tim.chen@xilinx.com

Abstract—An on-chip ring oscillator based process monitoring vehicle embedded within host automatic place and route digital blocks and accessed through design for testability (DFT) circuit is introduced and characterized. Within-wafer AC uniformity (ACU), performance and power consumption for the ring oscillator are analyzed in a 7 nanometer technology testchip. The design and analysis techniques described are suitable to monitor process variation, real-time power fluctuation and performance proxy of host digital blocks in products.

Keywords—ring oscillator, process variation, process monitoring, AC characterization, AC uniformity, test structure, design for testability

I. INTRODUCTION

Process-induced variation and model correlation to silicon have become major challenges for integrated circuit (IC) performance in advanced technology nodes [1]. Characterizing and monitoring process variation and model correlation are keys in design for manufacturability, product performance and process knob selection.

During the adoption of new advanced technology nodes, significant engineering efforts are done to characterize and understand sources of variability, to predict variation impact to circuit and product behaviors, and to develop design and layout strategies mitigating variability [2]. That generally involves measuring transistor parameters in scribe line test structures to correlate with product behaviors in various process corners. Measured process variations in device DC characteristics such as on-state current and threshold voltages are then categorized into within-wafer die-to-die and within-die local variations [3]. However, such test structures do not fully capture AC-related variations in the product [4]. The implementation and characterization of scribe line test structures are also costly and cannot be continuously monitored throughout product lifetime [2, 5].

Furthermore, layout environment and pattern density have prominent effects to transistor performance in advanced nodes, and the scribe line structures cannot mimic the diverse layout environments in different product blocks. Characterizing complex product behaviors alone can also be hard to pin point the process knob sensitivity. Thus, additional techniques for variation control and product correlation are needed for future nanometer-node VLSI circuits.

To address these challenges, we propose a ring oscillator (RO) based test structure embedded within various product host IP's for process monitoring and tuning. RO is commonly used in product circuits and its performance is representative of actual product applications especially in the AC domain [4, 6]. The embedded RO process monitor is built with standard cells as a hard macro. The testing of the embedded RO macro is enabled as part of on-chip design for testability (DFT) features and can be inserted into various host IP's through automatic place and route (APR) flow.

The first-order short-range layout dependent effect for transistors within the RO macro is thus similar across different host IP's. The RO serves as performance proxy of host circuit in its native environment, reticle location and neighboring circuits within a die. The RO is also field-measurable in real-time through DFT when the IP is in mission mode. Performance fluctuations throughout product lifetime and different operation mode can be monitored, unlike scribe line tests which can only be done before wafer dicing [7].

II. DESIGN AND MEASUREMENT

Figure 1 shows the block diagram of a RO module embedded within the DFT controller and host IP. Key components and signals of the DFT controller related to RO test structure control are also shown. Upon receiving test initiation signal and a reference clock frequency at f_{clock} from the host IP, the controller enables the embedded RO and two separate counters. The clock cycle counter records the number of reference clock cycles until reaching the pre-defined maximum clock count at which also sends the stop-count signal to the RO counter. The RO counter, which is an asynchronous ripple counter driven by RO output, therefore counts the number of RO cycles between the enable-high event and the stop-count-high event. The RO frequency can be measured as:

$$f_{RO} = f_{Clock} \times \frac{(RO\ Count)}{(Max\ Clock\ Count)}$$

The maximum clock count and the RO count are parts of the DFT scan chain and can be retrieved by the host IP. RO speed is controllable by assigning the speed selector bits S1/S0 at test initiation.

2018 INTERNATIONAL CONFERENCE ON MICROELECTRONIC TEST STRUCTURES, MARCH 19-22, 2018, AUSTIN, TEXAS, USA

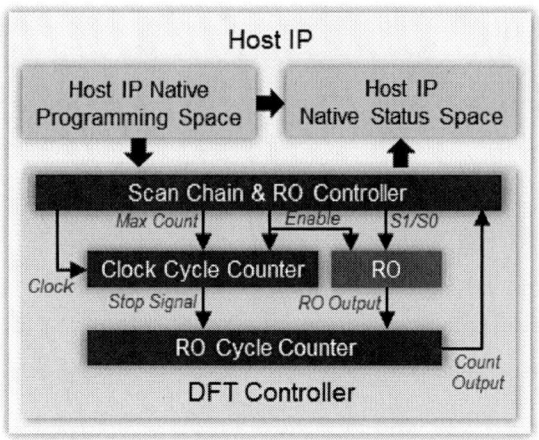

Fig. 1. Simplified block diagram for ring oscillator embedded in the DFT controller and host IP. S1/S0 input to RO selects its speed.

Figure 2 shows the design schematic, layout, and functions of the ring oscillator manufactured using 7 nm process technology. The RO is capable of three speed modes for different application scenarios. It is made of three inverter loops connected by two MUX with user-selectable S1/S0 speed selection pins to determine number of RO stages, as shown in Figure 2(a). Figure 2(b) illustrates the complete layout synthesized by APR flow for the DFT controller with RO embedded. The host IP is not included in this testchip, instead the DFT controller is interfaced with JTAG-compatible general-purpose IO circuits that connects to the test bench. The area of the RO and DFT logic controller is actually a small portion in Figure 2(b). Figure 2(c) shows the truth table of the frequency-selectable RO, with f0, f1, and f2 corresponding the frequencies in slow, mid, and fast speed modes. The slow mode incorporate inverter chains in all three loops and is designed to mimic the logic operation <1 GHz. The fast mode closes the inverter chain after Loop 1 and is used as a proxy for high-frequency operations.

Figure 3(a) illustrates the testing sequence and transitions between key controller states. The test program first resets all counters and the JTAG finite state machine in the DFT controller. The DFT controller sets S1/S0 and maximum clock count which are parts of the input stream to the controller. The controller then applies the trigger signal that enable both RO counter and reference clock counter until the stop-count-high event occurs when the reference clock counter reaches maximum clock count. The RO counter value is then retrieved as part of the scan chain readout and the controller returns to the idle state. Figure 3(b) shows a sample output waveform from the scan chain, with the maximum clock count and the RO count marked within the bit stream. The f0, f1, f2 frequency readouts for a sample RO can be calculated from the RO count readouts, as shown in Table I.

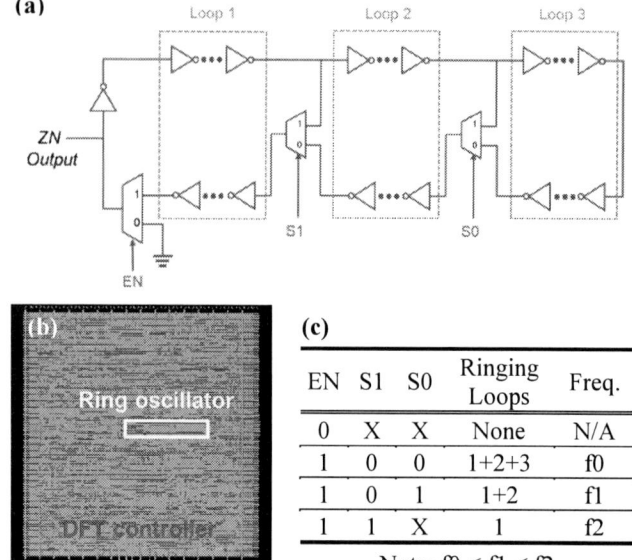

Fig. 2. (a) Schematics of the RO. EN is the enable signal, S1/S0 are the speed selector bits, and ZN is the output to counter. (b) Completed placed-and-routed layout showing RO embedded within the DFT controller. (c) Truth table showing RO frequency dependency on input.

EN	S1	S0	Ringing Loops	Freq.
0	X	X	None	N/A
1	0	0	1+2+3	f0
1	0	1	1+2	f1
1	1	X	1	f2

Note: f0 < f1 < f2

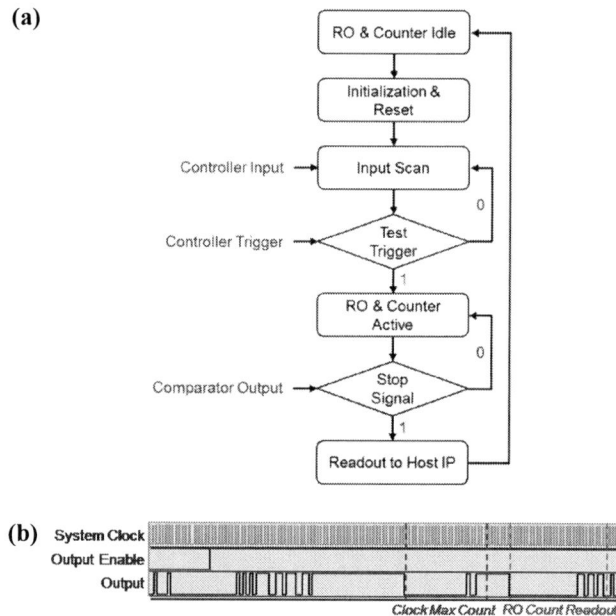

Fig. 3. (a) Simplified flow diagram illustrating key controller states and transitions. (b) Sample output bit stream from DFT controller after sampling. RO counter output and pre-defined maximum clock count can be observed from the output sequence.

TABLE I.   SAMPLE RO COUNT AND FREQUENCY READOUTS

Mode	Slow	Mid	Fast
S1/S0	0/0	0/1	1/1
RO Count	4779	8545	21605
Frequency [MHz]	f0=600.4	f1=1073.5	f2=2714.2

[a.] Reference clock frequency at 25 MHz and maximum clock count at 199

978-1-5386-5072-1/18 $31.00 © 2018 IEEE   118

2018 INTERNATIONAL CONFERENCE ON MICROELECTRONIC TEST STRUCTURES, MARCH 19-22, 2018, AUSTIN, TEXAS, USA

TABLE II.   MEASUREMENT RESULTS AT ROOM TEMPERATURE

Mode	Vcc [V]	Mean Freq. [MHz]	Std. Dev. [MHz]	3-Sigma AC Uniformity
Slow	0.55	295.7	10.3	10.4%
	0.675	484.4	13.9	8.6%
	0.75	585.1	16.1	8.3%
	0.96	812.6	21.8	8.0%
Mid	0.55	526.3	18.6	10.6%
	0.675	865.3	25.5	8.8%
	0.75	1046.2	29.2	8.4%
	0.96	1456.8	39.2	8.1%
Fast	0.55	1315.1	50.0	11.4%
	0.675	2169.8	65.4	9.0%
	0.75	2628.1	75.1	8.6%
	0.96	3661.4	99.6	8.2%

III. RESULTS AND DISCUSSION

Table II shows the list of measurements completed for RO across a wafer, with varying Vcc and three speed modes. The result validates the RO and DFT controller designs with very good measurement repeatability observed across a wide range of reference clock frequencies, thus the controller can be deployed through APR flow to various host IP's with different reference clocks available. The variability and AC characteristics data generated from each RO is a better performance proxy to host IP circuits than traditional transistor monitors placed far away on scribe line.

Data is collected in a checker board pattern of 37 dies to calculate the statistical mean and deviation for each measurement. The RO frequency across all three RO speed modes and different Vcc ranges from about 300 MHz to >3 GHz, representing typical operating frequencies of digital circuits. AC uniformity (ACU), defined as the 3-sigma deviation divided by wafer frequency mean, can be monitored from the test results. The within-die transistor variation is effectively minimized in the multiple-stage RO design. In the Vcc = 0.55 V experiments, ACU is slightly higher than measurements done at other Vcc since the overdrive voltage in excess of transistor threshold voltage is very small.

Figure 4 shows the wafer contour map for all 37 dies and their respective measured RO frequency, capturing the distribution of key AC performance indicator across the wafer. The contour distribution of AC performance, along with traditional DC process control monitors, can be used to monitor signatures of process-induced variation. The results can be compared across wafers in the same lot and monitor potential lot-to-lot process changes. With several RO's placed in multiple host IP's across a die, the local AC uniformity can also be measured to gauge process variability from factors such as mask reticle making and scanner exposure.

The RO measurement results of all 37 dies, along with the expected frequency vs. Vcc data from simulation, are plotted in Figure 5 for all three RO speed modes. The data shows good sensitivity of frequency vs. Vcc across typical digital circuit operating frequencies. This shows the test structure, by periodically issuing test initiation commands to the DFT controller and retrieving measured RO frequencies, can be used to monitor voltage fluctuations on connected power rail. The duration of power rail fluctuation must be longer than the sampling rate of the test structure in order to be captured

effectively. When embedded within the host IP, the test structure provides useful power rail voltage information at designated IP state, and can be analyzed in conjunction with other DFT features within the same host IP.

(a)

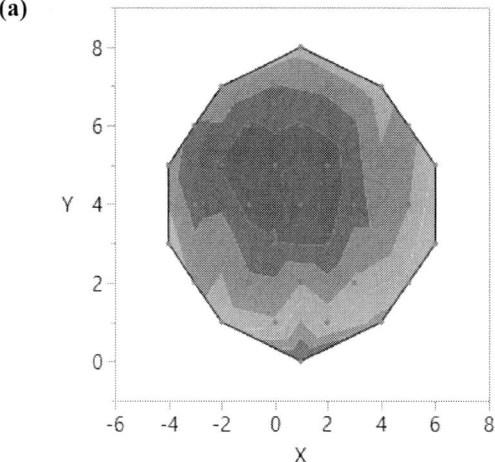

(b)

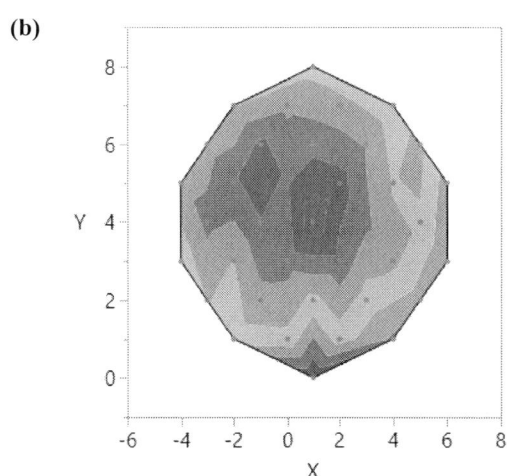

(c)

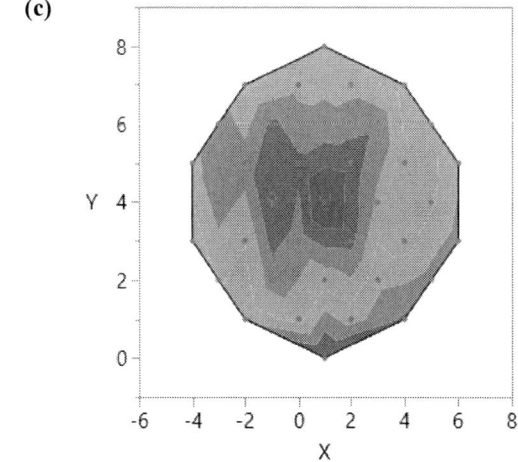

Fig. 4. Wafer contour map of RO frequency measurement results, for (a) slow, (b) mid, and (c) fast RO modes. RO is measured with Vcc = 0.75 V. Within-wafer AC variation is captured from this embedded test structure.

978-1-5386-5072-1/18 $31.00 © 2018 IEEE   119

2018 INTERNATIONAL CONFERENCE ON MICROELECTRONIC TEST STRUCTURES, MARCH 19-22, 2018, AUSTIN, TEXAS, USA

(a)

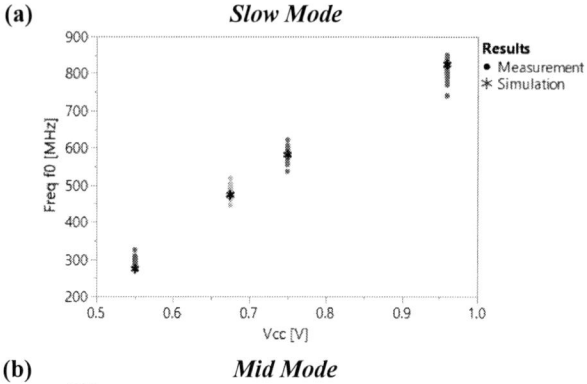

(b)

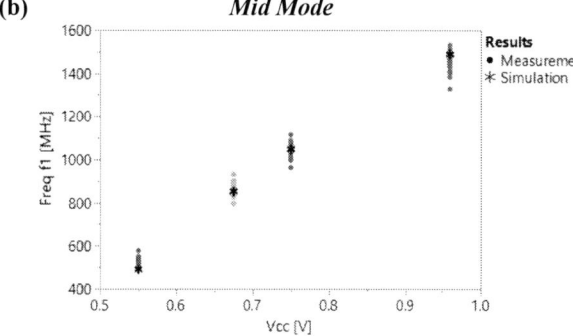

(c)

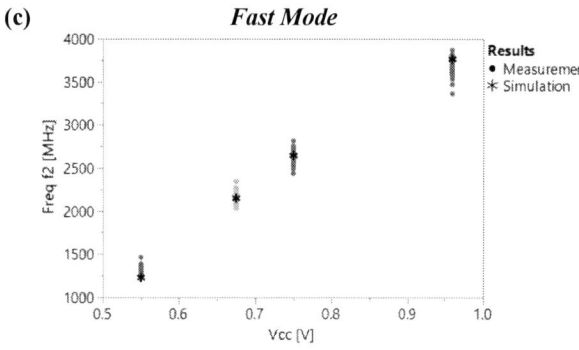

Fig. 5. Frequency vs. Vcc measurement results for slow, mid, and fast RO speed modes, respectively. Expected RO frequencies from simulation are shown along with measurement results.

The RO dynamic power consumption can also be monitored from the proposed test structure by measuring the Icc current draw by the RO. Icc can be calculated as the difference in total current consumption between two states in the measurement sequence, one with RO turned on and the other with RO off. All other components of the DFT controller remain the same between the two states. This differential mode measurement of Icc is effective when the RO is embedded within a standalone DFT test structure, such as this testchip, and when the host IP current draw is less than or comparable to Icc. Figures 6 and 7 shows the relationship between Icc vs. Vcc and Icc vs. f0, respectively, using slow RO mode. The expected Icc obtained from RO simulation are also plotted. Since the RO frequency is controlled in part by Vcc and f0-Vcc follows approximately a linear trend, the Icc can be fitted with Vcc^2 and $f0^2$ for both measured and simulated data. The AC capacitance can be extracted and monitored from the Icc measurement.

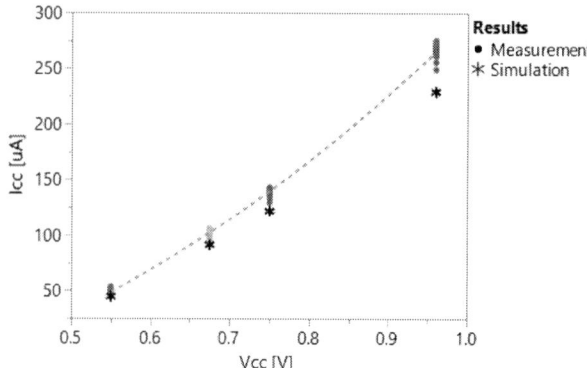

Fig. 6. Dynamic Icc vs. Vcc measurement and simulation results for all samples using slow RO mode. Measured Icc current can be fitted to Vcc^2 relationship. Similar results can be obtained with other RO speed modes.

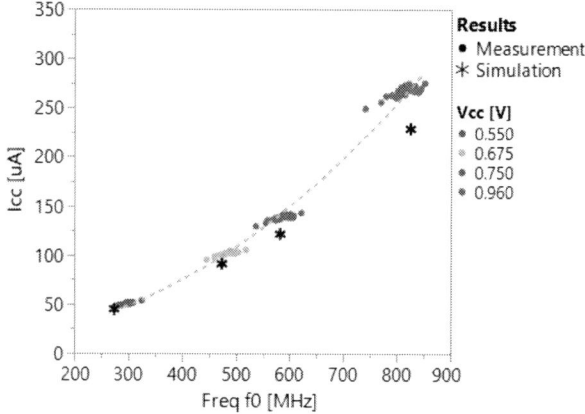

Fig. 7. Dynamic Icc vs. frequency measurement and simulation results using slow RO mode. Measured Icc current can be fitted to $f0^2$ relationship. Similar results can be obtained with other RO speed modes.

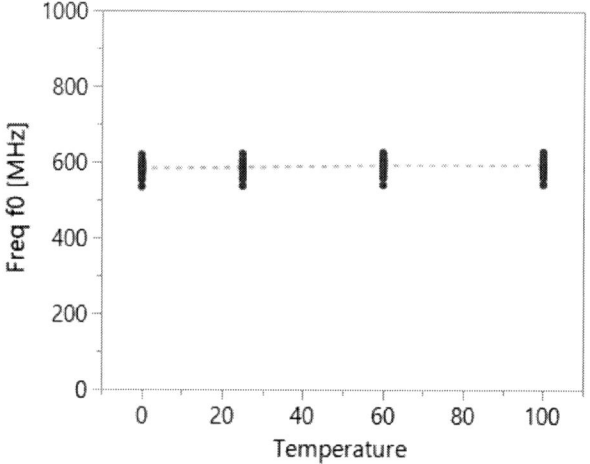

Fig. 8. Temperature dependency of measured RO frequency for slow RO mode. Result shows RO frequency is not affected by environment temperature across typical IC operating conditions. Simulation expectation is plotted in dashed lines and matches with measured data.

978-1-5386-5072-1/18 $31.00 © 2018 IEEE

Since the proposed RO test structure is embedded within the host IP, the native operation of the host IP can have significant impact on the environment temperature when RO is oscillating. We measured frequency variation across four typical IC operating temperatures and summarized the results in Figure 8. The results show little temperature sensitivity across typical operating temperature and frequency ranges. Therefore the proposed test structure is relatively immune to the temperature change resulted from host IP and DFT controller operations.

IV. CONCLUSION

The DFT-enabled ring oscillator validated here provides a reliable way to capture on-chip key process variations and serve as performance proxy throughout product life in 7 nm technology. RO frequency across wafer is measured and the AC uniformity is obtained through statistical analysis. Power consumption and temperature dependency are also measured and analyzed.

ACKNOWLEDGMENT

The authors would like to thank X. Wu and J. Wong for helpful discussions, S. Raman, P. Sripathi, F. Noorbasha for their design support, and H.C. Lin, C.C. Yeh, H. Chen for their testing support.

REFERENCES

[1] S. Borkar et al., "Parameter variations and impact on circuits and microarchitecture," in Proc. 2003 Design Automation Conf. (DAC), pp. 338-342, 2003.

[2] S. R. Nassif, "Process variability at the 65nm node and beyond," in Proc. IEEE 2008 Custom Intergrated Circuits Conference (CICC), p. 2-1-1, 2008.

[3] S. Fujimoto, A. Islam, T. Matsumoto, and H. Onodera, "Inhomogeneous ring oscillator for within-die variability and RTN characterization," IEEE Trans. Semiconductor Manufacturing, vol. 26, pp. 296-305, 2013.

[4] M. Bhushan, A. Gattiker, M. B. Ketchen, and K. K. Das, "Ring oscillators for CMOS process tuning and variability control," IEEE Trans. Semiconductor Manufacturing, vol. 19, pp. 10–18, 2006.

[5] M. Bhushan, M. B. Ketchen, M. Cai, and C. Kim, "Ring oscillator technique for MOSFET CV characterization," IEEE Trans. Semiconductor Manufacturing, vol. 21, pp. 180-185, 2008.

[6] M. Nourani and A. Radhakrishnan, "Testing on-die process variation in nanometer VLSI," IEEE Design & Test of Computers, vol. 23, pp. 438-451, 2006.

[7] D. Cheng, A. Majumdar, X. Wang, and N. Chong, "Field profiling & monitoring of payload transistors in FPGAs," in Proc. 2017 IEEE 23rd International Symposium on On-Line Testing and Robust System Design (IOLTS), pp. 180-185, 2017.

Versatile Chip-level Integrated Test Vehicle for Dynamic Thermal Evaluation

Suresh Parameswaran
Silicon Technology Group
Xilinx, Inc. 2100 Logic Drive,
San Jose, CA - 95124, USA
sureshp@xilinx.com

Saravanan Balakrishnan
Silicon Technology Group
Xilinx, Inc. 2100 Logic Drive,
San Jose, CA - 95124, USA

Boon Ang
Silicon Technology Group
Xilinx, Inc. 2100 Logic Drive,
San Jose, CA - 95124, USA

Abstract - Thermal management of semiconductor chips is becoming very important as the demand for chip performance increases. It is necessary to evaluate/manage the thermal aspects of a chip throughout the development cycle – starting from initial planning stage to deployment on customer board and beyond. In this paper, we present a versatile thermal evaluation vehicle that addresses the above requirements. This paper describes the circuit architecture/implementation, details of operation, programming aspects, usage model and various applications of a silicon chip that is successfully used as a thermal evaluation tool. The chip has 1600 sectors with programmable heat-generation and temperature-sensing capability – enabling it to generate up to 3W per mm^2 and has a temperature detection range of 30C to 125C with an accuracy of +/-2C. It has a simple implementation and is easy to program and test - yet has substantial thermal evaluation capabilities. It was fabricated in 0.18um technology and packaged as flip-chip. The chip has ability to do automated on-chip temperature measurements through a tester-friendly interface and has been successfully controlled through a simple and inexpensive test-platform. The ability to generate heat on-die and monitor spatial & temporal on-die temperature makes this chip suitable to emulate many different use cases of a product during the development stage ahead of product silicon availability. The capabilities of this test-vehicle make it a suitable candidate for demonstrating power-aware/thermal-aware testing. Silicon measurement data and comparison to simulation results based on numerical models are also presented in this paper.

Keywords— thermal management, thermal evaluation tool, on-die heating, on-die temperature sensing, package-level thermal evaluation, power-aware testing, thermal-aware testing, on-chip measurements, test-structures

1. INTRODUCTION

High die-temperature, if not properly controlled, can result in reliability issues, performance degradation, circuit malfunction and even thermal-runaway leading to possible destruction of the die. These issues highlight the need for special attention to die-level and package-level thermal aspects and cooling solutions starting from the chip planning stage. We have developed a hardware thermal evaluation system to evaluate thermal behavior of a chip under different power usage conditions. The heart of this system is a specially designed semiconductor chip whose on-chip heating and temperature measurement abilities make it easily programmable to emulate on-die thermal behavior under different customer use conditions. The tester used to program this chip can be relatively simple because the chip control complexity is absorbed into the chip design. In addition, creative digitization techniques have been used in this chip to further simplify the tester interface for fast data acquisition to enable temporal temperature measurement.

2. BACKGROUND

The features of a comprehensive thermal evaluation tool should include the following [5]:

- The ease of generating various heating patterns and heat concentrations within the die through software program
- The ability to monitor thermal gradient, dynamic temperature profile and static temperature profile within the die
- The capability to accurately sense and read out the die temperature at various locations within the die
- The flexibility to apply various cooling solutions for evaluation purpose

The difficulties we encountered were to design a chip architecture that meets the above requirements and to implement it in a simple yet very reliable method. Besides these, we also emphasized on the ease of programming and collecting data in using this evaluation system. The design covered in this paper attempts to satisfy all these requirements.

3. CHIP ARCHITECTURE & IMPLEMENTATION

The Sector, which is the building-block, has its own heating-elements and sensors (Fig. 1a). 20-sectors form a Row (Fig. 1b) and 20-rows form a Bank (Fig. 1c). Thus, each Bank is an array of 400 (20 X 20) Sectors. Heating-elements and sensors have sector-level programmability. The heating-elements in a Sector can be either turned on or off - the percentage turned-on is programmable. The Sensing/digitization block in each Sector, senses the local temperature, converts into a digital-code and sends out the code. Multiple Sectors can be sensed simultaneously - read-out of the digital code will be done sequentially, 1-sector at a time.

Bank has independent controls, this enables multiple Banks to be placed together to form a chip. The evaluation vehicle that we fabricated, packaged and tested has 4 Banks. The results are equally valid for a chip with a different number of Banks as well.

Heating-elements have a mix of Resistor heating (I^2R) and Ring-oscillator heating (CV^2f). The heating-elements are distributed uniformly as 4-quadrants within each Sector (Fig. 1a) and can heat up to 3W per mm^2

Sensing-system has 2 types of sensing-elements - digital and analog. The digital sensing-element is a Ring-oscillator (RO). The digitization-scheme produces a digital-code corresponding to the temperature-dependent frequency of the RO (Fig. 2a) ([1], [3]). The analog sensing-elements include a Resistor or a Diode – this is user selectable. The temperature dependent analog voltage from these sensing-elements ([2], [4]) is converted to a digital-code through a digitization-scheme. The implementation of the analog sensing-elements and the digitization scheme uses simple circuit blocks – a high-gain Comparator, a D-flipflop, a State-machine and a Digital to Analog Converter (DAC) (Fig. 2b). The DAC has 2 implementations – a Resistor-divider DAC and the R-2R DAC. The user has the ability to pick one of them. The sensing-element produces a temperature dependent voltage vout1 and the DAC produces a voltage vout2. The comparator determines if vout2 is less than or greater than vout1. This will trigger the state-machine code to move up or down by one binary-code. The system will make 1 step in the right direction with each clock-pulse and will eventually stabilize when vout1 = vout2. The analog sensing-scheme doesn't need any precision reference voltages – but, calibration is required to compensate for silicon variations. Temperature detection range is 30C to 125C with an accuracy of +/-2C.

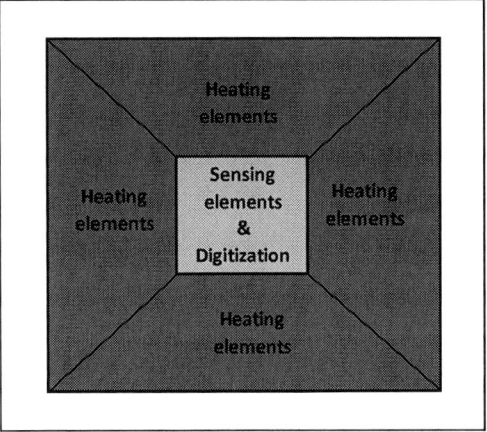

Fig 1a: Sector-level Architecture

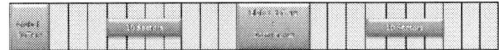

Fig. 1b. Row-level Architecture

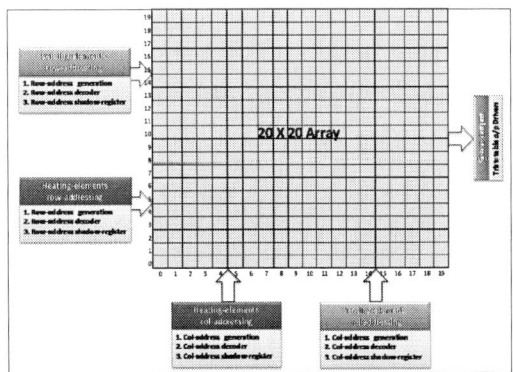

Fig. 1c. Bank-level Architecture

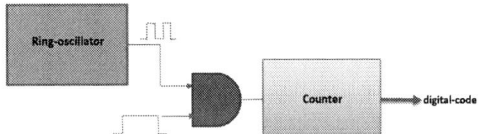

Fig. 2a. Digital sensing & digitization scheme

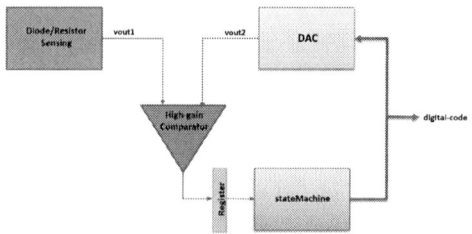

Fig. 2b. Analog sensing & digitization scheme

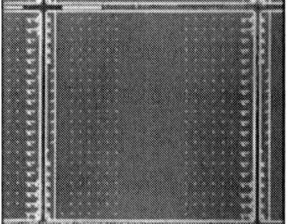

Fig. 3. Die photo (shows 1 Bank)

4. CHIP PROGRAMMING & USAGE

This chip is flexible, programmable and capable of near real-time temperature monitoring at all 1600 sectors (4-banks) in the chip. The programming and usage of the chip can be done through digital pattern files using a software test program (Fig. 4). These features make this chip very flexible as a thermal evaluation vehicle.

Heat-programming: Each Sector can be individually programmed to 0% through 100% of heating capacity. The following features make it possible to program real life use cases on this chip:

- Heat-programming can be done at Sector-level, Row-level or Bank-level.
- Concurrent heat-programming while heating is in progress based on previous settings is supported.
- While a Bank is being heat-programmed, the heat-programming in other 3 Banks is preserved.

The chip was heat-programmed in a slow & steady rate by adding 1 row at a time. A dedicated 3.3V supply domain powers the heating-elements.

Sensing: Sensing is done concurrently in 1 Sector within each of the 4 Banks, while read-out will be done for only 1 of the above 4 Sectors. Sensing sequence is totally flexible – binary-up, binary-down or random. Chip is capable of continued sensing in a Sector, Row or Bank for an indefinitely long period of time. Sensed Digital-code is preserved in a Sector till that Sector is accessed again for sensing. The sensed digital code can be read-out either in parallel or serial format. A

dedicated 1.8V supply domain powers the sensing circuitry.

The chip was characterized to over 150W of heating. Peak-currents during heat-programming and the DC-currents during heating need special attention. Realistic chip use conditions were mimicked. Temperatures within the package can rise very quickly – use of proper thermal interface material (TIM) at required interfaces and sufficient cooling are required.

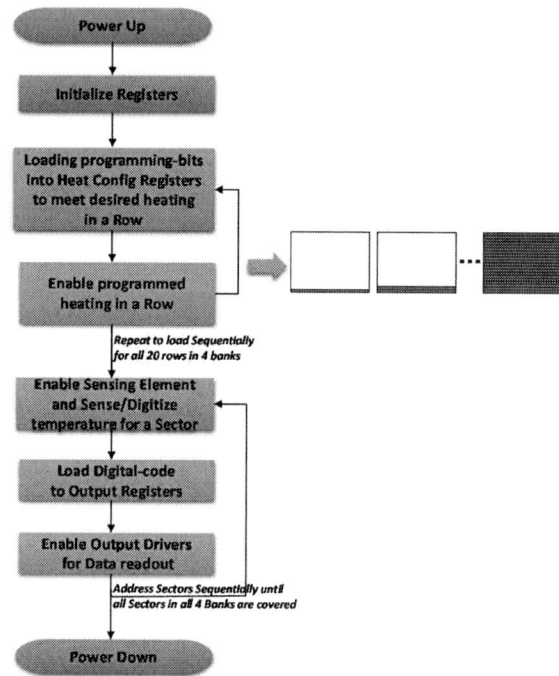

Fig. 4. Flow diagram of programming sequence

Calibration: is required at the package-level to cancel out die-to-die and within-die variations. The packaged unit is subjected to external-heating till thermal equilibrium is reached and the digitized temperature codes are read out from every Sector. A look-up table is created between the read-out digital-code and the corresponding package temperature. This look-up table is used in all subsequent sensing operations. As shown in Fig. 5, the digital-code for the Ring-oscillator based sensing moves quite linearly and monotonically wrt temperature. Similar behavior has been observed with the analog sensing schemes as well.

5. APPLICATIONS

Some of the applications of this thermal evaluation system are listed below:

Chip floor planning – How closely power-hungry circuits can be packed inside a chip without running into heat dissipation issues is an important information to have during the floor planning phase of any chip. We can use this evaluation system to provide this insight by programming heat maps mimicking that of the targeted floor plan into the evaluation test chip and measuring the temperature profile. In doing so, the most optimal distribution of power-hungry circuit blocks can be achieved.

Package & cooling solution planning - When packaged in different package types, this evaluation test chip can be used to collect on-die heat maps as a function of different package types, thermal interface materials and cooling methods (e.g. passive heat sink vs active cooling solutions). The results can be used to select the most optimal package type and cooling solutions of a product ahead of product silicon availability as long as the power specifications of the product is known.

Power-aware/thermal-aware testing - The concept of having distributed sensing elements in a chip and monitoring on-die temperature at various spots can be applied to enable thermal-aware testing. A typical application is as follows: during testing, temperature at critical on-die locations can be monitored and mitigation action can be taken on the fly when the temperature goes above the safe limit.

6. SILICON RESULTS

This thermal evaluation system was used to evaluate several use cases. This includes monitoring on-die temperature profile over time (i) for different wattages for the same cooling and (ii) for the same wattage with different cooling. Two such cases are shown on Fig 6 & Fig 7. In each case, the evaluation chip was programmed to meet the targeted spatial wattage distribution and on-die temperature was monitored over time. In Fig 6, the high heat density circuits are concentrated on the left side. Fig 6c shows the die temperature delta wrt ambient as a function of time till thermal equilibrium. Figs 6d, 6e & 6f display the temperatures measured at different locations of the chip at various points in time. Figs 6g & 6h display the temperature map with increased cooling applied to the packaged unit. In Fig 7, the power-hungry circuits are grouped symmetrically in the middle of each Bank.

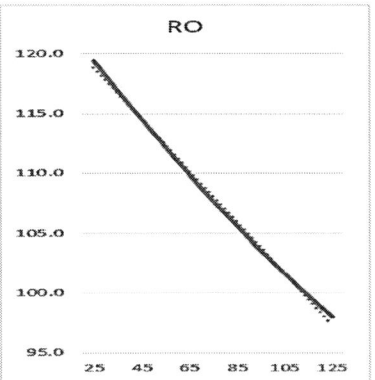

Fig. 5. Temperature digital code (Y) vs temperature (X)

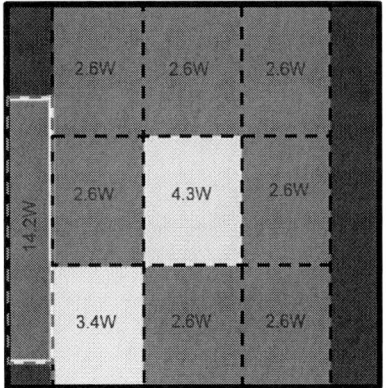

Fig. 6a. Power-map (programmed into the die)

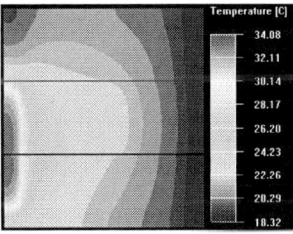

Fig. 6b. Simulated die temperature delta w.r.t. ambient

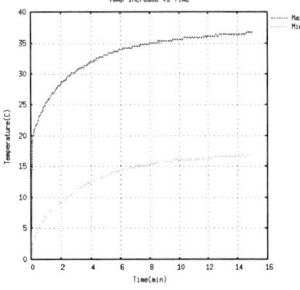

Fig. 6c. Measured die temperature delta w.r.t. ambient

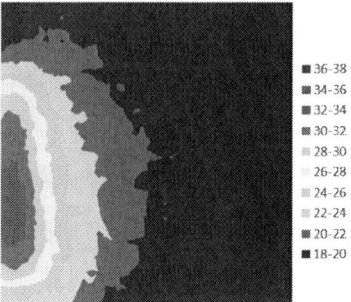

Fig. 6d. Heat-map of measured die temperature delta w.r.t ambient at 5 minutes

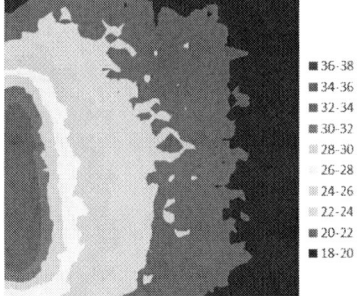

Fig. 6e. Heat-map of measured die temperature delta w.r.t ambient at 10 minutes

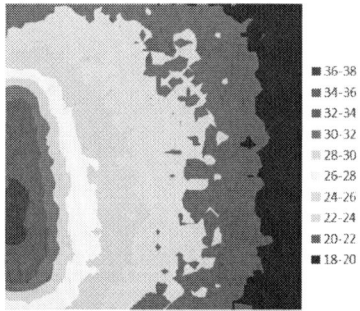

Fig. 6f. Heat-map of measured die temperature delta w.r.t ambient at 15 minutes

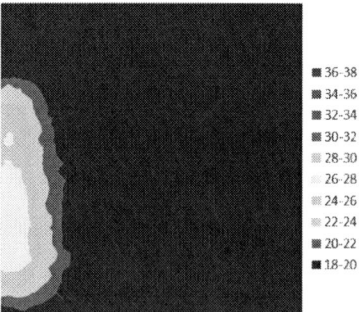

Fig. 6g. Heat-map of measured die temperature delta w.r.t ambient with increased cooling at 15 minutes

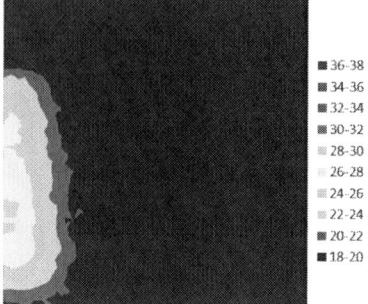

Fig. 6h. Heat-map of measured die temperature delta w.r.t ambient with increased cooling at 30 minutes

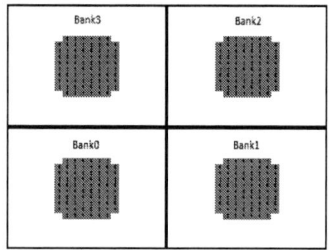

Fig. 7a. 65W power-map (programmed into the die)

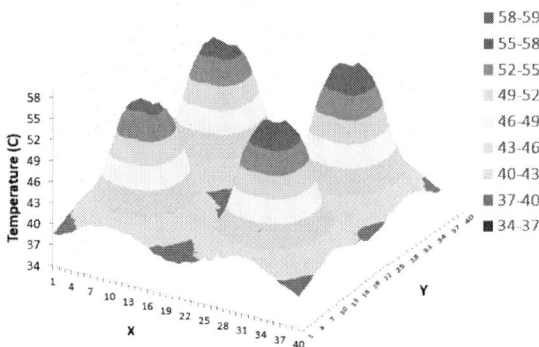

Fig. 7b. 65W heat-map of measured die temperature delta w.r.t ambient at 30 minutes

7. CONCLUSION

A highly flexible, programmable and versatile thermal hardware evaluation tool with software control is presented. This tool can be used to characterize package-level thermal management solutions by providing both spatial and near real-time on-die temperature information. Given the thermal issues faced by many advanced semiconductor chips, we think that this chip is a valuable tool. Some of the applications of this chip include chip floor-planning, package/cooling-solution planning and product use case evaluations. The concept is also applicable to the field of power-aware/thermal-aware testing. The chip's implementation details, usage model, application areas and silicon results are covered.

REFERENCES

[1] Shruti Suman et al, "Ring Oscillator Based CMOS Temperature Sensor Design" International Journal of Scientific & Technology Research Volume 1, Issue4, May 2012 pp.76-81

[2] Suhas Vishwasrao Shinde, "PVT Insensitive Reference Current Generation" Proceedings of the International MultiConference of Engineers and Computer Scientists 2014 Vol II

[3] Tejasvi Anand et al, "A VCO Based Highly Digital Temperature Sensor With 0.034C/mV Supply Sensitivity" IEEE JSSC, Vol.51, No.11, November 2016 pp.2651-2662

[4] Behzad Razavi "Design of Analog CMOS Integrated Circuits" ©2011 McGraw-Hill Higher Education

[5] Suresh Parameswaran et al, "On-die thermal evaluation system" 2017 IEEE

International Conference on

Microelectronic Systems Education (MSE) pp.55 - 58

2018 INTERNATIONAL CONFERENCE ON MICROELECTRONIC TEST STRUCTURES, MARCH 19-22, 2018, AUSTIN, TEXAS, USA

All-Digital On-Chip Heterogeneous Sensors for Tracking the Minimum Energy Point of Processors

Shu Hokimoto*, Jun Shiomi*, Tohru Ishihara* and Hidetoshi Onodera*

* Graduate School of Informatics, Kyoto University, Yoshida-honmachi, Sakyo-ku, Kyoto 606-8501, JAPAN

email: {s-hokimoto,shiomi-jun}@vlsi.kuee.kyoto-u.ac.jp, {ishihara,onodera}@i.kyoto-u.ac.jp

Tel: +81-75-753-5948, Fax: +81-75-753-5343

Abstract—Dynamically scaling the supply voltage (V_{DD}) and the threshold voltage (V_{TH}) is one of the most effective approaches for reducing the energy consumption of processors. However, since the best pair of V_{DD} and V_{TH}, which minimizes the energy consumption of processors is strongly dependent on the operating condition such as an activity factor and a performance required for the processor, it is not trivial to find the best pair of the voltages at runtime when the operating condition widely varies. With all-digital on-chip heterogeneous sensors, we propose a simple runtime method to accurately identify the best pair of V_{DD} and V_{TH}, which minimizes the energy consumption of a processor under a specific operating condition which is determined by a process variation, an activity factor, and a performance requirement for the processor. Measured results for a 32-bit RISC processor integrating the heterogeneous sensors show that the proposed method successfully tracks the minimum energy operating point (i.e. the best pair of V_{DD} and V_{TH}) of the processor even in a case that the operating condition widely varies.

I. INTRODUCTION

Dynamically scaling the supply voltage (V_{DD}) and the threshold voltage (V_{TH}) is one of the most effective approaches for reducing the energy consumption of processors. The V_{TH} of transistors can be dynamically changed by tuning the back-gate bias (V_{BB}). We refer to the pair of V_{DD} and V_{TH}, which minimizes the energy consumption of the processor under a given operating condition, as a minimum energy point (MEP in the following). Since the MEP heavily depends on the operating condition such as a process variation, an activity factor, and a performance required for the processor, it is not trivial to find the MEP at runtime under a wide range of the operating condition. This paper proposes a runtime method to accurately identify the MEP using all-digital on-chip heterogeneous sensors. Since the sensors are fully-digital circuits, they work very well over wide operating voltages ranging from 1.2 V down to 0.4 V. In addition to the wide operating range, their power overhead and the circuit footprint are very small since the structure of the sensors is very simple. We designed and fabricated a 32-bit RISC processor integrating the sensors with a 65 nm process technology. Fig. 1 shows a chip photograph of the processor.

II. RELATED WORK AND OUR CONTRIBUTION

In this paper, we assume to exploit an adaptive back-gate biasing technique [1], [2] which tunes the back-gate bias

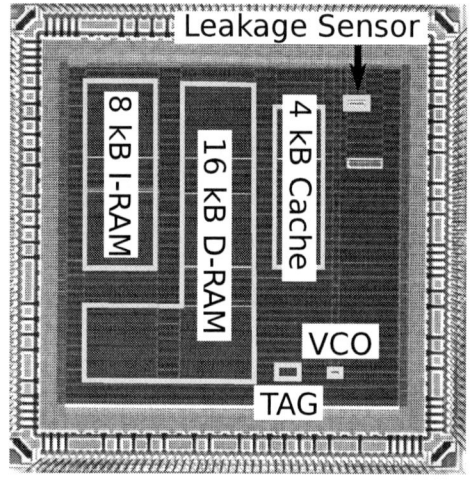

Fig. 1. Test chip structure fabricated in 65 nm process.

(V_{BB}) for dynamically changing V_{TH} of transistors. Since simultaneous scaling of V_{DD} and V_{BB} drastically reduces the energy consumption of processors, techniques for dynamically scaling V_{DD} and V_{BB} under dynamic workloads of processors are widely investigated over the past 15 years [1]–[3]. They derived the optimal pairs of V_{DD} and V_{BB} for different operating conditions analytically. However, since there are a lot of uncertain parameters in their analytical models such as parasitic capacitance in a circuit, it is hard to dynamically find an optimal pair of V_{DD} and V_{BB}, which minimizes the energy consumption using the analytical model only.

In [4], Nose et al. have shown that the total power consumption of a circuit is minimized when the P_D/P_{LEAK} ratio is maintained at a specific value, where P_D is switching power and P_{LEAK} is leakage power. It is not, however, very easy to maintain the P_D/P_{LEAK} ratio to the specific value under a wide range of operating conditions. In [5], Nomura et al. proposed a closed-loop system to automatically keep the P_D/P_{LEAK} ratio to the specific optimal value. They assume that the ratio of switching current to leakage current (i.e., I_{SW}/I_{LEAK}) is kept at a specific optimal value when the total power consumption of a circuit is minimized even if the operating conditions vary. However, the optimal I_{SW}/I_{LEAK} ratio depends on the operating conditions. If the operating

978-1-5386-5072-1/18 $31.00 © 2018 IEEE

2018 INTERNATIONAL CONFERENCE ON MICROELECTRONIC TEST STRUCTURES, MARCH 19-22, 2018, AUSTIN, TEXAS, USA

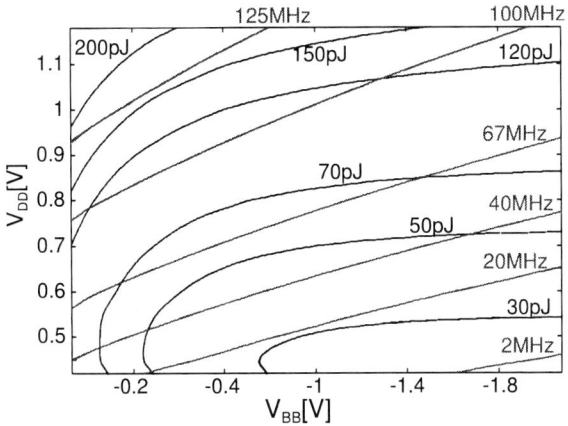

Fig. 2. Example of frequency contours, energy contours of the processor simulated for a temperature of 25 °C.

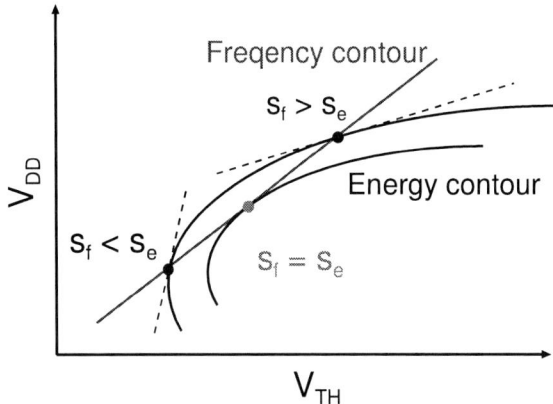

Fig. 3. Concept of minimum energy point identification.

conditions widely vary, the optimal I_{SW}/I_{LEAK} ratio also moves, which leads to a non-optimal setting of V_{DD} and V_{BB}, where the energy consumption of the circuit is not minimized. Our algorithm is based on more accurate guideline for closely tracking the MEP even in a case that the operating conditions widely vary.

In [6]–[8], a minimum energy point tracking algorithms which tune V_{DD} are proposed. Those algorithms incrementally change supply voltage V_{DD} to get close to the MEP for a given performance constraint. However, these algorithms tune only V_{DD}. Our algorithm tunes V_{DD} and V_{BB} simultaneously and it is more effective than the algorithms which tune V_{DD} only in terms of the reduction in total energy consumption.

In [9], a minimum energy point tracking algorithm which tunes V_{DD} and V_{BB} simultaneously is proposed. The algorithm is based on relative power values between two measurement iterations. Power monitor computes difference between power consumed in current step and the power consumed in previous step. Based on the sign of the difference (i.e., negative or positive), power monitor decides whether power is increasing or decreasing. The algorithm incrementally changes body bias V_{BB} by step size ΔV_{BB} based on the decision of power monitor. This differential operation reduces the measurement area and power overhead significantly. However, since there are several sources for the change in the power consumption, it is hard to decide a right direction of stepping V_{BB} using the difference between the power consumption values in two consecutive steps only. For example, the leakage power is exponentially related to the chip temperature which may change over time. If the switching activity increases and the temperature decreases simultaneously such that the total power is unchanged, the power monitor may lose the right direction of stepping V_{BB}. Unlike the technique presented in [9], our MEP tracking algorithm does not need to know the difference between the power consumption values in two consecutive steps. One-time estimation of dynamic power, static power and temperature is sufficient to identify whether V_{BB} should be increased or decreased. In addition, since our

algorithm identifies a right direction of stepping voltage based on a one-time estimation of the power consumption and the temperature, it does not lose the right direction of stepping voltage even if the total power consumption is unchanged due to the simultaneous impact of two or more factors as described above.

This paper first presents an evaluation function which indicates whether the processor is running at the minimum energy point (MEP) or not. The evaluation function uses the dynamic power, static power and the temperature of the processor as inputs. Then, it accurately tells whether V_{DD} and V_{BB} should be increased or decreased to get close to the MEP for a given operating condition. The evaluation function does not need to know the difference between the power consumption values in two consecutive steps. This paper next presents a simple but effective algorithm for tracking the MEP of the processor under a wide range of operating conditions. Even in a situation that the operating conditions widely vary, our algorithm can closely track the MEP. Finally, the paper proposes a MEP tracking system using all-digital on-chip heterogeneous sensors. The sensors include an activity sensor for estimating dynamic power of the targeting processor, a leakage sensor for estimating static power of the processor, and a temperature sensor.

III. MINIMUM ENERGY POINT TRACKING ALGORITHM

A method for tracking the MEP of a processor is proposed in [10]. It is based on the fact that the MEP is found at a tangent point of a constant energy contour and a constant frequency contour. Fig. 2 shows the energy and frequency contours obtained through gate-level post-layout simulation for the processor. Since the MEP is on the tangent point of the contours, the gradient of the frequency contour s_f and that of the energy contour s_e is equal to each other at the MEP as shown in Fig. 3. When $s_f < s_e$, both V_{DD} and V_{TH} should be stepped up toward the MEP while the voltages should be stepped down when $s_f > s_e$ to reach the MEP. Ref. [10] shows analytical expressions for s_f and s_e as presented in (1) and (2), respectively, where α is velocity saturation index. N_S is a product of thermal voltage and ideality factor of

978-1-5386-5072-1/18 $31.00 © 2018 IEEE

2018 INTERNATIONAL CONFERENCE ON MICROELECTRONIC TEST STRUCTURES, MARCH 19-22, 2018, AUSTIN, TEXAS, USA

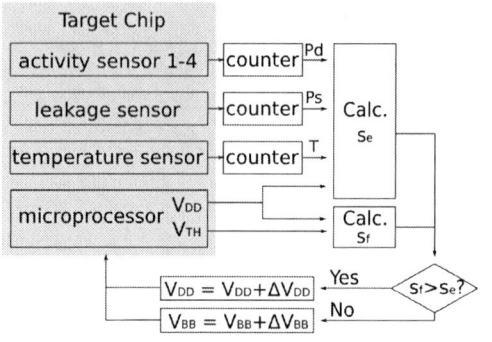

Fig. 4. Overview of the MEP tracking system.

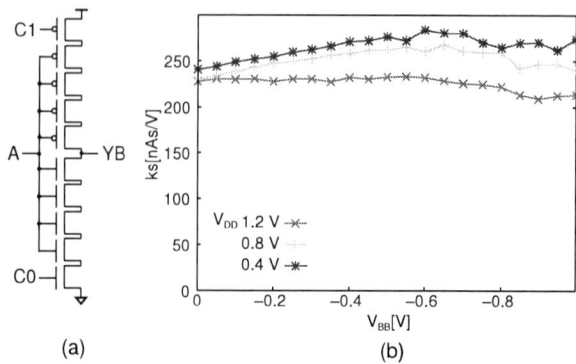

Fig. 5. (a) Reconfigurable leakage sensor cell. (b) Result of k_s for a 53-stage ring oscillator-based leakage sensor.

a MOSFET. P_d and P_s represent dynamic and static power consumption of the processor, respectively.

$$s_f = \frac{\alpha V_{DD}}{\alpha V_{DD} - (V_{DD} - V_{TH})} \qquad (1)$$

$$s_e = \frac{P_s V_{DD}}{(2P_d + P_s)N_s} \qquad (2)$$

Once we obtain the values of P_d and P_s at runtime, we can predict s_f and s_e. This makes it possible to identify the MEP at runtime even in case that the operating condition widely varies. Although a series resistor inserted on the power supply line may measure the runtime power consumption of the processor, it is not trivial to measure the dynamic and static power separately at runtime.

IV. MINIMUM ENERGY POINT TRACKING SYSTEM

Fig. 4 shows an overview of the MEP tracking system. The microprocessor and sensors are integrated on the same chip. Microprocessors typically employ cache memories and scratch-pad memories on the same die. All the sensors are based on an oscillation circuit. The activity sensors shown in Fig. 4 are designed so that each oscillation frequency is proportional to the switching activity of gates in the corresponding hardware module such as a pipelined data-path and a cache memory. Similarly, the oscillation frequencies of the leakage sensor and the temperature sensor are proportional to the subthreshold leakage current and exponentially proportional to the reciprocal of the absolute temperature in the processor, respectively. Once we count the oscillation frequencies of the sensors using dedicated counters as shown in Fig. 4, we can obtain the activity factor, the leakage and the temperature of the processor as digital values which can be handled by software program running on the processor. As explained in the previous section, the value of s_f can be calculated from (1) using dynamic power P_d, static power P_s, temperature T and V_{DD} of the processor. Note that the values of P_d and P_s can be accurately estimated by activities of representative hardware modules and subthreshold leakage in the processor, respectively. The value of s_f can be also calculated from (2) using V_{DD} and V_{TH}. These calculations can be done with either dedicated hardware or software running on the processor

since all the values needed for calculating s_f and s_e can be obtained as binary numbers through the dedicated counters. Therefore, sensors for estimating P_d, P_s and T are key components in the MEP tracking system. The rest of this paper explains the detail of individual on-chip sensors which always maintain the processor operating at the minimum energy point even in a wide range of operating condition. We use four types of activity sensors to estimate the runtime P_d, and a leakage sensor to measure the runtime P_s. The leakage sensor presented in [11] can be used not only as a static power sensor but also as a temperature sensor. However, we assume that the temperature of the processor is constant and we do not use the temperature sensor in this paper. Implementing a minimum energy point tracking system considering the temperature variation in the processor is our future work.

V. RUNTIME STATIC ENERGY ESTIMATION

We use a leakage sensor presented in [11] to measure the runtime static power. It is based on a 53-stage ring oscillator consisted of a standard cell inverter shown in Fig. 5 (a). This oscillator operates with V_{DD} and V_{TH} which are used for the processor to closely track the subthreshold leakage of the processor. When the nMOSFET "C0" is OFF and pMOSFET "C1" is ON, the frequency of the leakage sensor f_{leak} is proportional to the product of leakage current of nMOS $I_{leak,n}$ and V_{DD}. Therefore, the runtime static power of the processor can be estimated using (3), where k_s is a fitting coefficient.

$$P_s = I_{leak} V_{DD} = k_s f_{leak} V_{DD}^2 \qquad (3)$$

We suppose to find the value of k_s at test time. Once the leakage (I_{leak}) and the f_{leak} values are obtained, we can fit the value of k_s using (3). Fig. 5 (b) shows the fitting results for k_s. As can be seen from Fig. 5 (b), the value of k_s is almost constant. The maximum P_s estimation error is 24% using k_s at $V_{DD} = 0.8$ V and $V_{BB} = 0$ V. Therefore, we can accurately estimate P_s of the processor using k_s at $V_{BB} = 0$ V.

VI. RUNTIME DYNAMIC ENERGY ESTIMATION

We use four types of activity sensors to estimate P_d. The sensors measure the number of the following hardware events per second in the processor; (i) instruction execution, (ii)

978-1-5386-5072-1/18 $31.00 © 2018 IEEE

2018 INTERNATIONAL CONFERENCE ON MICROELECTRONIC TEST STRUCTURES, MARCH 19-22, 2018, AUSTIN, TEXAS, USA

TABLE I
AVERAGE FREQUENCIES OF INSTRUCTION EXECUTION AND MEMORY ACCESSES WHEN $V_{\mathrm{DD}} = 0.75$ V, $V_{\mathrm{BB}} = -0.5$ V AND $f = 50$ MHz.

Program	P_{d} [mW]	instruction execution [MHz]	cache [MHz]	I-RAM [MHz]	D-RAM [MHz]
1	3.06	10.3	8.90	0.215	4.15
2	2.86	10.3	0.137	8.75	4.05
3	1.52	0.950	0	0	0
4	1.54	0.975	0	8.00×10^{-3}	0.183
5	2.27	7.80	7.75	7.30×10^{-2}	0
6	3.20	10.1	7.90	0	0.900
7	2.96	10.1	5.00×10^{-4}	8.28	0.910
8	1.52	0.830	0	0	0
9	1.52	0.905	0	0	7.10×10^{-2}
10	2.37	7.62	6.50	0	0

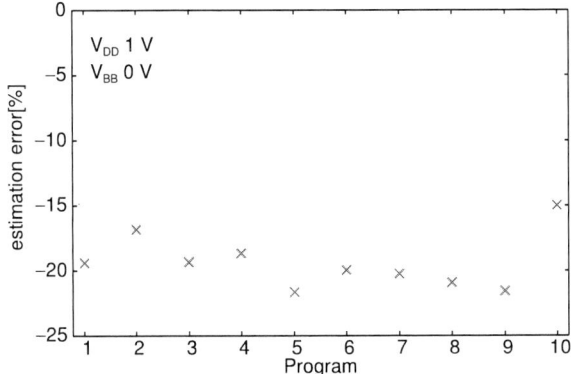

Fig. 6. Results of dynamic power consumption modeling with activity sensors for a supply voltage of 1 V and a back-gate bias of 0 V.

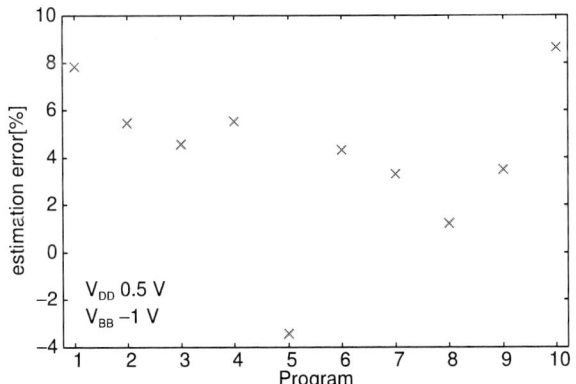

Fig. 7. Results of dynamic power consumption modeling with activity sensors for a supply voltage of 0.5 V and a back-gate bias of −1 V.

Fig. 8. Results of dynamic power consumption modeling with activity sensors for a supply voltage of 0.75 V and a back-gate bias of −0.5 V on another chip.

also shows that the power is dependent on the memory access pattern. This implies that the dynamic power consumption of embedded processors can be estimated using the number of instructions executed per second and the number of memory accesses per second. The activity sensors in our system are simple counters which are composed of normal standard cells and integrated with the processor. We run several training programs on the processor, which induce different switching activities of the processor. We then fit a linear approximation model presented in (4) to the measured P_{d} values through linear regression analysis. f in (4) is frequency of the processor. Coefficients A, B, C, D, and E in (4) are variables to be determined through the regression analysis. We assume that the model fitting through the regression analysis is done once at test time.

cache access, (iii) instruction RAM (I-RAM) access, and (iv) data RAM (D-RAM) access. Our idea here is estimating the dynamic power consumption of processors using weighted linear summation of above four activity factor values. This strategy is based on the model presented in [12]. It empirically shows that the variation in the dynamic power consumption across instructions in embedded processors is quite small. It

$$
\begin{aligned}
P_{\mathrm{d}} = \ & (A \times [\text{No. of instructions executed per second}] \\
& + B \times [\text{No. of cache accesses per second}] \\
& + C \times [\text{No. of I-RAM accesses per second}] \\
& + D \times [\text{No. of D-RAM accesses per second}] \\
& + E \times f) V_{\mathrm{DD}}{}^2
\end{aligned}
\qquad (4)
$$

978-1-5386-5072-1/18 $31.00 © 2018 IEEE   131

2018 INTERNATIONAL CONFERENCE ON MICROELECTRONIC TEST STRUCTURES, MARCH 19-22, 2018, AUSTIN, TEXAS, USA

TABLE II
RESULT OF PARAMETER FITTING FOR DYNAMIC AND STATIC POWER ESTIMATION MODEL.

A	B	C	D	E	Chip1 k_{s}	Chip2 k_{s}
7.18×10^{-10}	-4.7×10^{-10}	-5.05×10^{-10}	6.25×10^{-11}	4.07×10^{-11}	2.29×10^{-7}	2.37×10^{-7}

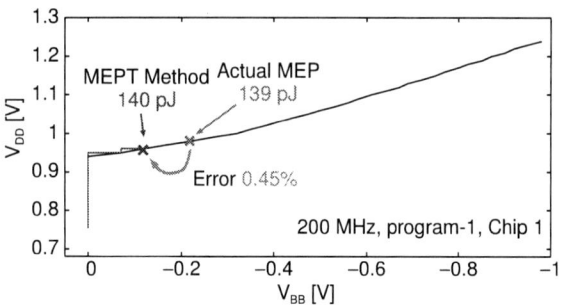

Fig. 9. Result of MEP tracking method for the performance of 200 MHz and a temperature of 24 °C running *program*-1.

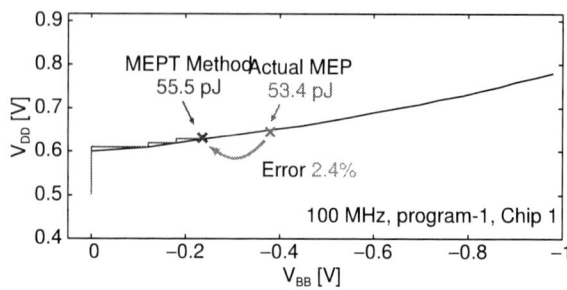

Fig. 11. Result of MEP tracking method for the performance of 100 MHz and a temperature of 24 °C running *program*-1.

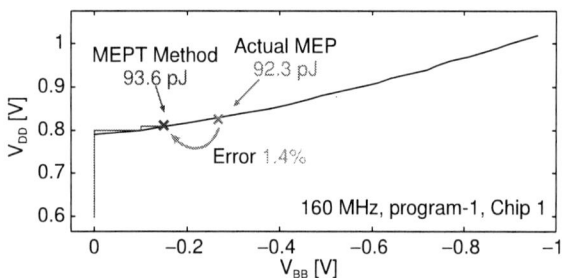

Fig. 10. Result of MEP tracking method for the performance of 160 MHz and a temperature of 24 °C running *program*-1.

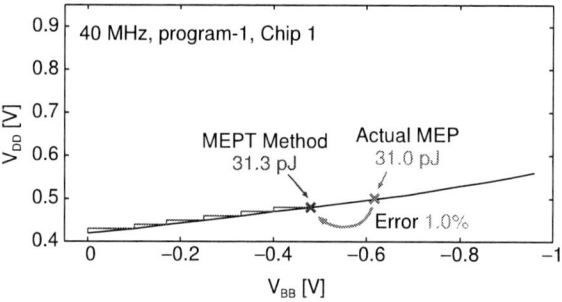

Fig. 12. Result of MEP tracking method for the performance of 40 MHz and a temperature of 24 °C running *program*-1.

We determine A, B, C, D and E by least squares method. TABLE I shows average frequencies for instruction execution, cache, I-RAM and D-RAM accesses and P_{d} when $V_{\mathrm{DD}} = 0.75$ V, $V_{\mathrm{BB}} = -0.5$ V and $f = 50$ MHz. Figs. 6 and 7 show the results of P_{d} estimated for different V_{DD} and V_{BB}; $V_{\mathrm{DD}} = 1$ V and $V_{\mathrm{BB}} = 0$ V in Fig. 6, and $V_{\mathrm{DD}} = 0.5$ V and $V_{\mathrm{BB}} = -1$ V in Fig. 7. As can be seen from Figs. 6 and 7, the maximum modeling error of P_{d} is 22%. Fig. 8 shows the result of P_{d} estimated for another chip using the same parameters as used in Figs. 6 and 7, when $V_{\mathrm{DD}} = 0.75$ V and $V_{\mathrm{BB}} = -0.5$ V. As can be seen from Fig. 6, the maximum modeling error of P_{d} is 4.5%. Therefore, once we obtain the values of A, B, C, D and E at one chip, we can estimate P_{d} of all chips.

VII. EXPERIMENTAL RESULTS

A. Experimental Setup

We use the 32-bit, 5-stage pipelined RISC procesor designed with a 65 nm process technology. Fig. 1 shows a chip photograph of the processor. The following on-chip memories consisted only of standard cells are implemented to the processor.

- 4 kB instruction cache
- 8 kB instruction RAM (I-RAM)

- 16 kB data RAM (D-RAM)

V_{DD} and V_{BB} can be arbitrarily adjusted. The values of N_{s} and α in the target 65 nm process at 25 °C are about 36 mV and 1.4 respectively. In addition, we used a 10 mV for the value of ΔV_{DD} and ΔV_{BB} in this experiment. P_{s} is estimated using (3), where the value of k_{s} is obtained by fitting the model to the measured P_{s} once at a condition of $V_{\mathrm{DD}} = 0.8$ V, $V_{\mathrm{BB}} = 0$ V, and a temperature of 25 °C. P_{d} is estimated by (4), where the values of A, B, C, D and E are obtained by fitting the model to the measured P_{d} once at $V_{\mathrm{DD}} = 0.75$ V and $V_{\mathrm{BB}} = -0.5$ V. The values of the fitting parameters A, B, C, D, E and k_{s} are shown in TABLE II.

B. Results of Minimum Energy Point Tracking

We evaluated the MEP tracking method described above. Fig. 9 to Fig. 14 show the results of the MEP tracking method applied to the processor chips. The black line shows the delay contour which corresponds to the performance requirement. We use F_{max} test to obtain the constant delay contour. The blue line is a locus of V_{DD} and V_{BB} which are tracked by our method. The actual MEP is found by thoroughly measuring the power consumption of the processor at different V_{DD}s and V_{TH}s. A red point is the actual MEP and a blue cross point is the MEP obtained by the MEP tracking method. Fig. 9

978-1-5386-5072-1/18 $31.00 © 2018 IEEE

2018 INTERNATIONAL CONFERENCE ON MICROELECTRONIC TEST STRUCTURES, MARCH 19-22, 2018, AUSTIN, TEXAS, USA

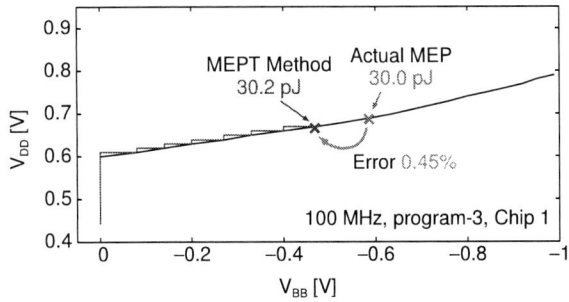

Fig. 13. Result of MEP tracking method for the performance of 100 MHz and a temperature of 24 °C running *program*-3.

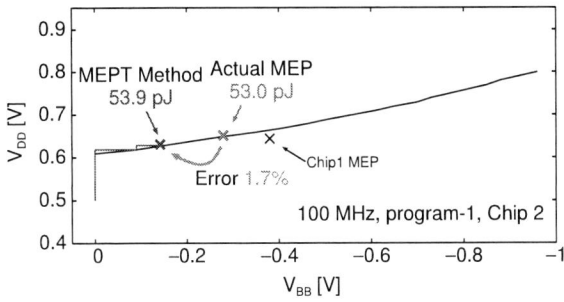

Fig. 14. Result of MEP tracking method for the performance of 100 MHz and a temperature of 24 °C running *program*-1 for another chip.

to Fig. 12 show the results of the MEP obtained for the performance requirements of 200 MHz, 160 MHz, 100 MHz and 40 MHz, respectively. The temperature is 24 °C and a program running on the processor is *program*-1 for all cases. Fig. 13 shows the result when *program*-3 is running on the processor with a performance requirement of 100 MHz at a temperature of 24°C. Fig. 14 shows the result of a different chip. The temperature is 24 °C, the performance requirement is 100 MHz and the program running on the processor is *program*-1 in this case. The energy loss introduced by the method is less than 2.4%. Our method closely tracks the MEP even when the operating condition is widely changed. Thus, the proposed method using the on-chip sensors are useful to maintain the minimum energy operation of the processor.

VIII. CONCLUSION

With all-digital on-chip heterogeneous sensors, we propose a simple runtime method to accurately identify the best pair of V_{DD} and V_{TH}, which minimizes the energy consumption of a processor under a specific operating condition determined by a process variation, an activity factor, and a performance requirement for the processor. In the method, we use four types of activity sensors to estimate the runtime dynamic energy of the processor. The maximum estimation error of dynamic energy is 21%. We use the leakage monitor proposed in [11] to estimate the runtime static energy of the processor. The maximum estimation error of static energy is 24%. Measured results for a 32-bit RISC processor integrating the heterogeneous sensors show that the proposed method successfully tracks the minimum energy operating point of the processor

within the error of 2.4% even in a case that the operating condition widely varies. One of our future work is expanding functionality of the current MEP tracking system with the on-chip temperature sensor so that the MEP can be closely tracked even in case that the temperature of the chip widely changes.

ACKNOWLEDGMENT

This work has been partly supported by KAKENHI 26540021. The authors acknowledge the support of VLSI Design and Education Center (VDEC), the University of Tokyo.

REFERENCES

[1] L. Yan, J. Luo, and N. K. Jha, "Joint Dynamic Voltage Scaling and Adaptive Body Biasing for Heterogeneous Distributed Real-time Embedded SSstems," *IEEE Transactions on Computer-Aided Design of Integrated Circuits and Systems*, vol. 24, no. 7, pp. 1030–1041, July 2005.

[2] S. M. Martin, K. Flautner, T. Mudge, and D. Blaauw, "Combined Dynamic Voltage Scaling and Adaptive Body Biasing for Lower Power Microprocessors under Dynamic Workloads," in *Computer Aided Design, 2002. ICCAD 2002. IEEE/ACM International Conference on*, Nov 2002, pp. 721–725.

[3] A. Basu, S.-C. Lin, V. Wason, and A. Mehrotrat, "Simultaneous Optimization of Supply and Threshold Voltages for Low-Power and High-Performance Circuits in the Leakage Dominant Era," in *Proceedings of the 41st Design Automation Conference*, 7 2004, pp. 884–887.

[4] K. Nose and T. Sakurai, "Optimization of VDD and VTH for Low-power and High Speed Applications," in *Proceedings of Asia and South Pacific Design Automation Conference*, ser. ASP-DAC '00. New York, NY, USA: ACM, 2000, pp. 469–474.

[5] M. Nomura, Y. Ikenaga, K. Takeda, Y. Nakazawa, Y. Aimoto, and Y. Hagihara, "Delay and Power Monitoring Schemes for Minimizing Power Consumption by Means of Supply and Threshold Voltage Control in Active and Standby Modes," *IEEE Journal of Solid-State Circuits*, vol. 41, no. 4, pp. 805–814, April 2006.

[6] N. Mehta and K. A. A. Makinwa, "Minimum energy point tracking for sub-threshold digital CMOS circuits using an in-situ energy sensor," in *2013 IEEE International Symposium on Circuits and Systems (ISCAS2013)*, May 2013, pp. 570–573.

[7] C. Ababei and C. Tamma, "Distributed minimum energy point tracking for systems-on-chip," in *IEEE International Conference on Electro/Information Technology*, June 2014, pp. 246–251.

[8] S. V. Gubbi and B. Amrutur, "All Digital Energy Sensing for Minimum Energy Tracking," *IEEE Transactions on Very Large Scale Integration (VLSI) Systems*, vol. 23, no. 4, pp. 796–800, April 2015.

[9] N. Mehta and B. Amrutur, "Dynamic Supply and Threshold Voltage Scaling for CMOS Digital Circuits Using In-Situ Power Monitor," *IEEE Transactions on Very Large Scale Integration Systems*, vol. 20, no. 5, pp. 892–901, May 2012.

[10] S. Hokimoto, T. Ishihara, and H. Onodera, "Minimum Energy Point Tracking Using Combined Dynamic Voltage Scaling and Adaptive Body Biasing," in *2016 29th IEEE International System-on-Chip Conference (SOCC)*, Sept 2016, pp. 1–6.

[11] I. Mahfuzul, J. Shiomi, T. Ishihara, and H. Onodera, "Wide-Supply-Range All-Digital Leakage Variation Sensor for On-Chip Process and Temperature Monitoring," *IEEE Journal of Solid-State Circuits*, vol. 50, no. 11, pp. 2475–2490, Nov 2015.

[12] A. Sinha, N. Ickes, and A. P. Chandrakasan, "Instruction Level and Operating System Profiling for Energy Exposed Software," *IEEE Transactions on Very Large Scale Integration (VLSI) Systems*, vol. 11, no. 6, pp. 1044–1057, Dec 2003.

978-1-5386-5072-1/18 $31.00 © 2018 IEEE

978-1-5386-5072-1/18 $31.00 © 2018 IEEE

Session 7

Test Parallelism

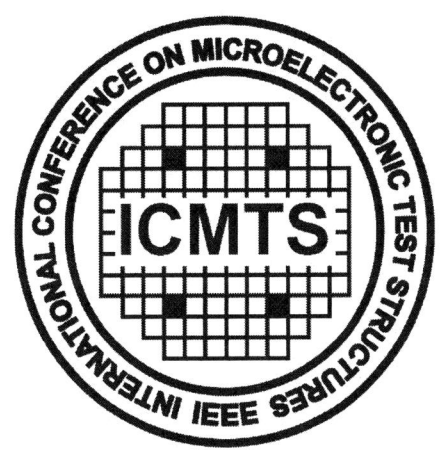

March 21, 2018

14:00-14:40

Chair:

Bill VERZI, Keysight Technologies, USA

978-1-5386-5072-1/18 $31.00 © 2018 IEEE

2018 INTERNATIONAL CONFERENCE ON MICROELECTRONIC TEST STRUCTURES, MARCH 19-22, 2018, AUSTIN, TEXAS, USA

Addressable test structure design enabling parallel testing of reliability devices

Lee DeBruler[1], Dennis Pretti[2], Mike Violette[3], Dave Peterson[4], Salil Mujumdar[5], Xia Li[6], and Ken Marr[7]

[1-7] Micron Technology Inc. Boise, Idaho 83707-0006

Abstract— This new design enabled an efficient layout of breakdown test devices for parametric and reliability testing. These reliability circuits consisted of interlocking combs of routing layers with varying widths and spaces that were representative of the design rules. These were accessed through multiplexer controlled pass gates. All addresses could be simultaneously enabled for stress biasing and addressed individually for failure detection. Once a breakdown was detected, as current leakage of the comb, each device could be addressed sequentially to find the failing structure. This was an improvement over previous designs which either grouped many devices in parallel, but could not electrically identify which device was failing, or only had a single device enabled, but suffered from poor pad efficiency. The grouping of these devices allows for simultaneous parallel stressing of each force and ground pad pair on the parametric testers. Electrical measurement showed that the same breakdown voltage values measured on this mux design were the same as standalone devices.

Keywords— *CMOS integrated circuits, parallel test parametric structures, reliability*

I. INTRODUCTION

Increasing die size and process complexity have driven a need to build scribe line test structures that utilize the available space as effectively as possible [1]. The use of multiplexed control logic or addressable arrays in test structures have been frequently used to add more devices but still utilize the same amount of scribe space [2][3][4]. Test structure efficiency in the scribe is an important consideration when designing new test modules. As these test structures become more complex there is commonly a trade-off with what the device-under-test (DUT) can effectively measure and the amount of space in the scribe that it utilizes [5]. Furthermore, enabling parallel testing of DUT's in the scribe required novel approaches to how the devices were wired and the parametric pads were shared [1]. Enabling parallel testing has shown to decrease the time on the tester thereby reducing the testing cost. Reliability structures further confound the tradeoffs already mentioned; these devices are typically tested to failure with high voltages and long test times [6]. Additionally, there are many DUT's required as they typically test all routing levels in the product.

We have developed a test structure design which effectively stressed and measured reliability structures in a multiplexed controlled scribe line test module. This design significantly increased the number of structures that were

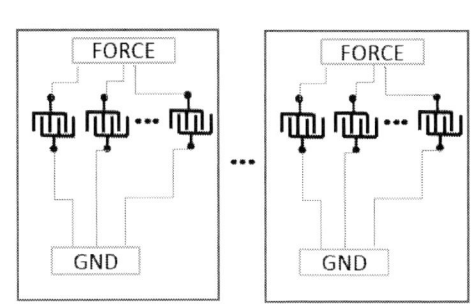

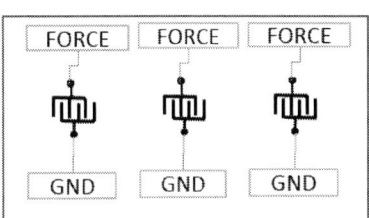

Fig. 1 Example of previous structures used for testing reliability devices. Each group of devices shared force and ground pads. 1A) has multiple DUT's connected to each Force and GND pad, 1B) has a single device for each Force and GND pad.

placed in the module compared to placing one structure per two pads. It also enabled parallel testing of many devices simultaneously while still retaining its ability to isolate a DUT failure.

II. DESIGN METHODOLOGY

There were multiple versions of test modules on the previous designs. The version shown in Fig. 1A had no way to access each device individually. A selection of the DUT's, grouped by voltage conditions, shared a common force and ground pad, Fig. 1A. Each force and ground pad grouping of DUTs were then stressed simultaneously until a failure or pass condition was detected. Fig. 2 shows a representative I-V curve of the devices tested in Fig. 1A. If a failure condition was detected there was no way to electrically determine which device failed or to continue testing that group. Fig. 1B shows

978-1-5386-5072-1/18 $31.00 © 2018 IEEE   137

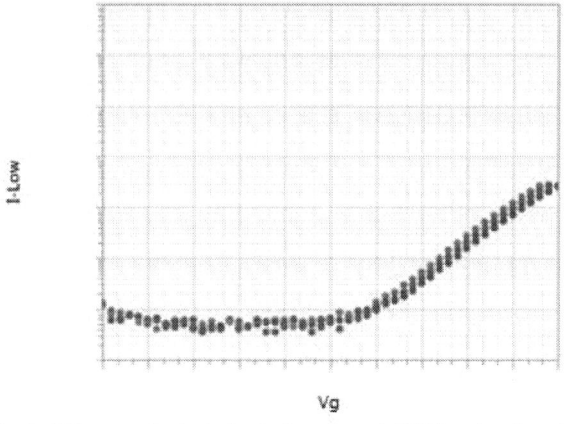

Fig. 2 I-V curve of a typical set of comb-comb DUT's using the original design, see Fig. 1A. Multiple sites across a wafer are shown.

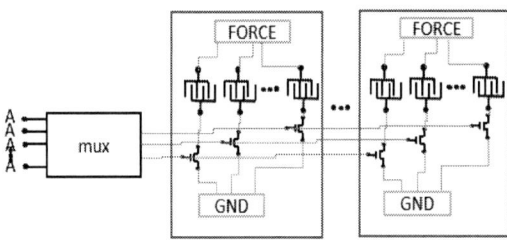

Fig. 3 Schematic representation of the multiplexed controlled reliability devices. Each set of force and ground pads were distinct and each pass gate was controlled by the mux.

the alternate layout which had one DUT per force/ground pad pair. While this design allowed for individual stressing of a single DUT, its poor pad efficiency restricted the number of DUT's that could be effectively placed in the scribe.

Each DUT, in this new design, were pairs of interlocking combs of metal routing layers. The widths and spaces of these lines were representative of the design rules defined for the manufacturing process. These were the same DUT's that were placed in the original designs. This new test structure design used multiplex addressable logic to access a pass gate array, Fig. 3. Each pass gate was a single NMOS transistor which was used to connect force and ground through the DUT. The logic was designed to allow all addresses to be simultaneously enabled as well as to sequence each single address. This was done so that during stress all DUT's could be accessed in parallel, accelerating the test. Once the stress was completed each device could be addressed individually to determine which device failed. If more granularity was required, each device could be stressed individually.

There were multiple force and ground pad pairs that were simultaneously controlled by the mux, this enabled parallel testing on each of these banks. This provided the ability to optimize the module configuration for the number of devices and for the test conditions. Equation (1) calculates the total number of DUT's that can be tested using this design. This optimization could be based on test/stress conditions, voltage requirements, or the sensitivity to failure conditions of the

pass gate and DUT. The data collected for this paper used a module configuration with 4 address pads, 5 pairs of force and ground pads, and had 64 DUT's. If desired, the configuration of the mux could be changed to enable a gate on and gate off voltage so that more isolation of the DUTs could be achieved. Other configurations could yield more structures, for example sharing the ground pad for all force pads would significantly increase the number of devices, but could possibly result in reduced sensitivity in parallel testing due to pad sharing.

$$\#DUT = (F*2^{P-F-G-N})-F \qquad (1)$$

Where: F: # of force pads
P: total pads
G: # ground pads
N: # of logic force & ground pads

Having a single address that enabled all DUT's had several benefits. This prevented a floating node between the DUT and the pass gate during stress. This floating node would couple to the stress voltage which could damage the circuit or give false readings. It also accelerated the testing of the module, allowing for multiple devices to be stressed and tested simultaneously. If a failure was detected during parallel stressing, discrete addressing of the multiplexer allowed the test to identify the failing structure. Stressing all devices simultaneously only identified first to failure on the FRC-GND pair, the remaining FRC-GND pairs continue to function. Discretely addressing each DUT allowed for single point failure detection of any DUT as long as all pass gates isolated stress voltage and ground. The pass gate transistor, as shown in Fig. 3, was placed on the ground side of the DUT to help isolate it from the high potentials used for stressing. Since no force and ground pads were shared in this design it would be possible to switch the force and ground pad if desired.

III. ELECTRICAL SIMULATION

A DC HSPICE simulation was done on the DUT and pass gate circuit to determine if the pass gates would be sufficiently isolated from the stress bias conditions. This isolation was required so that the pass gate would not be electrically damaged resulting in an inability to address the devices. It was also necessary to determine if a measurable signal would be generated from a failing device. Fig. 4A shows a picture of the schematic which simulated the voltages on the floating node during Vforce ramp (stress voltage). The voltage on the floating node was dependent on the capacitance of the DUT

2018 INTERNATIONAL CONFERENCE ON MICROELECTRONIC TEST STRUCTURES, MARCH 19-22, 2018, AUSTIN, TEXAS, USA

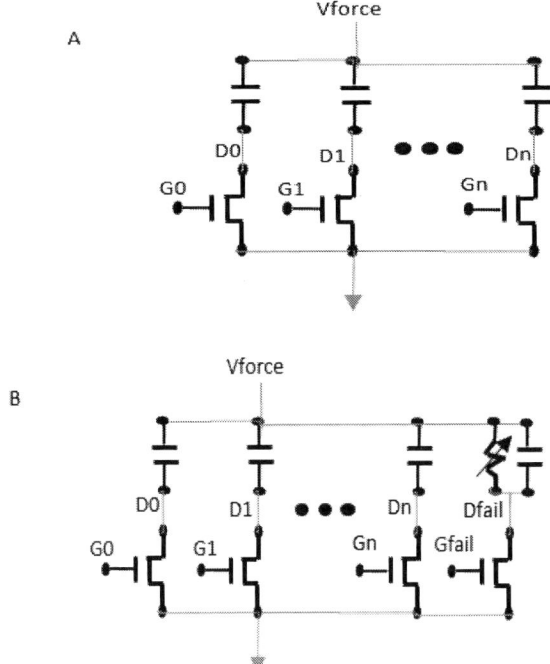

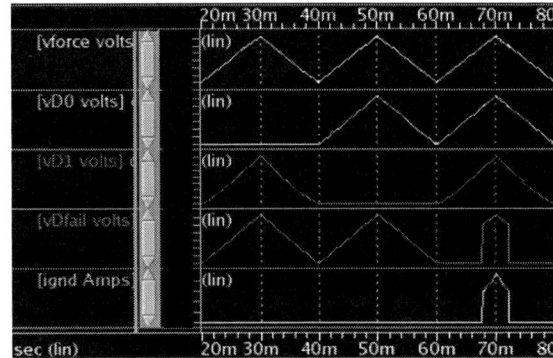

Fig. 6 Simulated waveform of Fig 4B. Similar bias conditions as Fig. 5 with the additional failing device (Dfail). The failing device showed a signal of drain voltage (vDfail) and gnd current tracking with Vforce. The ignd spike at 70mS was the signal that the test program was designed to detect as this represents a failing device.

Fig. 4 Schematic representation of circuits simulated. The DUT was represented by a capacitor which was connected to Vforce (stress voltage) and a connection to ground through an addressable pass gate. A. shows all reliability combs as ideal capacitors with no failing DUTs. B. has a voltage controlled resistor in parallel with one of the capacitors to represent a failing DUT, which was the "fail" device. The voltage controlled resistor behaved as an open until Vforce ramped above Vfail. Each select transistor (G1 – Gn and Gfail) has a separate gate bias to sequentially connect the DUT between Vforce and GND.

when the respective pass gate was turned off. The capacitive values were chosen so they would range around the anticipated capacitance of the existing DUTs. Typically, these DUT's on silicon were in the .5pF to 100pF range, the simulated capacitances were in the same ranges. Fig. 5 shows the floating node voltage (D0, D1, etc.) when cycling Vforce for each pass gate. For 20-40mS G0 was enabled, 40-60mS G1 was enabled, etc. The simulation was done this way to show the response of the selected device as well as the unselected devices. When the pass gate was on, the stress voltage (Vforce – GND) was across the DUT and when the pass gate was off, the stress voltage was across the pass gate. Since the voltage on the floating node coupled with Vforce and resulted in a voltage drop across the pass gate, this restricted the maximum stress voltage to a potential less than the breakdown of the pass gate. This was the benefit of addressing all devices simultaneously, the voltage drop was always across the DUT and not the pass gate.

Fig. 6 shows the simulated curve with an additional device added to the circuit, Gfail. For this DUT, a voltage controlled resistor was placed in parallel with the capacitor to represent a resistive or shorted DUT. The simulation showed ground current (ignd) representing device breakdown. The good devices (D0 – Dn) show no comparable ground current when they were accessed. The failing DUT (Dfail) showed a similar current signature to the devices in Fig. 2 with the current across the DUT increasing with Vforce. This simulation indicated the circuitry could identify the failing DUT.

This simulation was limited to the use of ideal capacitors and resistors which were a simulated representation of a comb structure during stress and failure. Additional work was done, through measurement of the DUT's on Silicon, to determine if the simulated results accurately represent a failure condition.

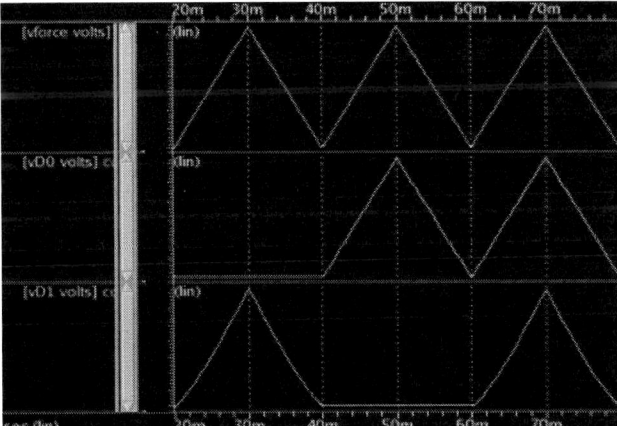

Fig. 5 Simulated waveform of the schematic shown in Fig. 4A with Vforce, VD0, and VD1 plotted over time. Each cycle of Vforce, a different pass gate was turned on. The first peak VG0 was enabled, the second peak VG1, etc. This showed the floating node (D0 and D1) coupled to the Vforce voltage when the gate was turned off but was pulled to ground by the pass gate when it was turned on.

IV. ELECTRICAL RESULTS

The test module contained 5 different banks of structures. Each bank contained structures with different materials and varying line space patterns. The first part of the electrical testing we kept the stress voltage less than the avalanche breakdown (BVDSS) of the pass gate. Fig. 7 shows the

978-1-5386-5072-1/18 $31.00 © 2018 IEEE   139

2018 INTERNATIONAL CONFERENCE ON MICROELECTRONIC TEST STRUCTURES, MARCH 19-22, 2018, AUSTIN, TEXAS, USA

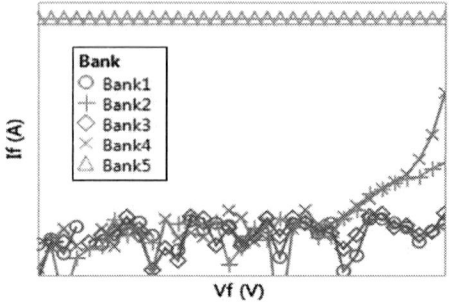

Fig. 7 Parallel electrical stress of all banks with all DUT's enabled. Bank 5 contains the electrically shorted device which was for diagnostic and to mimic a failed DUT.

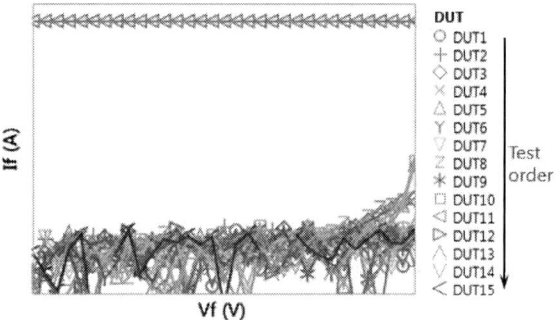

Fig. 8 Discrete addressing of the DUTs in bank 5. DUT11 in Bank 5 contains the DUT with a designed short for diagnostic and failure detection. This DUT, as expected, shows high levels of current, mimicking a failure, at all vforce levels.

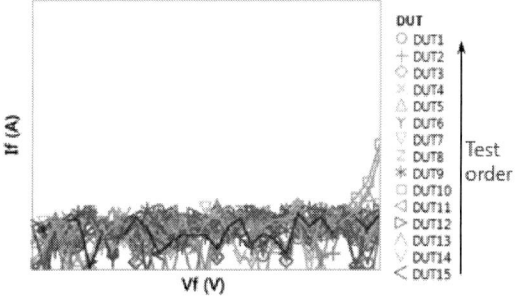

Fig. 9 Discrete addressing of DUT's in Bank 4. DUT 4 and DUT 10 showed increases in leakage with increased Vforce. The remaining DUTs were not showing any appreciable leakage current.

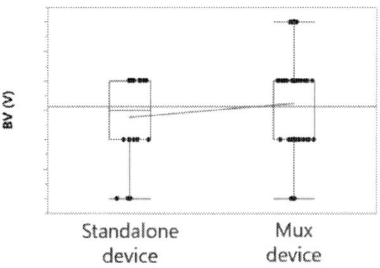

Fig. 10 Boxplot of electrical data comparing breakdown voltage measurements between a standalone DUT with isolated force and ground pads and a mux controlled DUT.

measurements from the testing of this module. All banks were stressed with all DUT's enabled. Banks 2, 4, and 5 show indications of devices starting to breakdown. Bank 5 contains an intentionally shorted device for diagnostic and failure detection. As expected this shows appreciable currents immediately during the stress. Bank 2 and 4 show an increase in the leakage current as the stress voltage increases, while the remaining banks continue to show currents in the noise level of the tool.

Now that failing banks have been identified, the measurements can focus on those failing banks to identify the failing DUT. Fig. 8 shows the sweeps for bank 5. DUT 11, which corresponds to the design shorted DUT, immediately shows high current while the other devices which haven't broke down show low level (noise) current. Fig. 9 shows similar curves for Bank 4. DUT 5 and DUT 10 had identical dimensions, DUT 10 had a 90-degree rotation. All other DUT's in this bank have larger line space patterns so it was appropriate that DUT 5 and 10 were the failing DUT's. When the breakdown voltage (BV) measurements were compared between DUT 5 and 10 to the standalone structures shown in Fig. 1B, similar breakdown voltage values were measured, see results in Fig. 10.

The previous measurement values only stressed up to the BVDSS of the pass gate. The majority of these DUT's had a breakdown voltage higher than the BVDSS. Additional work was done to determine if a BV could be measured above BVDSS. We measured Bank 3 devices on several sites with varying current clamp conditions beyond the BVDSS limit. Fig. 11 shows all DUT's enabled in bank 3. The spikes in the curve, "A", were suspected to be due to DUT's failing. When the DUT's were discretely addressed with these new test conditions some failing devices were detected. DUT 2 and 10, which were the tightest dimensions, showed the earliest increase in current. This indicated that the circuit could identify a failing device, see Fig. 12. A new site was measured with a higher current limit (>100X).

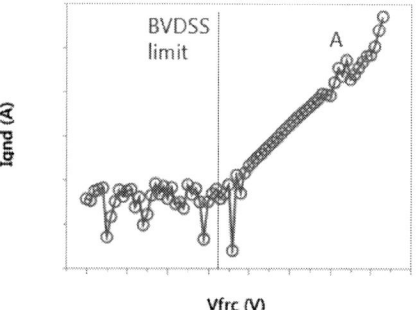

Fig. 11 Sweep of bank 3 with all devices enabled beyond the BVDSS of the pass gate. Site "A" was suspected to be due to a DUT failing causing a spike in the curve.

978-1-5386-5072-1/18 $31.00 © 2018 IEEE

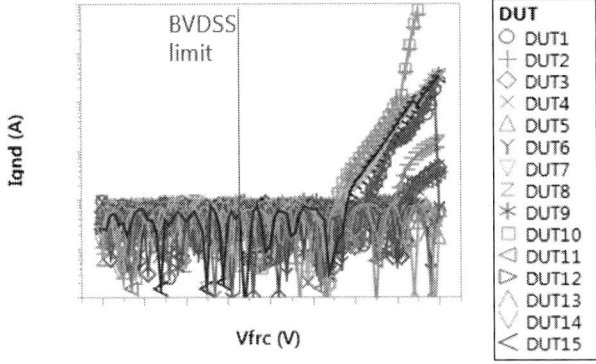

DUT	
○	DUT1
+	DUT2
◇	DUT3
×	DUT4
△	DUT5
Y	DUT6
▽	DUT7
Z	DUT8
*	DUT9
□	DUT10
◁	DUT11
▷	DUT12
∧	DUT13
∨	DUT14
<	DUT15

Fig. 12 Discretely accessing each DUT in bank 3 to determine which DUT had failed. DUT 2 and 10 show the lowest BV for the bank.

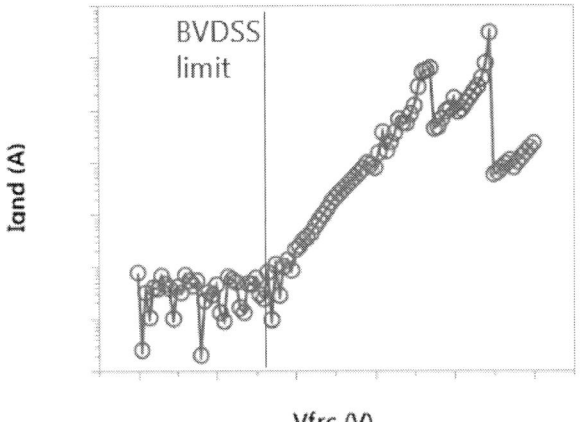

Fig. 13 Sweep of a new site in bank 3 with all devices enabled but with the current limit over 100 times higher than the previous sweep shown in Fig. 11.

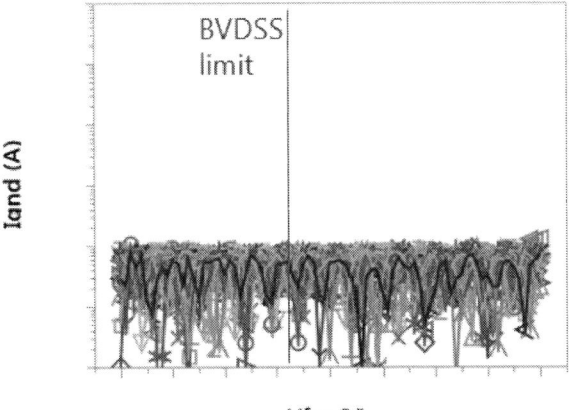

Fig. 14 Discretely addressing each DUT for the site shown in Fig. 13. No devices showed an increase in current, it was suspected that with the higher max current value that the circuit had been damaged.

Spikes in the curve were seen when all devices were enabled. However, when individual devices were discretely addressed we did not see any indication of device failure as seen in Fig. 13 and 14 respectively. It was suspected that either the pass gate or related circuits were damaged by the higher current causing the structure to lose resolution. This indicated that the maximum Vforce could exceed the BVDSS of the pass gate at the risk of damaging the control circuitry.

V. CONCLUSION

We have shown through electrical simulation that this test structure design stressed and identified failing sites. The use of multiplex logic and pass gates can test many reliability structures in parallel which will be advantageous during high volume manufacturing parametric test. We have also shown through electrical test of the circuit that DUT's can be stressed and identified up to the BVDSS of the pass gate. Testing can continue beyond that voltage limit but there was a risk to the circuit if the stress conditions were not managed correctly. Further work will need to be done to explore the characteristics of this limit. The simulation and measurement of this new design has shown the ability to develop a reliable product by detecting process failures.

ACKNOWLEDGMENT

Thanks to Will Otteson, Joonwoo Nam, and Warren Parsons for their help with the simulation. Thanks to Keerti Kalia and the rest of the scribe design team for getting this module design placed on a reticle set for manufacturing. Thanks to Tim Owens, John Smythe, and Pratap Murali for their reliability insights.

REFERENCES

[1] G Moore et al, "Accelerating 14nm device learning and yield ramp using Parallel test structures as part of a new inline parametric test strategy" in International Conference on Microelectronic Test Structures (ICMTS), 2015 pp 44 -49.

[2] C. Chen et al, "A compact test structure for characterizing transistor variability beyond 3σ," *IEEE Transactions on Semiconductor Manufacturing*, vol 28, issue 3, pp 329 – 336, Aug 2015.

[3] H Lin, J Segal, and K. McGaughey, "An addressable test structure for geometrical design rules characterization," *in IEEE International Symposium on Semiconductor Manufacturing*, 2006, pp 25 – 28.

[4] K Doong, J. Cheng, and C. Hsu, "Design and simulation of addressable failure site test structure for IC process control monitor," *in International Symposium on VLSI Technology, Systems, and Applications*, 1999, pp 219 – 222.

[5] S. Chitrashekaraiah, et al, "Addressable test structure for MOSFET variability analysis," in *IEEE Internation Conference on Microelectronic Test Structures (ICMTS)*, 2012, pp 31 – 35.

[6] M. Karthikeyan, et al, "32nm yield learning using efficient parallel-test structures," *in Advanced Semiconductor Manufacturing Conference (ASMC)*, 2010, pp 1-6.

Algorithm Based Adaptive Parametric Testing for Outlier Detection and Test Time Reduction

Veenadhar Katragadda, Martin Muthee, Arthur Gasasira, Frank Seelmann, Jiun-Hsin Liao
GLOBALFOUNDRIES,
400 Stone Break Rd Extension
Malta, New York
USA – 12065
Email: veenadhar.katragadda@globalfoundries.com, martin.muthee@globalfoundries.com

Abstract—Parallel test capability, enabled by numerous independent measurement channels has significantly increased throughput in parametric testing. It involves testing of numerous devices simultaneously synchronously or asynchronously. The number of devices tested for a given pad layout is increased by using higher dimensional arrays, the hallmark of which is pad sharing. Parallel testing of multiple devices with shared pads is vulnerable to device fails, where a failing device adversely affects measurement of all other devices. Information about this failing device or compromised measurement would only be evident at post analysis where a retest with a recipe change can then be ordered. In some cases retest is impossible as wafers would have already moved on to subsequent processing steps, thereby losing valuable learning opportunity. On the other hand, having to wait for post analysis requires time. Ideally failure detection and subsequent re-measure is done dynamically while the device is under test. This would require that decision making capability to be implemented in an automated tester equipment. In this work, we will discuss an algorithm based approach to adaptively change the test program allowing testing or skipping devices based on data collected real time while device is under test. The adaptive algorithm is also extended to aid in test time efficiency by eliminating tests based on measurement results of preceding tests.

Keywords—Adaptive algorithm, Parallel testing, Parametric testing, Outlier Detection, Automated Test Equipment (ATE), Keysight P9000 MPPT.

I. INTRODUCTION

While scaling on semiconductor integration continues, more advanced and complex structures demand comprehensive parametric testing driving test time and increased data-rate for post-test yield/device analysis. With increased test content per structure, providing quality data for quicker yield learning, while maintaining optimal manufacturing test time is a challenge [1].

To cope with these challenges, an algorithm that adapts to the results of a test furthermore dynamically changes the test recipe is demonstrated [6]. The implementation allows users to choose methods ranging from limit based to several real time calculations to command the algorithm adaptivity. The implementation of the algorithm allows test recipe program setup to be less complex as decision is made by the algorithm and not test program. This design is scalable to add calculations/routines for future use.

II. ALGORITHM DESCRIPTION

An Automated Test Equipment (ATE) system uses several measurement routines designed to perform a particular test as described in the test recipe. A few adaptive test implementations involve performing the analysis and decision making within the test plan. This works for simpler adaptive test and is not advisable for analysis involving multiple data points and calculations that would make the test-plan setup convoluted. Also adaptive test involving outlier detection and re-measure in parallel test of multiple DUTs cannot be possible with test plan approach.

As an ATE environment has the maximum control between, reading the user defined content in test plan to measure and output data, the adaptive algorithm is implemented as a pseudo measurement routine to work in conjunction with other routines. This choice makes the adaptive implementation call the measurement algorithm when needed to re-measure as described in section III-B. As a pseudo measurement routine by itself the algorithm can be called by the test plan recipe for suitable applications. This hybrid implementation of adaptive algorithm is suited for several applications that dynamically adjust test program and test sequences locally on the tester [2].

The data required for the adaptive algorithm is inputted in two ways. One way is the feed forward method where the ATE system (environment) allows previous data to be fed to the adaptive algorithm, as described in the test plan recipe. The other way is passing data by reference which is suitable for applications where re-measure is required. Here the measurement routine feeds the measured data to adaptive algorithm for analysis. A re-measure is followed if the analysis results in a fail. All operations needed for adaptive test are implemented and precompiled in ATE code. Section III describes several applications in which a single adaptive implementation can be used for different criteria. The scalable design of the algorithm allows addition of new operations and

978-1-5386-5072-1/18 $31.00 © 2018 IEEE

2018 INTERNATIONAL CONFERENCE ON MICROELECTRONIC TEST STRUCTURES, MARCH 19-22, 2018, AUSTIN, TEXAS, USA

thus is future ready. Figure 1 shows the high-level algorithm flow. The setup consists of a test recipe detailing regular test sequences for a Device under Test (DUT) along with adaptive operation information. DUT starts testing per sequence, data is collected and an operation (calculation, if any) is performed per the adaptive setup. The adaptive algorithm then analyzes the distribution of data. The analysis result along with calculation (if any) is fed to the test sequence. A 0 indicates the DUT performing out of spec and will be rejected from its additional test sequences. The test will move onto next set of DUTs and will test per its setup sequence. A 1 indicates a normal test sequence. Once the DUT completes all of its test sequences, test will move to the next DUT. For Parallel DUT group testing, the algorithm rejects DUTs that failed the adaptive analysis, other DUTs in group will move onto their next test sequence.

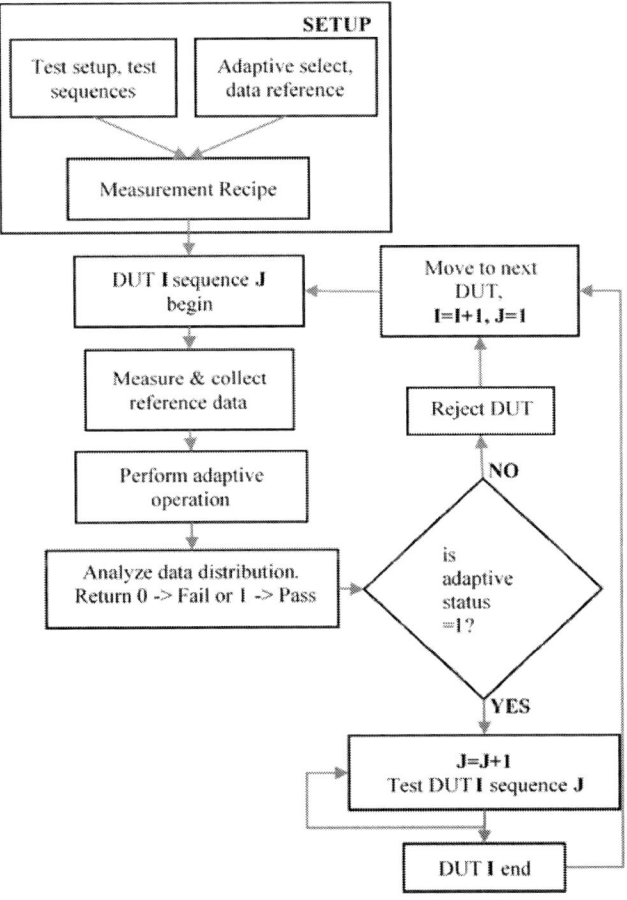

Figure 1: High level implementation of adaptive algorithm

III. APPLICATIONS

A. Dynamic Recipe Tuning

An ATE follows the sequence of tests as setup in the test recipe, where bias conditions are fixed. Retest for the purpose of test condition optimization, which demands additional setup time, can be executed with an adaptive test algorithm to dynamically tune recipes as shown in figure 3(a).

The test recipe is setup with all information needed for recipe tuning for a given test. A measurement routine sends its measured data point/s to adaptive algorithm routine. Here a pre-selected calculation is performed and analyzed. Depending on the analysis, re-measure can be performed with tuned parameters. The operation repeats itself until the adaptive analysis is satisfied or for a pre-defined number of times, whichever occurs first.

Dynamic recipe tuning can also be applied to switch measurement units having different measuring range capability. One such test is measuring capacitance on *Keysight massively parallel parametric test (MPPT) system [4]* where two different kinds of capacitance measurement units (CMU): A parallel per-pin CMU where each pin has its own measurement resource on the test head. The second kind is the external unit which is shared among all pins. Our systems are configured with two of these external units. The per-pin CMU although gives a high throughput, is limited by the conductance range of 1uS outside of which the capacitance measurement becomes unreliable. A cap measurement routine performs the measurement and reads back the conductance from per-pin resource and if beyond the resource range, re-measures from external CMU. This ultimately can increase the measurement time as a single DUT gets measured on both kinds of the CMU units. Instead, if conductance range of DUT is measured prior to capacitance measurement, one can ultimately avoid measuring cap on both measure units. One good indicator of conductance range can be a simple test measuring leakage current through the cap DUT.

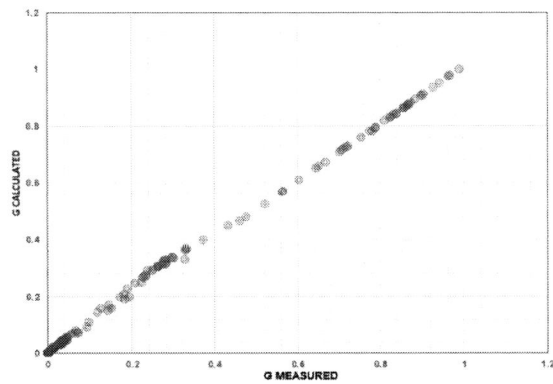

Figure 2: Normalized conductance calculated from leakage test vs. conductance measured from cap measurement unit.

Figure 2 shows good correlation between conductance calculated by leakage measurement vs. conductance measured from CMU unit. Here adaptive algorithm can be used to analyze the leakage measurement for conductance range and tune the cap measure routine to measure from suitable resource. In typical applications, the leakage test is part of the test plan and is executed before the capacitance measurements.

978-1-5386-5072-1/18 $31.00 © 2018 IEEE   143

2018 INTERNATIONAL CONFERENCE ON MICROELECTRONIC TEST STRUCTURES, MARCH 19-22, 2018, AUSTIN, TEXAS, USA

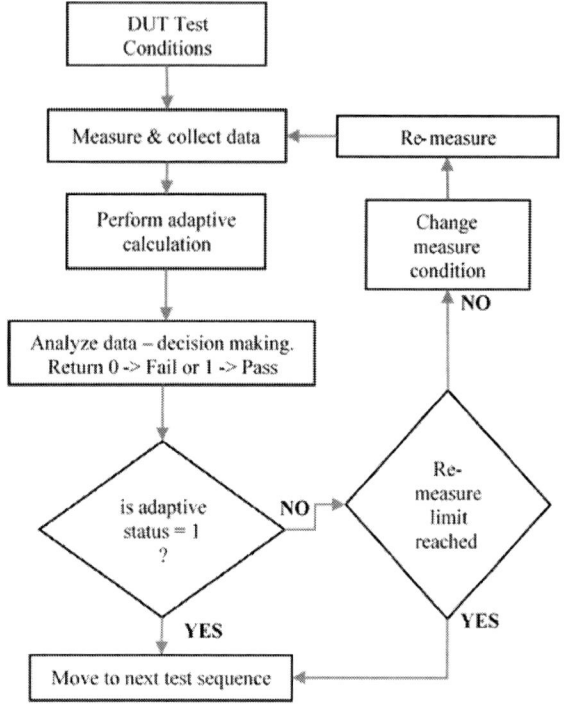

Figure 3(a): Adaptive algorithm flow for Dynamic Recipe Tuning.

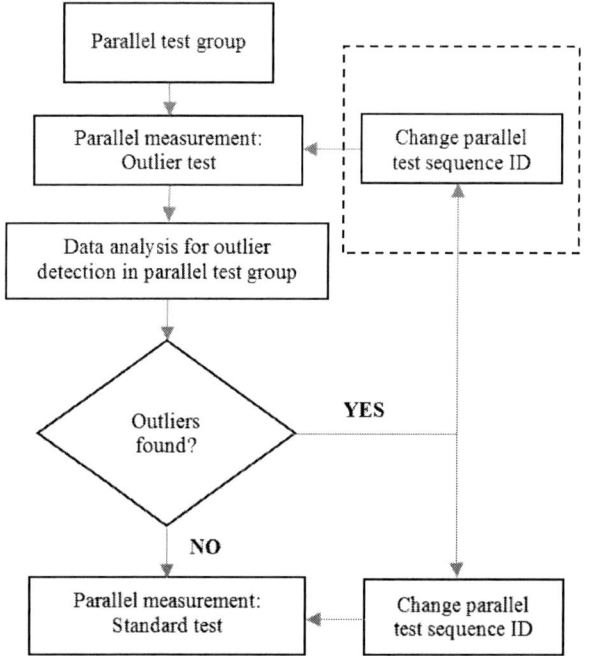

Figure 3(b): Adaptive algorithm flow for outlier detection

B. Outlier Detection

The macro design approach commonly referred to as design for parallel test is based on maximizing the number of devices tested for a given pad layout, normally accomplished by using higher dimensional arrays with pad sharing. Figure 2 shows

examples of some pad sharing schemes [1],[5]. While high dimensional arrays help to increase throughput, pad sharing is particularly vulnerable to parasitic wire resistance and measurement uncertainty due device failure.

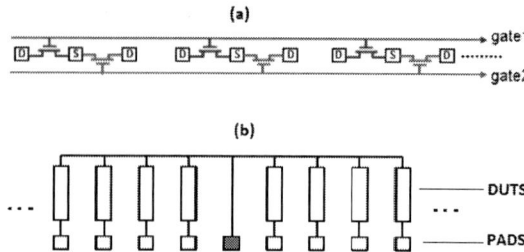

Figure 4: (a) FET array with shared Gates and Source design. (b) Short monitor array with a common bias pad shown in red

In the event of a hard fail, the measurement for all or a handful of devices can be severely compromised. This is especially the case when a short occurs on a shared pad used to apply bias. To mitigate the effect of failing devices we propose using an adaptive algorithm method to detect outlying data in a first pass of measurement, we call the outlier test, to test for failures primarily in the shared pads. Based on a given criteria the algorithm analyzes the data and removes any outlying devices (data points) from the parallel test sequence. The measurement is then repeated with the outliers either tested separately, serially or ignored altogether.

In a typical parallel test system, the test plan contains information that controls the execution of tests, serially or in parallel, synchronously or asynchronously. In the *Keysight P9000 MPPT* system [4], the execution of serial versus parallel tests is controlled by a *sequence* id, pre-assigned before actual test execution. In the adaptive framework, the algorithm adaptively changes the measurement sequence in the event of outlier detection. Fig 3(b) shows an implementation flow where a group of DUTs sharing the same sequence id is measured in parallel. Moreover, the group of DUTs has one or more shared pads. The first pass test is the outlier test, specifically designed to capture failures in the shared pad, the result of which is analyzed for outliers in distribution. If any outliers are found, their sequence id is changed for the succeeding second pass test or alternatively re-measure of the outlier test (highlighted). The result of this change is that the outliers will not be tested in parallel with the standard group.

Two cases are illustrated: A FET array as shown in figure 4(a) with shared gates and sources, and an array of short monitor devices [5] as illustrated in figure 4(b). In the FET array, parallel testing of all devices from the non-shared drain pads to measure *Idlin/Idsat* for example, can be completed in one pass per gate. In the event of a gate-drain or gate-source short in any of the devices, the entire measurement would be compromised and the results invalid. The outlier test for the FET array therefore is most suitable one that probes the gate-drain or gate-source integrity. The gate is biased at 0V while Drain and Source at *Vdd*. Current measurements are then

978-1-5386-5072-1/18 $31.00 © 2018 IEEE

2018 INTERNATIONAL CONFERENCE ON MICROELECTRONIC TEST STRUCTURES, MARCH 19-22, 2018, AUSTIN, TEXAS, USA

made from all three terminals simultaneously. Note that this particular outlier test is not optimized for test time. Figure 5 shows the outlier test results, normalized currents for each FET tested in parallel. From the Source and Gate results, some clear outlying data indicate a short mode failure in the Gate-Source for FETs 5, 11 and 14. The adaptive algorithm then responds by removing those FETs from the parallel test group.

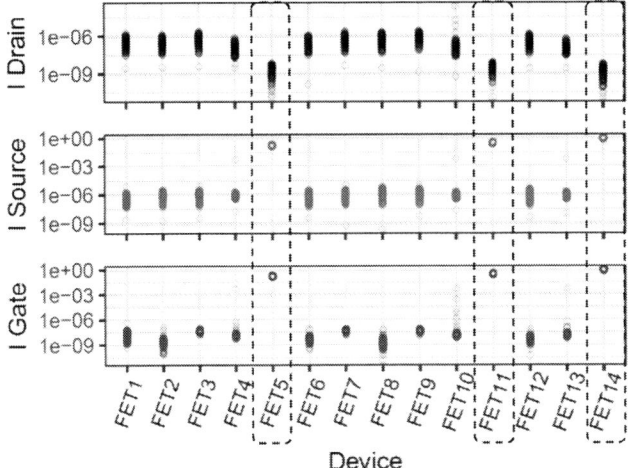

Figure 5: Normalized Gate, Source and Drain currents from outlier test performed on a 14 FET array with design as figure 4(a).

To show the effect of the failing FETs an Idsat measurement result is shown in figure 6 where devices adversely affected by the failing ones recover in the adaptive test which excludes the failing devices.

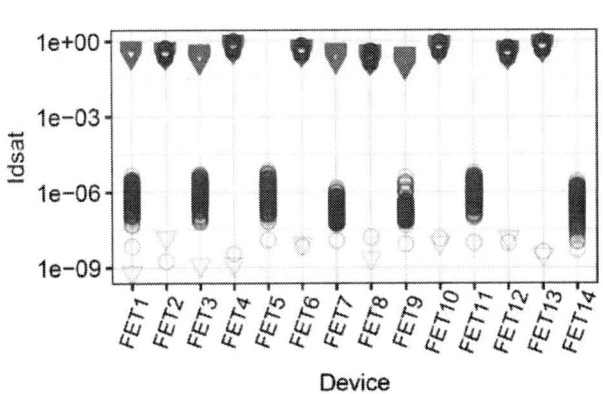

Figure 6: Normalized Idsat measurement results from an adaptive and non-adaptive test.

For the short monitor array case, in this design, all devices share a common bias pad with unique measurement pads. Any shorted device would grossly affect the measurement by providing a path to ground. Figure 7 shows some simulated data of short monitors showing the effect of a failing device during parallel measurement of a group sharing a common bias

pad. In figure 7(a) is the expected result while 7(bottom) shows the effect of having a shorted device (DUT12). The takeaway is that any one short considerably affects the measurement result for all.

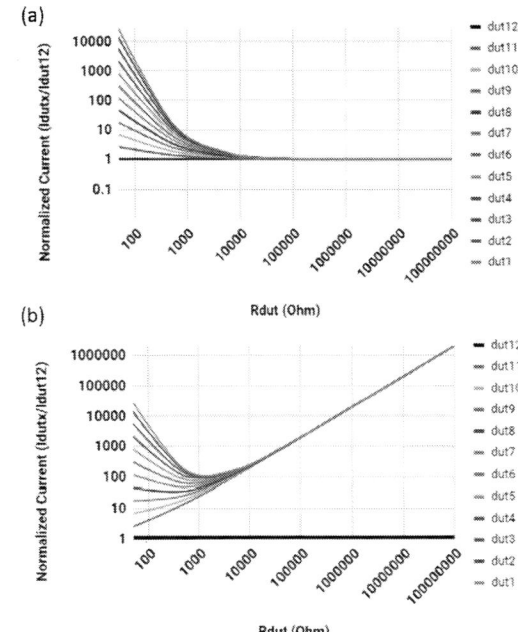

Figure 7: Simulation results from an array of 12 DUTs with design as shown in figure 4(b). (a) All DUT resistances are approximately equal. (b) DUT12 shorted while all others are approximately equal.

In the adaptive framework, the actual Short test can be used as both the 1st and 2nd pass test. Any shorted devices would draw the highest currents compared to the standard DUTs. Figure 8 shows the result of an adaptive (red) and non-adaptive test (blue). DUTs 1, 2, 6, 7, 11, 12, 18 and 19 show the typical short signature and the mitigating adaptive algorithm shows the standard group results clearly differing with the more accurate adaptive data.

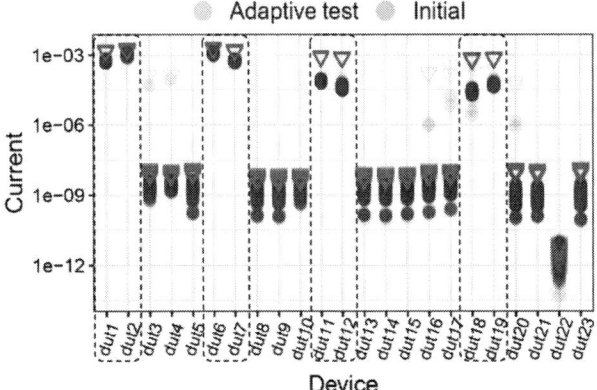

Figure 8: Normalized measured current results from an array of 23 short monitors tested adaptively and non-adaptively.

978-1-5386-5072-1/18 $31.00 © 2018 IEEE   145

C. Test Time Efficiency

Adaptive test for test time efficiency is an application with significant impact. As the number of devices per structure and test content per device is increasing, a significant growth in test time is observed. Each DUT goes through a series of tests where a single test itself might not be time consuming but adds up quickly for a whole test suite. Devices fail tests, especially early in the technology development cycle.

A whole suite of tests on a failing DUT can be redundant especially when subsequent tests are not needed for device learning. These redundant tests on a failing DUT can take a significant amount of time yielding no valuable information. An adaptive algorithm can save test time when a DUT fails a preliminary test that gives the information on device functionality. If a DUT passes this preliminary test/s, subsequent suite of tests can be applied. If a DUT fails the preliminary test, then the subsequent tests will be rejected from measuring and an error code can be applied to the output data indicating failing DUT. The implementation is similar to the flow shown in figure 1.

For demonstration purposes we tested 4557 FETs that each underwent a suite of tests, a test time of approximately 580ms. As part of preliminary tests, we chose Ion and Ioff tests taking approximately 80ms combined. As Ion/Ioff ratio should be large for good device functionality, a number 100 is chosen for adaptive criteria. DUTs that have Ion/Ioff ratio to be less than 100 are considered fails (0 as per the adaptive algorithm figure 1a) and above 100 are considered to be passes of the preliminary test (1 per adaptive algorithm implementation). Figure 9 shows Vt vs. Ion/Ioff ratio, where above 100 valid Vt measurements are reported and below, an error code is reported. Table 1 gives stats on total test time and adaptive test time benefits. Out of 4557 DUTs tested, our preliminary test failed on 2363 devices, which are adaptively removed from successive tests saving around 19.6 minutes per wafer. It is advised to note that these time benefits are contingent upon a large number of failing devices and number of tests a DUT undergoes.

# of DUTs	4557
Time for preliminary tests/DUT	80 (ms)
Time for all successive tests / DUT	500 (ms)
Time for all successive Tests all DUTs	37.9 (mins)
Fail # of DUTs	2363
Adaptive Test Savings	19.6 (mins)

Table 1: Table showing adaptive test-time savings

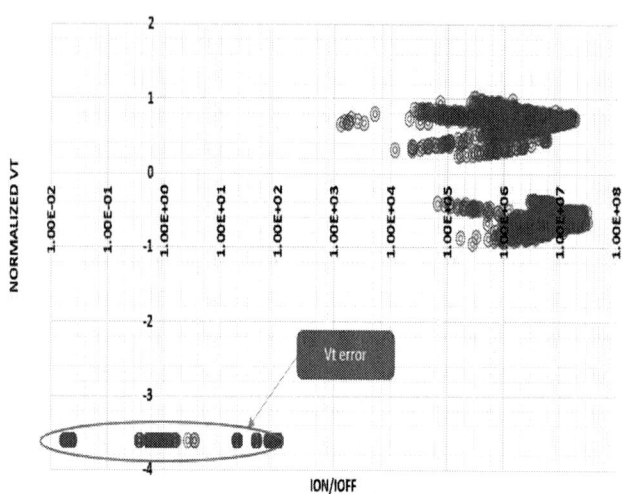

Figure 9: Normalized Vt vs Ion/Ioff ratio showing higher ratios having valid Vt while lower ratios have Vt errors.

REFERENCES

[1] G. Moore, J. H. Liao, S. McDade and B. Verzi, "Accelerating 14nm device learning and yield ramp using parallel test structures as part of a new inline parametric test strategy," Proceedings of the 2015 International Conference on Microelectronic Test Structures, Tempe, AZ, 2015, pp. 44-49.

[2] S. Benner and O. Boroffice, "Optimal production test times through adaptive test programming," Proceedings International Test Conference 2001 (Cat. No.01CH37260), Baltimore, MD, 2001, pp. 908-915.

[3] M. Bhushan, M. B. Ketchen, Microelectronic Test Structures for CMOS Technology. Springer 2011, pp. 11-64

[4] Keysight Technologies: 'P9001A Massively Parallel Parametric Test System' [Online]. Available: https://www.keysight.com/en/pdx-2881433-pn-P9001A/massively-parallel-parametric-test-system

[5] M. Lauderdale and B. Smith, "A versatile defectivity monitor designed for efficient test and failure analysis," 2011 IEEE ICMTS International Conference on Microelectronic Test Structures, Amsterdam, 2011, pp. 25-30.

[6] N. Kupp and Y. Makris, "Applying the Model-View-Controller Paradigm to Adaptive Test," in IEEE Design & Test of Computers, vol. 29, no. 1, pp. 28-35, Feb. 2012.

Session 8

Device Characterization

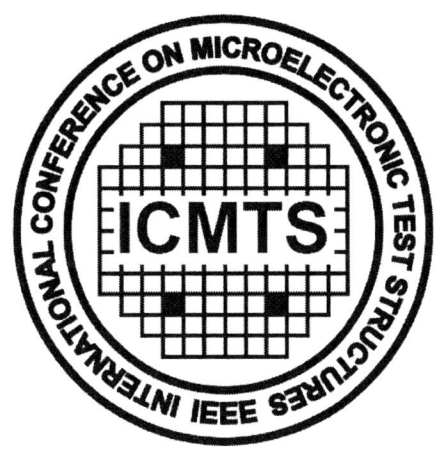

March 21, 2018

15:20-17:00

Co-Chairs:

Kjell JEPPSON, Chalmers University of Technology, Sweden

Mark POULTER, Texas Instruments, USA

978-1-5386-5072-1/18 $31.00 © 2018 IEEE

2018 INTERNATIONAL CONFERENCE ON MICROELECTRONIC TEST STRUCTURES, MARCH 19-22, 2018, AUSTIN, TEXAS, USA

Evaluation of Qss on SOI Back Si/SiO₂ Interface by Newly Designed Charge Pumping Method-TEG

Kazuma Takeda[1], Jiro Ida[1], Takayuki Mori[1] and Yasuo Arai[2]

[1] Division of Electrical Engineering, Kanazawa Institute of Technology, Ishikawa, Japan,
E-mail: ida@neptune.kanazawa-it.ac.jp
[2] High Energy Accelerator Research Org., (KEK), Tsukuba, Japan

Abstract—The surface state density (Qss) of SOI back Si/SiO2 interfcae was evaluated by newly desiged Charge Pumping (CP) method-TEG. The CP method was also re-examied to apply to the thick oxide MOS. It was noted that the high volatge aplitude and attention on the slope on the gate pulse are nesessary to evaluate the Qss of SOI back interface made of the thick oxide. It was founded out that the Qss of SOI back interface (bonded wafer interface) is comparabe to that of the thermal oxidation interface and also that the Qss of Floating Zone (FZ) wafer is larger than that of Czochralski (CZ) wafer, and the Qss of FZ wafer varies from lot to lot.

Keywords—Charge pumping, SOI, Surface state density

I. INTRODUCTION

We have developed "SOI pixel sensor", shown in Fig. 1 [1], where the PN photo diode is fabricated in the Si substrate. The surface state density (Qss) on the SOI back Si/SiO2 interface (indicated in Fig. 1) determines the sensitivity of this sensor. Therefore, it is important to evaluate the Qss on the SOI back interface. We have used the bonded SOI wafer (SOITEC smartcut wafer). The SOI back interface is the wafer bonded interface, not the thermal oxidation interface. Therefore, the evaluation of the Qss on the SOI back interface is also interesting to know the Qss of the wafer bonded interface. However, the evaluation reports of the Qss on the SOI back interface have been limited because the C-V and the charge pumping (CP) method [2] are difficult to apply due to the existence of the buried oxide in SOI [3].

We have already developed the buried contact (indicated in Fig. 1) process to the substrate [1]. In this study, we have newly designed the MOSFET structure TEG on the SOI back interface with the buried contact and evaluated the CP currents and the Qss on the SOI back interface by applying the CP method [2]. We also re-considered the CP method for applying it to the thick gate oxide MOS made by the buried oxide of the SOI. In order to detect the high energy photon, the high resistivity substrate has been also applied on our sensor, using the wafer made by Czochralski (CZ) method and also the wafer made by Floating Zone (FZ) method. Therefore, comparisons of the Qss between the CZ wafer and the FZ wafer were also shown, for the first time, in this study.

II. NEWLY DESIGNED TEG STRUCTURE

Fig. 2 and 3 show the plane view and the cross sectional image of the newly designed TEG, respectively. We fabricated the MOSFET structure with the highly doped active layer of the SOI as a gate electrode of the MOSFET and the buried oxide (BOX) as a gate oxide of the MOSFET. The buried P-well layer (BPW) of the implantation layer was also used as a source/drain overlapped to the gate (highly doped active layer) and it also contacted to the metal with the buried contact. We made the several L/W MOSFETs, shown in Fig. 2. The buried oxide thickness is 200nm and the Si thickness of SOI (corresponds to the gate electrode) is 50nm. We also used the CZ wafer (resistivity:700Ω, Magnetic field applied CZ wafer used, here) or the FZ wafer (resistivity:>2KΩ) as a substrate of SOI.

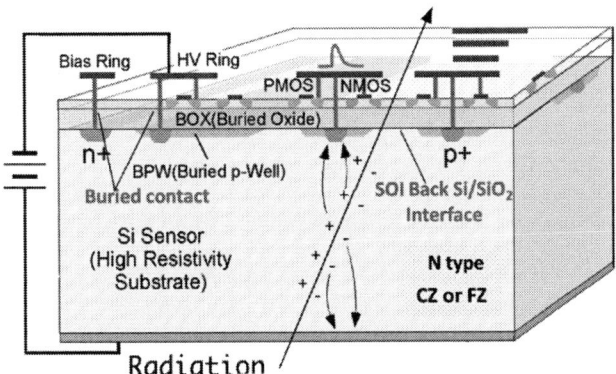

Fig. 1 SOI Pixel Sensor we developed [1].

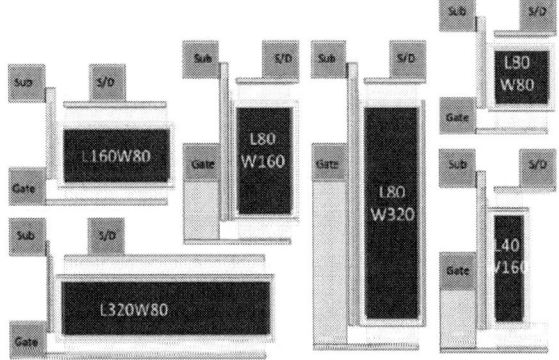

Fig. 2 Plan view of newly designed CP-TEG.

978-1-5386-5072-1/18 $31.00 © 2018 IEEE

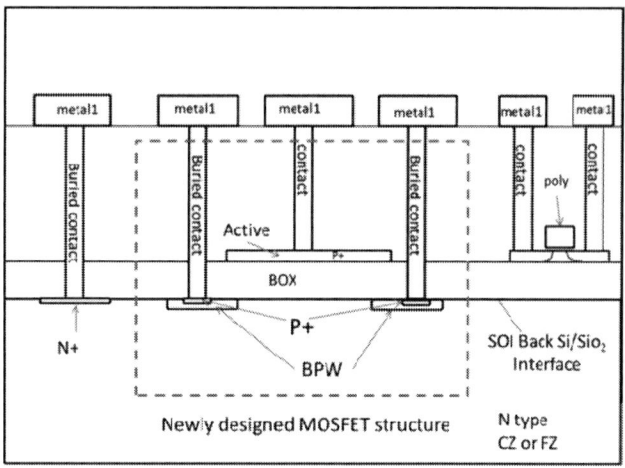

Fig. 3 Cross sectional image of newly designed CP-TEG

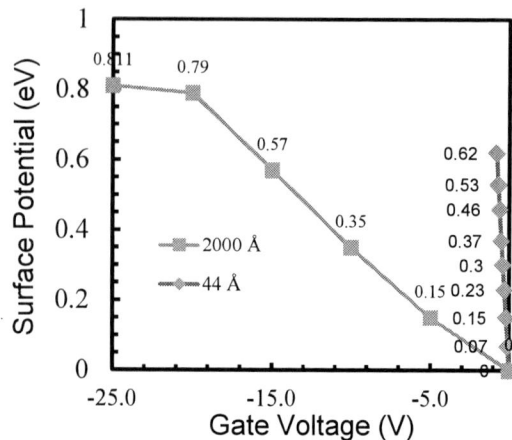

Fig. 5 Simulated surface potential shit with Tox=200nm MOS and Tox=4.4nm MOS

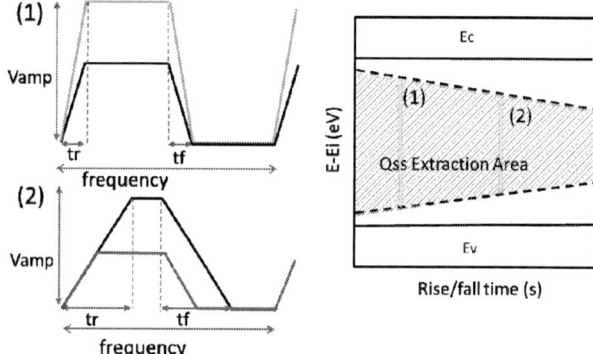

Fig. 6 Gate pulse amplitude increases by Fixed Rise/fall time (1) and Fixed slope (2) and image of Qss extraction area in the forbidden band with slope (Rise/fall time) of gate pulse.

III.   MEASUREMENT RESULTS AND DISCUSSIONS

Fig. 4 shows the measurement setup of the CP method. The base sweep method was used in this study. In order to apply the CP method to the thick gate oxide FET (Tox=200nm with BOX), it should be noted that the voltage amplitude of the gate pulse must be reconsidered. The MOS structures were set and simulated with the device simulator "HyENEXSS" [4]. Fig. 5 shows the simulated results of the shift on the surface potential with both Tox=200nm and the conventional thin oxide of Tox=4.4nm. It indicates that the near 20V of the voltage amplitude is necessary for enough surface potential change, compared with 1V on the thin Tox. Moreover, the slope of the gate pulse should be also noted when increasing the voltage amplitude. When the rise/fall time is fixed, the slope of the gate pulse increases with increasing the voltage amplitude. It results in Qss extraction from the wide area on the forbidden band, illustrated in Fig. 6 [2].

Therefore, at first, we checked the CP current dependence on the voltage amplitude with both the fixed rise/fall time and the fixed slope. Fig.7 shows the measured CP currents dependent on the pulse voltage amplitude with both slopes. Fig.8 shows the maximum CP current extracted from Fig.7. The slopes of the gate pulse are also indicated in Fig.8. Those results show that the fixed rise/fall time method of the 20V shows the largest CP current, which corresponds to Qss extraction from the wide area, and the small increase and saturation with the fixed slope are assumed to correspond to the surface potential change. We chose the 20V fixed rise/fall time in order to evaluate the maximum Qss, in this study. Fig. 9 shows the maximum CP current as a function of the measurement frequency. The linear dependency was confirmed, which is coincide with the CP current equation, as shown in Fig.4. Fig. 10 shows the maximum CP current with different L/W TEG's, normalized with the unit area. The long gate length shows the large CP current, which is expected to come from "geometric component" of CP method [3]. Therefore, we chose the L/W=80/80um.

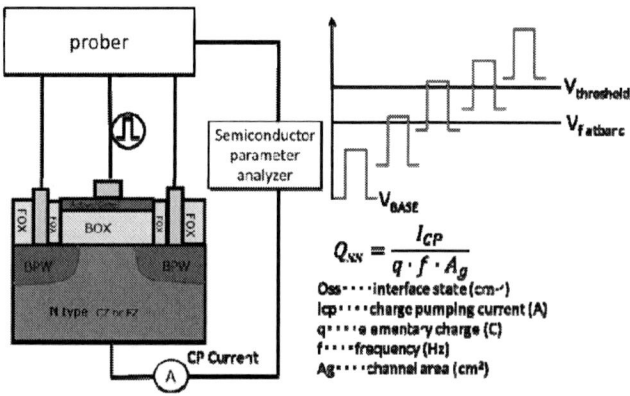

$$Q_{ss} = \frac{I_{CP}}{q \cdot f \cdot A_g}$$

Oss···· interface state (cm⁻)
Icp···· charge pumping current (A)
q···· elementary charge (C)
f···· frequency (Hz)
Ag···· channel area (cm²)

Fig. 4 Measurement setup of CP method in this study

978-1-5386-5072-1/18 $31.00 © 2018 IEEE

2018 INTERNATIONAL CONFERENCE ON MICROELECTRONIC TEST STRUCTURES, MARCH 19-22, 2018, AUSTIN, TEXAS, USA

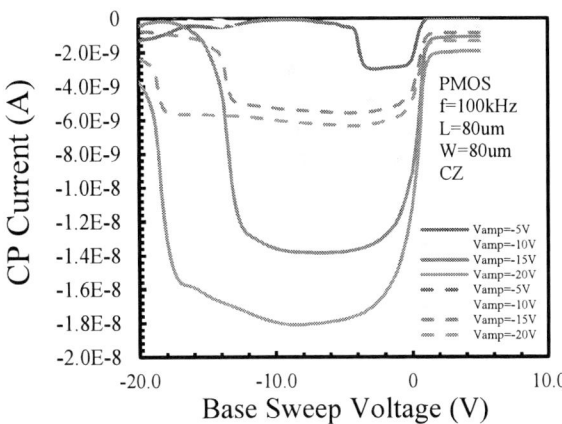

Fig. 7 Measured CP currents dependent on the gate pulse voltage amplitude. Line: Fixed rise/fall time, Dotted line: Fixed slope.

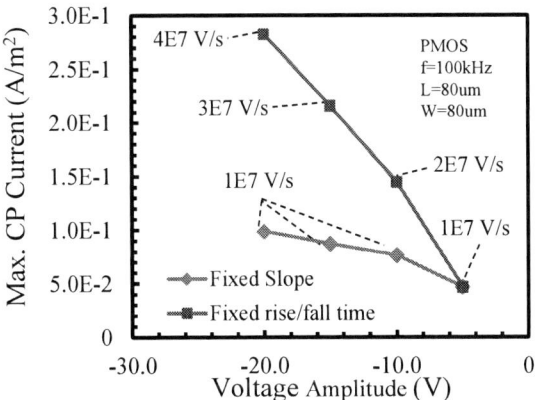

Fig. 8 Maximum CP currents extracted from Fig. 6. Slopes of gate pulse are also indicated.

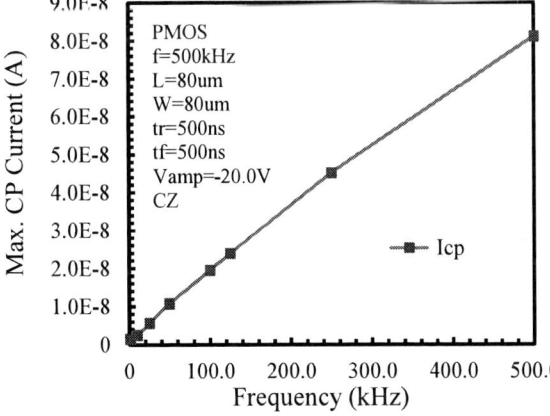

Fig. 9 Maximum CP currents dependent on Frequency.

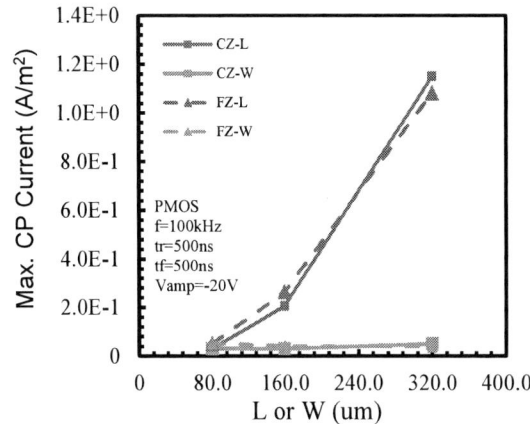

Fig. 10 Maximum CP currents with different L/W TEG's, normalized with the unit area. CZ-L and FZ-L indicate variable L and the constant W=80um, CZ-W and FZ-W indicates variable W and the constant L=80um.

Based on those pre-checking, we evaluated the CP currents and the Qss on the SOI back interface with the CZ wafer and the FZ wafer. The typical CP currents on the both wafers dependent on the voltage amplitude are shown in Fig. 11. Fig. 12 shows samples of the measured results from different three lots where both the FZ and the CZ wafer are used. Both figures including Fig. 7 indicate that the CP currents shows the typical flat shape of the base sweep CP method. Fig. 13 summarizes the calculated Qss from the maximum CP current with 4 different chips from 3 different lots. Table 1 shows the average Qss on 3 lots. Those results indicate that the Qss of the FZ wafer is larger than that of CZ wafer, and the Qss of the FZ wafer varies from lot to lot. On the contrary, the Qss of CZ wafer does not vary. It is assumed to be reasonable because the CZ wafer has the oxide impurity which acts as an intrinsic gettering, although the FZ wafer does not have it. It is also founded out that the Qss of 2 to 5E10/cm2 was obtained on the SOI back interface. It means that the Qss of the bonded wafer interface is comparable to that of the thermal oxide interface.

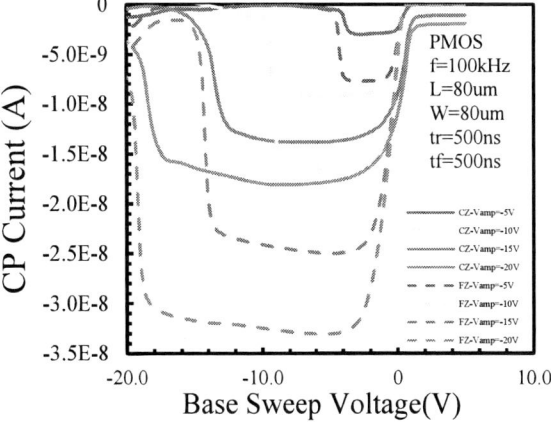

Fig. 11 CP currents dependent on the gate pulse voltage amplitude with CZ wafer and FZ wafer.

978-1-5386-5072-1/18 $31.00 © 2018 IEEE   151

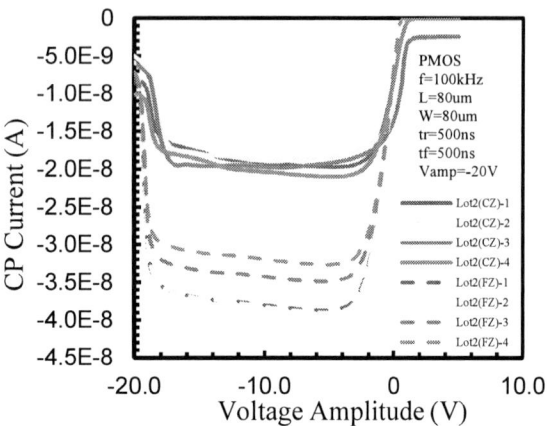

Fig. 12 Samples of CP currents from different lots with CZ wafer and FZ wafer.

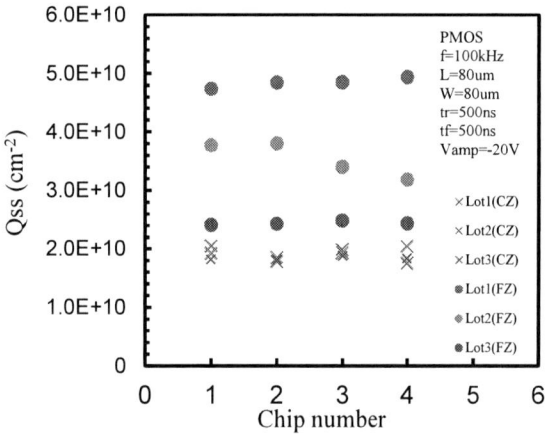

Fig. 13 Extracted Qss from 4 different chips of 3 lots with CZ wafer and FZ wafer.

Table 1 Extracted average Qss from 3 lots with CZ wafer and FZ wafer.

	CZ(cm^{-2})	FZ(cm^{-2})
Lot1	1.90E+10	4.85E+10
Lot2	1.94E+10	3.54E+10
Lot3	1.84E+10	2.45E+10

IV. CONCULUTIONS

The surface state density (Qss) of SOI back Si/SiO2 interfcae was evaluated by newly desiged Charge Pumping method-TEG. The MOS FET was fabricated with the highly doped active layer of the SOI as a gate electrode, the buried oxide (BOX) as a gate oxide and of the MOSFET and using the buried contact to the substrate of the SOI. The CP method was also re-examied to apply to the thick oxide MOS made by the bruied oxide of the SOI. It was noted that the high volatge aplitude and attention on the slope on the gate pulse are nesessary to evaluate the Qss of SOI back interface.

It was founded out that the Qss of SOI back interface (bonded wafer interface) is comparabe to that of the thermal oxidation interface and also that the Qss of Floating Zone (FZ) wafer is larger than that of Czochralski (CZ) wafer, and the Qss of FZ wafer varies from lot to lot.

ACKNOWLEDGMENT

This work was supported by MEXT KAKENHI Grand Number 25109002 and JST-CREST Grand Number JPMJCR16Q1, Japan. This work is also supported by VLSI Design and Education Center (VDEC), the University of Tokyo in collaboration with Cadence Design Systems, Inc. and Mentor Graphics, Inc.

The fabrication of this work is the results of collaboration with LAPIS Semiconductor Co., Ltd.

REFERENCES

[1] Arai, et.al, "Development of SOI pixel process technology", Nuclear Instruments and Methods in Physics Research, Vol.636, pp S31-36, 2011

[2] G. Groeseneken, H. Maes, N. Beltran, and R. F. D. Keersmaecker, "A reliable approach to charge-pumping measurements in MOS transistors", IEEE Trans. Electron Devices, 31, pp42, 1984

[3] W. Vandendaele, C. Malaquin, A. Ghorbel, M. Cassé, F. Allibert, G. Reimbold, "Novel CV/GV technique for top and bottom BOX interfaces traps density extraction on FDSOI wafer", IEEE S3S Conf., session11.1, 2017

[4] HyENEXSS™ ver5.5, Selete, 2011.

Quantitative Model of CMOS Inverter Chain Ring Oscillator's Effective Capacitance and Its Improvements in 14nm FinFET Technology

Seong Yeol Mun[1], J. Cho[2], B. Zhu[1], P. Agnihotri[1], C.Y. Wong[1], T.J. Lee[1], V. Mahajan[1], B.W. Liu[1], Y.J. Shi[1], W. Hong[1], J. Ciavatti[1], J.G. Lee[1], S. B. Samavedam[1] and D.K. Sohn[1]

[1]ATD 14NM Device
Globalfoundries, Malta, NY 12020
Email: seongyeol.mun@globalfoundries.com
[2]Global TCAD, Santa Clara, CA 95054
Email: Jin.Cho@globalfoundries.com

Abstract— **The quantitative model of effective total capacitance, Ceff, of a CMOS ring oscillator (R/O) inverter chain in a 14nm node FinFET 3D structure using advanced Replacement Metal Gate (RMG) is successfully extracted using all the unit capacitance components comprising the R/O, such as inverter, fan-out (F/O) MOSCAP, and metal routing. The extracted Ceff model is well validated by perfect matching to the measured Si Ceff in the R/O. This paper provides a concise and clear Ceff quantitative model of inverter R/O chain using individual transistor capacitance components such as channel capacitance (Cgc), overlap capacitance (Cov), junction capacitance (Cj) and metal wire capacitance (Cwire) considering the R/O layout and its operation mechanism, which has never been reported before. Furthermore, Cov is decomposed with the gate to contact capacitance (Cmol), EPI source-drain (S/D) to gate on Fin top (Cft), EPI S/D to gate on Fin sidewall (Cfb) and intrinsic gate to S/D overlap capacitance (Cdo) with Si data and simulation. Contribution to Ceff by all the capacitor components from Cgc, Cmol, Cj, Cwire, Cft, Cfb and Cdo is extracted with Si validation. Cov reduction without DC performance degradation is also provided in this paper.**

Keywords --- Ceff, CMOS ring oscillator, FinFET, RMG, Cgc, Cov, Cwire, Cft, Cfb, Cdo

I. INTRODUCTION

Capacitance in logic circuitry is one of the most critical parameters affecting its switching speed by Resistance Capacitance (RC) delay, which becomes even more critical as a technology node advances. The resistive component of the R/O, represented by Reff, is well understood where good correlation has been obtained with effective DC current (Ieff). However, the capacitive component of the R/O represented by Ceff, measured directly from R/O delay, is not well understood. In this paper, the Ceff of the CMOS R/O inverter chain representing the logic circuitry is investigated and its model is extracted using every capacitor component comprising the R/O with Si validation. For the precise modeling, detail analysis on the R/O's layout and the behaviors, during the oscillator toggling, of the device components comprising the R/O was done.

Each capacitance component and its contribution to Ceff are also well extracted and validated with Si and simulations. This effort is becoming critical because capacitor components are getting added in FinFET technology using advanced RMG due to its 3D structure [1, 2].

II. TECHNOLOGY OVERVIEW

A. Quantitaive model extraction of Ceff of CMOS Inverter Chain Ring Oscillator

The fan-out 3 (F/O-3) CMOS inverter chain R/O is composed of inverter part and two fan-out MOSCAPs as seen in Fig.1 of layout. The gates of fan-out MOSCAPs are connected to Vout of the inverter as described in the Fig. 2 schematics. R/O's total capacitance, Ceff, is determined by the sum of the inverter capacitance (Cinverter), two fan-out MOSCAPs and Cwire, the metal routing, as shown in Fig. 3 and equation (1).

Cinverter is composed of Cgc (channel capacitance= total gate capacitance Cgg-2Cov), Cov and Cj as described in Figs. 4 and 5, where Cgg and Cj are averages of NFET's and PFET's. The inverter part of the R/O has a capacitance load by NFET Cgcn, Cgsn+Cgdn and Cgdp when Vin= "H", and PFET Cgcp and Cgsp+Cgdp and Cgdn when Vin= "L" contributing to Ceff as highlighted in Figs. 4 and 5. In terms of Cj, P+/NW junction of the PFET plays a role when Vin= "H", and N+/PW junction of the NFET does when Vin= "L" as seen in Figs. 4 and 5. Channel capacitance, Cgc is not affected by "Miller effect" because the bulk is not connected to output, but rather, is connected to GND in NFET, or VDD in PFET. While, Cgd is connected to output, as seen in Figs.4 and 5, it should be subject to the "Miller effect" and become double the original capacitance [4]. This results in adherence to the Cinverter equation (2), where the total 5Cov components is detailed in equation (3).

MOSCAP consists of Cgc+2Cov like Cinverter; however, Cj doesn't play a role, because it is connected to GND in NMOSCAP or VDD in PMOSCAP without potential difference between junction diode terminals as shown in Figs. 6 and 7. NMOSCAP functions as a capacitor when Vout is "H", and PMOSCAP does when Vout is "L" as described in Figs. 6 and 7. MOSCAP has no "Miller effect" involved as shown in Fig. 6 for NMOSCAP and Fig. 7 for PMOSCAP, because the drain terminal is not connected to output. Fan-out MOSCAP is described as following equation (4).

Finally, the total capacitance, Ceff, for fan-out 3 inverter chain R/O, considering fin number (nfin= 4) and active width (Weff= 0.075um) can be calculated by summing the Cinverter, MOSCAP and Cwire as shown in equation (5). For fan-out 1 inverter chain R/O without two fanout MOSCAPs from fan-out 3, the Ceff is modeled by equation (6)

$$Ceff = Cinverter + 2*MOSCAP + Cwire \quad (1)$$

$$Cinverter = Cgc + 5Cov + Cj \quad (2)$$

$$5Cov = Cgs + 2(Cgdn + Cgdp) \quad (3)$$

, where Cov is average of Cgsn, Cgdn, Cgsp, and Cgdp.

$$Fan\text{-}out\ MOSCAP = Cgc + 2Cov \quad (4)$$

$$Ceff\ (F) = nfin*Weff*(3Cgc + 9Cov + Cj) + Cwire \quad (5)$$

, where the units of Cgc, Cov and Cj are Farad per unit width (F/um), and Cwire simulated using Parasitic Extraction (PEX) model is Farad (F).

$$Ceff\ (F) = nfin*Weff*(Cgc + 5Cov + Cj) + Cwire \quad (6)$$

B. Si Validation of the Ceff model

Calculated Ceff using equation (5) shows perfect matching to Si Ceff measured using massive wafers as shown in Fig. 9, well validating the Ceff model. Ceff in Si is extracted from the R/O by dividing the R/O's delay by ACReff as described in equation (7), where ACReff is the effective R/O's resistance during toggling and extracted from Idda and Iddq by equation (8). Idda is the total current flowing when the R/O is oscillating extracted with root mean square (rms), and Iddq is the leakage when the R/O is not oscillating.

$$Measured\ Ceff = delay/ACReff \quad (7)$$

$$ACReff = VDD/(Idda - Iddq) \quad (8)$$

From this mode, the Ceff contribution by each capacitance component, Cgc, Cov, Cj and Cwire, is extracted as seen in Table 1. The Cov contribution to Ceff is highest with 54.7% and higher than Cgc, 35.3%, mainly due to a higher contribution factor with 9Cov compared to 3Cgc as described in equation (5). The "Miller effect" of Cov is one of the reasons for the higher Cov contribution compared to Cgc's that is not affected by the "Miller effect". The Cov contribution to Ceff is well validated using Si by ΔCov/Cov(%) vs ΔCeff/Ceff(%) correlation as shown in Fig. 8, which shows quite good matching with 55.0% to the calculated contribution, 54.7%. Fig. 10 shows ΔCgg/Cgg(%) vs ΔCeff/Ceff(%) correlation, where the Cgg (Cgc+Cov) contribution to Ceff is 90.0% by the correlation equation. Cgc is 35.0% (=90.0% by Cgg-55.0% by Cov) proving perfect matching to the calculated Cgc contribution ratio, 35.3%. Cwire is also not small with 9.4% of Ceff in 14nm FinFET technology. The impact of Cwire becomes bigger as the feature size decreases, as it is well known. [3]

C. Cov decomposition

Cov is composed of Cmol, Cft, Cfb and Cdo of gate to S/D intrinsic overlap capacitance as described in Fig. 11, 12. From the gate height (GH) modulated by Chemical mechanical polishing (CMP) process vs Cov correlation, Cmol+Cft and Cdo+Cfb can be decomposed, because CMP can modulate the GH on Fin top only without affecting GH on Fin side wall as highlighted in Fig. 12. So, when the GH is 0nm on Fin top, the Cov is contributed by only Cfb and Cdo only as seen in equation (9).

$$Cov = (Cft + Cmol)*GH\ scaling + Cfb + Cdo \quad (9)$$

This concept is well explained in Fig. 13 and the actual correlation is plotted using Si data as shown in Fig. 14. Yint, Cfb+Cdo is confirmed to be 70.8% of total Cov in the plot and Cdo is 46.8% by subtracting TCAD simulated Cfb, 24.0% and the remaining is Cmol+Cft with 29.2% of the slope affected by GH scaling in Fig. 14. Cft is just 0.6% on Technology Computer-aided Design (TCAD) simulation, and the left capacitance component, Cmol, is 28.6% as seen in Fig.15. Gate to S/D intrinsic overlap capacitance is major contributor to total Cov, followed by Cmol and Cfb as shown in Fig. 14. Stringent control of gate to S/D overlap is required by tight process control such as EPI cavity depth, EPI cavity proximity, gate spacer thickness or thermal budget affecting the gate to S/D overlap influencing Cdo. For Cmol control, gate low-k spacer thickness or contact (CONT) critical dimension (CD) or gate height is a critical parameter to control. Cfb should be sensitive to gate spacer thickness or S/D EPI volume.

Cov can be reduced by 2.9% with 10% GH reduction by lowering the Cmol component, which has no side effect on DC performance, as shown in Fig. 14. By sizing down the CONT CD by 12%, the average Cov of NFET and PFET is reduced by 2.9%, lowering the Ceff by 1.5% as seen in Figs. 16 and 17, which results in R/O performance improvement by1.5% as seen in Fig. 18.

III. Conclusion

Fan-out 3 CMOS inverter chain R/O's effective capacitance quantitative model is successfully extracted and well validated by perfect matching to the Si measured Ceff in 14nm technology using FinFET RMG process. Using this proven model, R/O's Ceff can be obtained from the unit device level capacitances with high accuracy. Fan-out 1 or 3 R/O inverter chain's Ceff can be predicted or assessed with high exactness when design parameters, such as gate length or nfin, change or process changes using this model based on known process sensitivity to each capacitance component. Finally, this model is quite useful along all the technologies for the deep dive analysis to find root causes affecting the final R/O's Ceff or estimate the final R/O's speed when combined with DC performance, Ieff, without R/O testing.

Acknowledgement

The authors gratefully acknowledge the contributions of the Globalfoundries development and operations in Fab-8 to process the wafers and data analysis.

References

[1] K. P. Pradhan; S. K. Saha; P. K. Sahu; Priyanka, "Impact of Fin Height and Fin Angle Variation on the Performance Matrix of Hybrid FinFETs" IEEE Trans. on Elec. Devices, pp. 52-57, 2017

[2] E. Y. Jeong; J. S. Yoon; C. K. Baek; Y. R. Kim; J. H. Hong; J. S. Lee; R. H. Baek; Y. .H. Jeong, "Investigation of RC parasitics considering middle-of-the-line in si-bulk FinFETs for Sub-14nm node logic applications", IEEE Trans. on Elec. Devices, pp. 3441-3444, 2015

[3] I. Ciofi; A. Contino; P. J. Roussel; R. Baert; V. H. Vega-Gonzalez, "Impact of Wire Geometry on Interconnect RC and Circuit Delay" IEEE Trans. on Elec. Devices, pp. 2488 – 2496, 2016

[4] W. Li; H. Liu; S. Wang; S. Chen, "Reduced Miller Capacitance in U-Shaped Channel Tunneling FET by Introducing Heterogeneous Gate Dielectric" IEEE Elec. Device Letters, pp. 403-406, 2017

2018 INTERNATIONAL CONFERENCE ON MICROELECTRONIC TEST STRUCTURES, MARCH 19-22, 2018, AUSTIN, TEXAS, USA

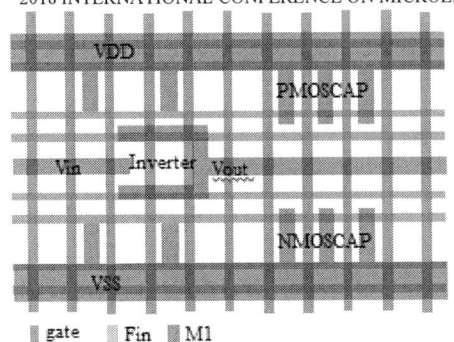

Fig. 1. Fan-out 3 (F/O-3) CMOS inverter chain R/O's layout in 14nm technology using FinFET with RMG process, gate length= 14nm, nfin= 4

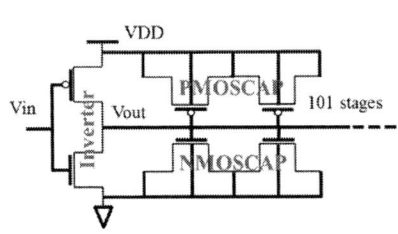

Fig. 2. Fan-out 3 (F/O-3) CMOS inverter chain R/O's schematics

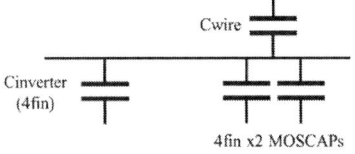

Fig. 3. Fan-out 3 (F/O-3) CMOS inverter chain R/O's schematics with capacitor components

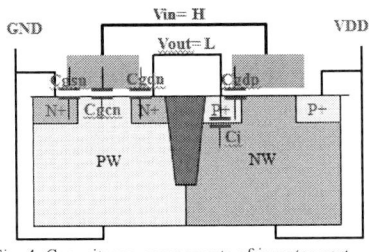

Fig. 4. Capacitance components of inverter part when Vin= H. Total Cinverter= Cgcn+Cov (Cgsn+Cgdn) of NFET, Cov (Cgdp) of PFET and PFET Cj (P+/NW)

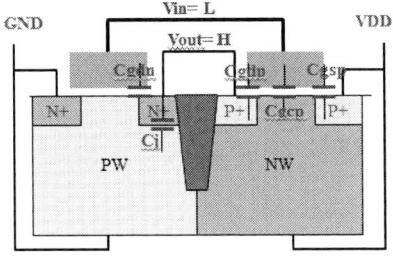

Fig. 5. Capacitance components of inverter part when Vin= L. Total Cinverter= Cgcp+Cov (Cgsp+Cgdp) of PFET, Cov (Cgdn) of NFET, and NFET Cj (N+/PW)

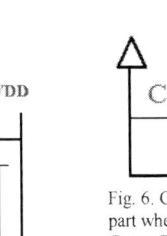

Fig. 6. Capacitance components of MOSACP part when Vout= H. Total fan out MOSCAP= Cgcn+Cov (Cgsn+Cgdn) of NFET. No Cj components involved

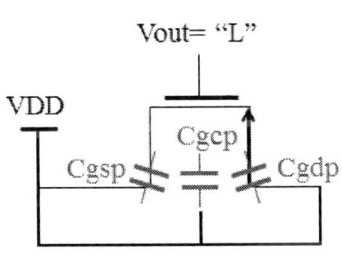

Fig. 7. Capacitance components of MOSACP part when Vout= L. Total fan out MOSCAP= Cgcp+Cov (Cgsp+Cgdp) of PFET. No Cj components involved

	Cgc	35.3%	
Ceff	Cov	54.7%	100.0%
	Cj	0.6%	
	Cwire	9.4%	

Table 1. Ceff contribution by each capacitance component in Fan-out 3 inverter chain R/O

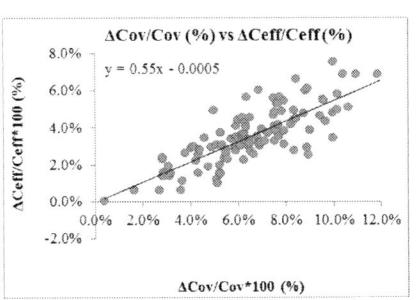

Fig. 8. Cov change ratio vs. Ceff change ratio correlation at Si. Cov contribution to Ceff is 55%

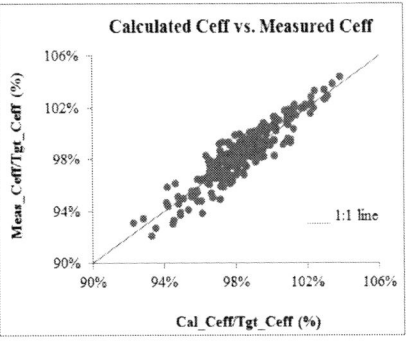

Fig. 9. Calculated Ceff/target Ceff vs. Si measured Ceff/target Ceff to see matching for fan-out 3 inverter chain R/O

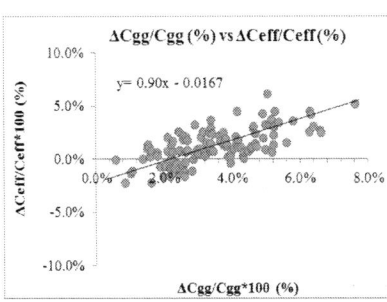

Fig. 10. Cgg (=Cgc+2*Cov) change ratio vs. Ceff change ratio correlation at Si. Cgg contribution to Ceff is 90%

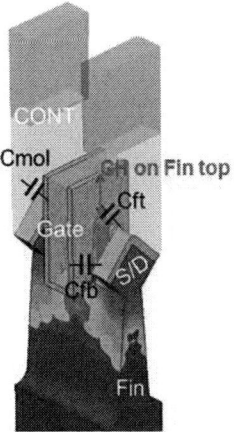

Fig. 11. TCAD simulation profile for Cmol (gate to CONT capacitance), Cft (capacitance by S/D to gate on Fin top), Cfb (capacitance by S/D to gate on Fin side wall)

978-1-5386-5072-1/18 $31.00 © 2018 IEEE   155

2018 INTERNATIONAL CONFERENCE ON MICROELECTRONIC TEST STRUCTURES, MARCH 19-22, 2018, AUSTIN, TEXAS, USA

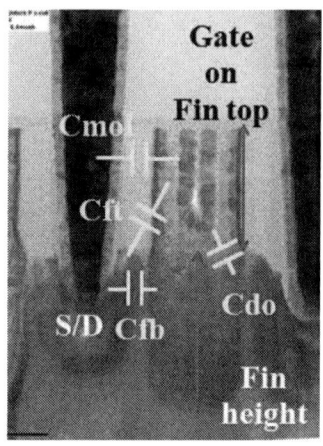

Fig. 12. TEM profile for Cmol (gate to CONT capacitance), Cft (capacitance by S/D to gate on Fin top), Cfb (capacitance by S/D to gate on Fin side wall)

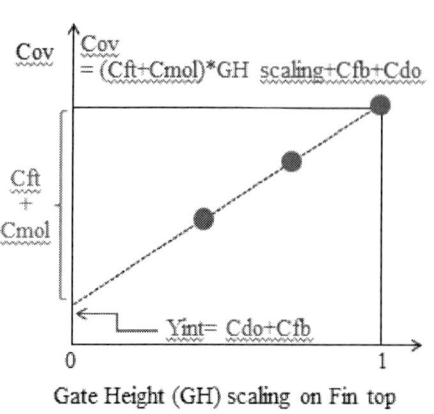

Fig. 13. Gate height scaling on Fin top vs Cov correlation to distinguish Cdo+Cfb and Cmol+Cft from total Cov

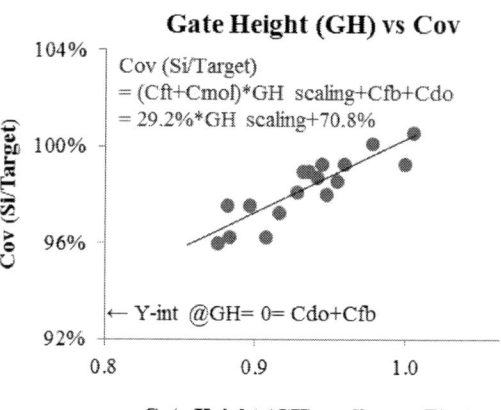

Fig. 14. Gate height (GH) scaling vs Cov correlation at Si. Yint is Cdo+Cfb. Cft+Cmol: 29.2% of Cov, Cdo+Cfb: 70.8% of Cov

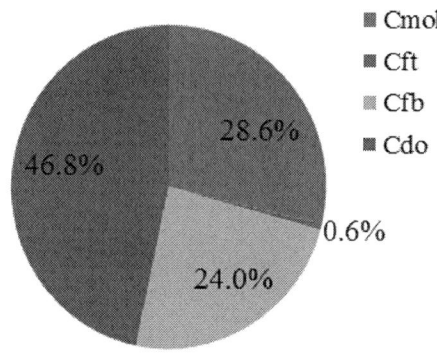

Fig. 15. Cov contribution ratio by each capacitance component at fan-out 3 R/O inverter chain

						100.0%
Ceff	Cov	Cgc		35.3%		
		Cf	Cmol	15.6%	54.7%	
			Cft	0.3%		
			Cfb	13.2%		
		Cdo		25.6%		
	Cj			0.6%		
	Cwire			9.40%		

Table 2. Fan-out 3 R/O inverter chain's total capacitance, Ceff, contribution ratio by each capacitance component

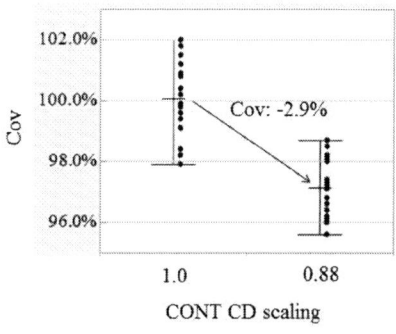

Fig. 16. Cov (average of NFET & PFET) reduction by CONT CD sizing down by 12%

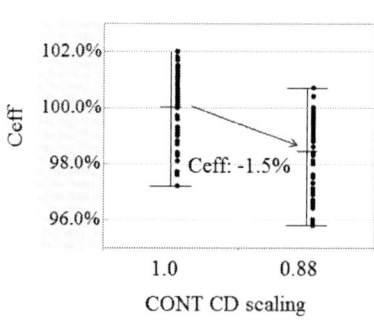

Fig. 17. Fan-out 3 R/O inverter chain's Ceff reduction by CONT CD sizing down by 12%

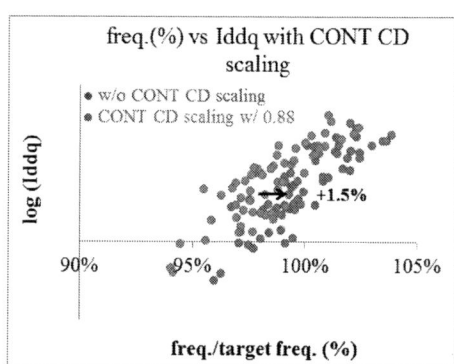

Fig. 18. Fan-out 3 R/O inverter chain's performance improvement by CONT CD sizing down by 12%

978-1-5386-5072-1/18 $31.00 © 2018 IEEE

2018 INTERNATIONAL CONFERENCE ON MICROELECTRONIC TEST STRUCTURES, MARCH 19-22, 2018, AUSTIN, TEXAS, USA

Measurement of IGBT trench MOS-gated region characteristics using short turn-around-time MOSFET test structures

Kiyoshi Takeuchi, Munetoshi Fukui, Takuya Saraya,
Kazuo Itou, Shinichi Suzuki, Toshihiko Takakura, and Toshiro Hiramoto
Institute of Industrial Science
the University of Tokyo
Tokyo, Japan
takeuchi@nano.iis.u-tokyo.ac.jp

Abstract— **Trench MOSFET test structures were fabricated for evaluating IGBT MOS-gated region performance. It was found that the test structures can be used for measuring saturation and sub-threshold current, though accurate estimation of linear resistance is difficult. Charge pumping measurement can be used to evaluate the oxide/substrate interface quality, for possible application to process optimization.**

Keywords—test structure; IGBT; trench MOSFET

I. INTRODUCTION

Today, Silicon Insulated Gate Bipolar Transistors (Si-IGBTs) are widely used in the market, and their improvement is still explored [1]. A modern IGBT includes vertical trench MOSFETs as electron emitters. Since they are one of the key building blocks of an IGBT, it is desirable that their performance can be separately measured and optimized. In this work, trench MOSFET test structures fabricated by a short turn-around-time flow, which omits many IGBT process steps that are not relevant to the MOSFET portion, are evaluated using a standard probe station and parametric tester. The effectiveness and limitations of the test structures will be discussed.

II. TEST STRUCTURE

Fig.1 shows the test structure used in this work. It is fabricated on an n-type substrate, whose thickness is 0.38mm. It comprises nine pairs of trench gates, and the central pair forms the target n-channel FET (called x1 FET hereafter). The rest of the pairs, four pairs on each side of the x1 FET, are dummy trenches, which also form an additional dummy FET (x8 FET) by parallel connection. To mimic a real device, the pitch of the pairs is sparse (16μm). This design is intended to control the hole sinking efficiency at the front side of an IGBT [2]. While the source electrodes Vs_x1 and Vs_x8 are separate, a single gate electrode (Vg) is shared by the x1 and x8 FETs. The N-substrate, which corresponds to the voltage supporting region of an IGBT, forms a lightly doped drain, which is also shared by the x1 and x8 FETs. The n+ drain electrode (Vd) is formed on the top surface to eliminate any back side processes. Since each pair of trenches are tied at the ends as shown in the top view schematic, the p-doped FET channel regions (p-body)

surrounded by the trench pairs are electrically isolated from the one outside the trench pairs (named p-float, but normally grounded for I-V measurements).

III. DRAIN CURRENT MEASUREMENT

To measure the I-V curves of the x1 FET, there are two possible ways; connecting both Vs_x1 and Vs_x8 to ground, or connecting only Vs_x1 to ground, leaving Vs_x8 floating. Fig.2a shows I_{DS}-V_{DS} curves for both the cases. Note that, since p-body and source are tied together, and gate current is negligible, $I_D=I_S$. If Vs_x8 is grounded, x1 I_{DS} does not saturate even at V_{DS}=10V, if V_{GS} is 15V or higher. However, if x8 is floated, I_{DS} rises much more steeply and fully saturates. If the x8 FET is connected, larger current flowing through the lightly doped n region causes significant voltage drop, lowering the intrinsic drain voltage of the x1 FET, hindering saturation. However, by floating Vs_x8, it was confirmed that the saturation drain current of the x1 FET can be stably measured. Fig.2b shows I_{DS}-V_{GS} curves of the same transistor. I_{DS} with and without Vs_x8 connection coincide, as long as V_{GS} is low enough. This shows that sub-threshold current can be also measured without interference of the n-substrate resistance.

During the on-state of a real IGBT, high carrier injection reduces the n region resistance, and the MOSFET normally operates in the linear region. Therefore, it is desirable that the linear resistance excluding the n region access resistance can be also estimated. To obtain information about the resistive voltage drop in the n region, the floating Vs_x8 can be used as a voltage sense terminal. Fig.3 shows floating Vs_x8 voltage vs V_{DS}, which was measured simultaneously with Fig.1. It can be seen that the Vs_x8 voltage is lower than V_{DS} by up to 1V or more, and the difference increases as I_{DS} increases, which is as expected. The inflections of the curves for V_{GS}=3V and 6V are caused by the turn off of the x8 FETs, which electrically disconnects Vs_x8 from the n region. Fig.4 shows I_{DS}-V_{DS} curves, where Vs_x8 voltage is used as V_{DS}, instead of the raw measured values. As can be seen, this compensation removes a significant portion of the n region voltage drop. However, since the target FET and dummy FET are still distant, it is also believed that non-negligible portion of the access voltage drop would still remain. To further remove the voltage drop to

978-1-5386-5072-1/18 $31.00 © 2018 IEEE   157

obtain a credible linear resistance value, improvement of the sense terminal design, combined with detailed TCAD analyses would be necessary.

Next, another approach for estimating the linear resistance is discussed. Since the saturation current (well defined, no short channel effect) is measurable, device parameters relevant to the linear drain current (I_{DLIN}) could be derived from the measured saturation current (I_{DSAT}) using some model. In the long channel limit,

$$I_{DSAT} = \frac{W\mu C_{OX}(V_{GS} - V_{TH})^2}{2mL} \quad (1)$$

and

$$I_{DLIN} = \frac{W\mu C_{OX}(V_G - V_{TH})V_{DS}}{L}. \quad (2)$$

If these two equations hold, for example, it would be possible to estimate I_{DLIN} from I_{DSAT}, only assuming the body effect parameter m. Unfortunately, it was found that this is not the case, in spite of the relatively long channel (3μm) used. As can be seen from Fig.2, saturated I_{DS} vs V_{GS} relationship is almost linear, suggesting strong velocity saturation. To confirm this, measured I_{DS} vs V_{GS} curves were fitted to a simple model. It is reported in [3] that the dependence of I_{DSAT} on channel length L and gate overdrive $V_{GT}=V_{GS}-V_{TH}$ is empirically given by

$$\frac{V_{GT}}{I_{DSAT}} = \frac{2m(L - L_{SAT})}{W\mu C_{OX}V_{GT}} + \frac{1}{WC_{OX}v_{SAT}}, \quad (3)$$

where L_{SAT} is an effective velocity-saturated (or pinch-off) region length, and v_{SAT} an effective saturation velocity. L_{SAT} and v_{SAT} can be extracted as constants by plotting V_{GT}/I_{DSAT} vs L straight lines for multiple V_{GT} values, and determining a point where the lines converge [3]. However, here, L dependence information is not available. Therefore, instead, V_{GT}/I_{DSAT} was plotted against $m/\mu V_{GT}$, as shown in Fig.5. To do so, it was assumed that the effective mobility μ equals the "universal mobility" [4], and m was estimated from the measured V_{TH} and known gate oxide thickness (100nm). Good linear relationship corresponding to Eq. (3) was obtained by slightly adjusting the V_{TH} value. From the slope and intercept, v_{SAT} and L-$L_{SAT} \equiv L_C$ can be estimated to be 7.3×10^4m/s and 0.76μm, respectively. The results suggest that the carrier reaches a high velocity v_{SAT} at only 0.76μm from the source, which is much shorter than $L=3$μm. In this situation, correlation between the linear and saturation drain current becomes unclear. Therefore, it was concluded that linear resistance estimation by simply using a drain current model is also difficult.

IV. CHARGE PUMPING MEASUREMENT

In addition to the I-V measurements, charge pumping (CP) measurements [5] were performed to evaluate the gate oxide quality, for possible application to process optimization. Since the source and p-body of the test structure are tied together, which is inherited from the original IGBT design, CP of the target MOSFETs is not possible. However, CP is still possible by using parasitic p-channel FETs formed between Vf and Vs, as shown in Fig.6a. Typical I_{DS}-V_{GS} curves for the parasitic pFET are shown in Fig.6b. Fig.7 shows CP current (I_{CP}) vs V_{INV}

characteristics for various frequencies, where V_{INV} is the low (inversion) side pulse level. The high (accumulation) side pulse level V_{BASE} is kept constant. V_{INV} at which I_{CP} starts to rise is consistent with the parasitic pFET threshold voltage seen in Fig.6b. Here, Vs_x1 and Vs_x8 were connected together to measure CP current of both x1 and x8 FETs altogether. Unlike the I-V case, measuring I_{CP} of the x1 or x8 FET alone is not possible, since the gate electrode is shared, and all the trench pFET channels are temporarily turned on by the CP gate pulse signal. Fig.8 shows I_{CP} vs pulse frequency. Linear frequency response was confirmed, as expected. Fig.9 shows I_{CP} vs V_{BASE}, where V_{INV} is fixed at -15V. I_{CP} gradually increases as V_{BASE} is raised. However, when V_{BASE} reached 4V, the measurement became unstable because of the onset of nFET turn-on. This shows that the nFET threshold voltage should be high enough to realize full pFET CP swing between inversion and accumulation, and that fixed V_{BASE} measurements are preferable to avoid complication by nFET turn-on.

I_{CP} thus measured contains contribution from the extra edge portions of the trenches (see top view schematic in Fig.1). To translate I_{CP} into surface state density, this edge component must be subtracted. For this purpose, I_{CP} was measured by varying trench length, as shown in Fig.10. By excluding the edge component, surface state density Nit of the gate oxide in contact with the n substrate was determined to be 4.5×10^9cm^{-2}, using the same pulse condition as Fig.7. This value is lower than the values usually reported for ordinary logic MOSFETs. For example, from the data in [5] for an nFET with similar gate oxide thickness (80nm) and pulse conditions to this work, Nit is calculated to be 1.3×10^{10}cm^{-2}. On the other hand, low Nit values similar to this work are reported for power VDMOS devices [5,6], which were obtained by measuring parasitic gated diodes. What seems to be common with this work and [5,6] is that CP is applied to a non-implanted substrate/oxide interface; i.e. the substrate receives no impurity implantation before the gate oxidation. Therefore, the authors currently consider that the obtained Nit value is valid, and there is a possibility that ion implantation to the substrate increases surface states after oxidation.

V. CONCLUSION

The short turn-around-time trench FET test structure can be used for measuring MOSFET saturation and sub-threshold current, though accurate estimation of linear resistance is not easy. Charge pumping measurements of parasitic FETs can be used to evaluate the oxide/n substrate interface quality.

ACKNOWLEDGMENT

This work was supported by the New Energy and Industrial Technology Development Organization (NEDO).

REFERENCES

[1] K. Kakushima et al., IEDM, p.268, 2016.
[2] M. Kitagawa et al., IEDM, p.679, 1993.
[3] K. Takeuchi et al., IEEE Trans. ED, 41, p.1623, 1994.
[4] S. Takagi et al., IEEE Trans. ED, 41, p.2357, 1994.
[5] G. Groeseneken et al., IEEE Trans. ED, 31, p.42, 1984.
[6] P. M. Igic, Int. Conf. Microelectronics (MIEL), 1, p.255, 1995.
[7] P. Habas et al., IEEE Trans.ED, 43, p.2197, 1996.

2018 INTERNATIONAL CONFERENCE ON MICROELECTRONIC TEST STRUCTURES, MARCH 19-22, 2018, AUSTIN, TEXAS, USA

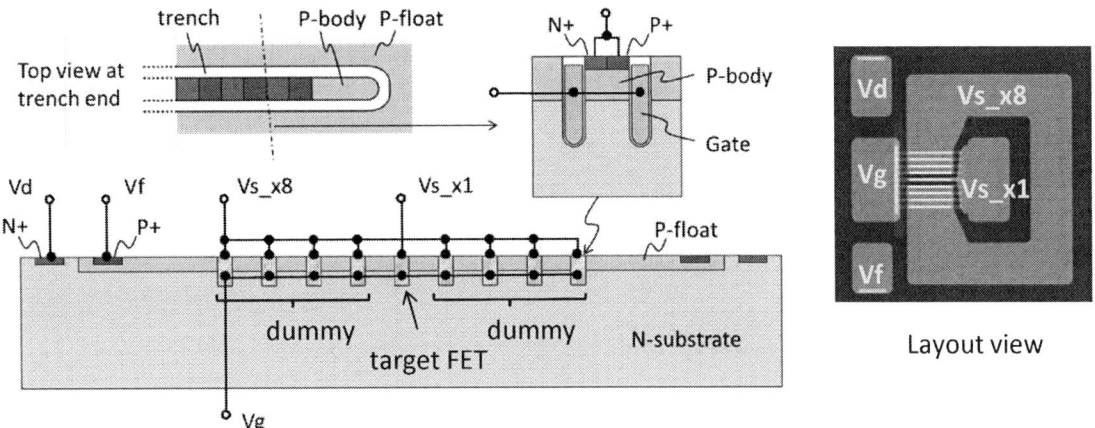

Fig.1 Test structure used for this work. Paired trenches are sparsely arranged to mimic electron emitters for an injection enhanced IGBT [2]. P-body and P-float regions are electrically isolated by trenches.

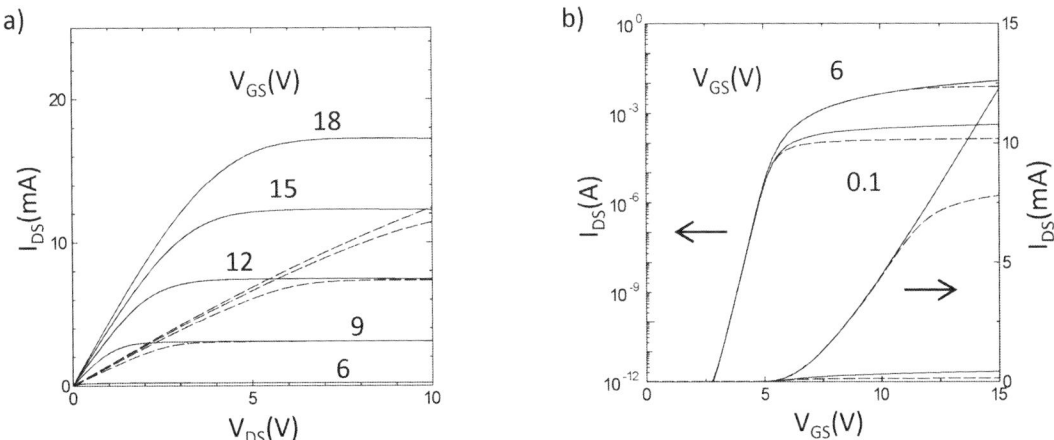

Fig.2 Current-voltage characteristics of x1 FET, with Vs_x8 grounded (dashed line) or floated (solid line). I_{DS} is measured at Vs_x1 terminal.

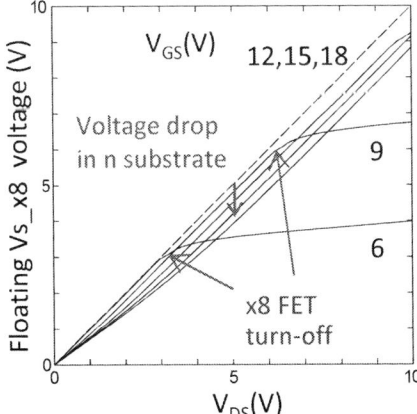

Fig.3 Floating voltage of Vs_x8 terminal. Large voltage drop in n substrate is detected.

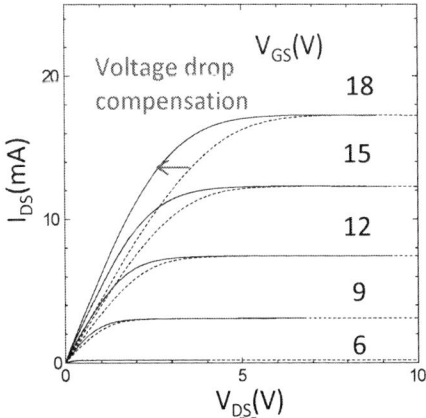

Fig.4 Corrected I-V using Fig.3. Since Vs_x8 is distant from x1 FET, parasitic resistance should still remain.

978-1-5386-5072-1/18 $31.00 © 2018 IEEE   159

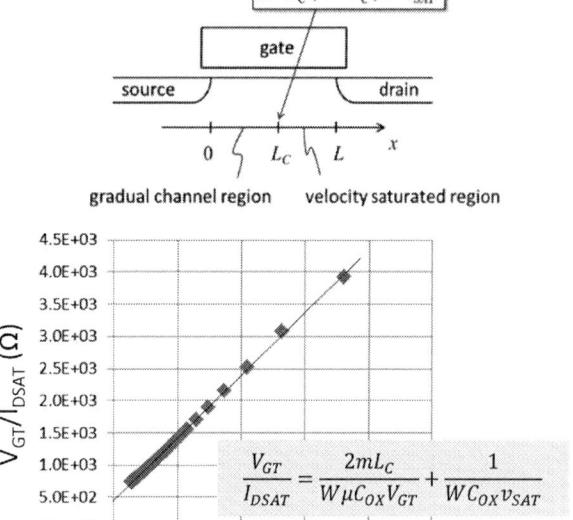

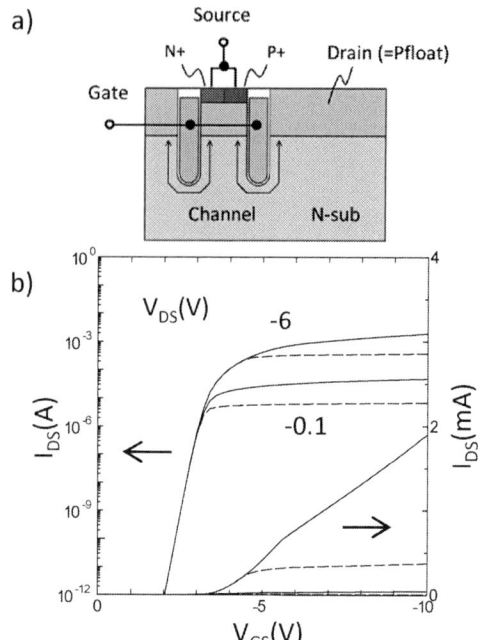

Fig.5 Comparison of measured I-V with simple drain current model. Results suggest that gradual channel length L_C is much smaller than entire channel length L.

Fig.6 a) Parasitic pFET structure and b) its I-V curves. I_{DS} is measured at Vs_x1. Vs_x8 is either grounded (dashed line) or floated (solid line).

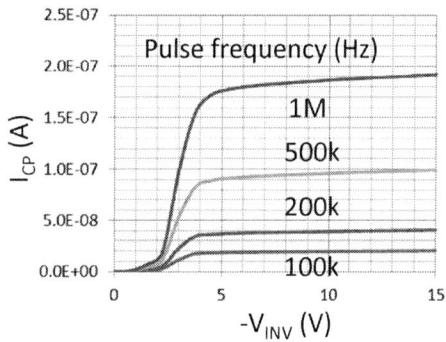

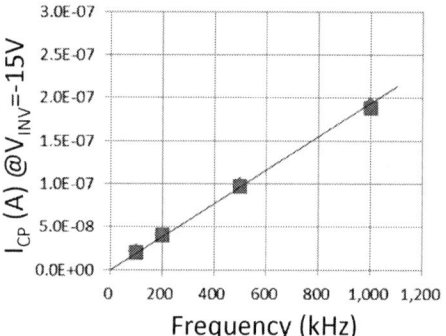

Fig.7 Charge pumping current I_{CP} vs low-side pulse voltage V_{INV}. High-side voltage is fixed at 2.25V.

Fig.8 I_{CP} @V_{INV}=-15V vs frequency. Linear response shows charge pumping current is measured.

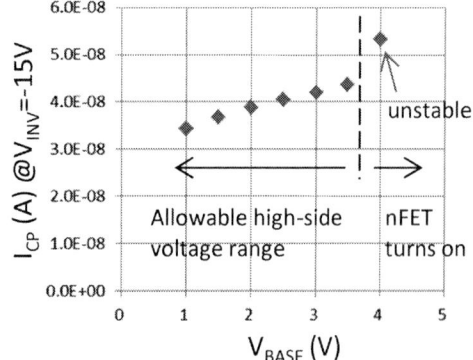

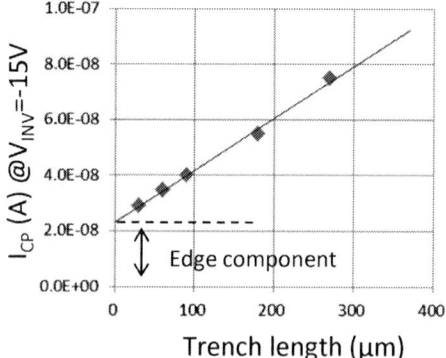

Fig.9 Dependence of I_{CP} on high-side pulse voltage V_{BASE}. To avoid nFET turn-on, fixed V_{BASE} measurements are preferred.

Fig.10 Trench length dependence of I_{CP}. Edge component must be subtracted for surface state density (Nit) extraction.

978-1-5386-5072-1/18 $31.00 © 2018 IEEE

2018 INTERNATIONAL CONFERENCE ON MICROELECTRONIC TEST STRUCTURES, MARCH 19-22, 2018, AUSTIN, TEXAS, USA

Sensitivity of High-k Encapsulated MoS$_2$ Transistors to I-V Measurement Execution Time

Pavel Bolshakov, Ava Khosravi, Peng Zhao,
Robert M. Wallace, and Chadwin D. Young
Department of Materials Science and Engineering,
The University of Texas at Dallas,
800 W Campbell Rd, Richardson, TX 75080, USA
E-mail: chadwin.young@utdallas.edu

Paul K. Hurley

Tyndall National Institute,
University College Cork,
Lee Maltings, Dyke Parade,
Cork, Ireland

Abstract – **High-k encapsulated MoS$_2$ field-effect-transistors were fabricated and electrically characterized. Comparison between HfO$_2$ and Al$_2$O$_3$ encapsulated MoS$_2$ FETs and their I-V response to execution time are shown. Changes in gate voltage step and integration time demonstrate that electrical characterization parameters can significantly impact device parameters such as the subthreshold swing and the threshold voltage.**

I. Introduction

Transition metal dichalcogenides (TMDs) are being investigated for their potential integration in high mobility, fast-switching devices. Significant effort has been devoted to studying the interaction of these 2D materials with contact metals [1]–[3] and high-k dielectric [4]–[6], with progress being made on both fronts. MoS$_2$ has been the most investigated TMD, with multiple studies on fabricated transistors and optoelectronic devices [7]–[9]. However,

there is a dearth of information regarding the electrical measurement parameters used in characterizing MoS$_2$ transistors and their sensitivity to such factors as voltage step and measurement speed, which can affect execution time. Using photolithography in conjunction with a functionalization treatment [10], [11], we fabricated MoS$_2$ transistors encapsulated in a high-k dielectric [12], [13] and studied the effects of gate voltage step and measurement speed on the transfer curve of the transistor. The results suggest that changes in electrical characterization parameters can significantly affect the subthreshold slope (SS) and the threshold voltage (V$_T$).

II. Fabrication and Characterization

Using a highly doped p-Si wafer, 27 nm of Al$_2$O$_3$ was deposited using atomic layer deposition (ALD) at 250 °C. Afterwards, the substrate was subjected to a 400 °C forming gas (5%H$_2$/95%N$_2$) anneal for 1 hour [14], [15] prior to any

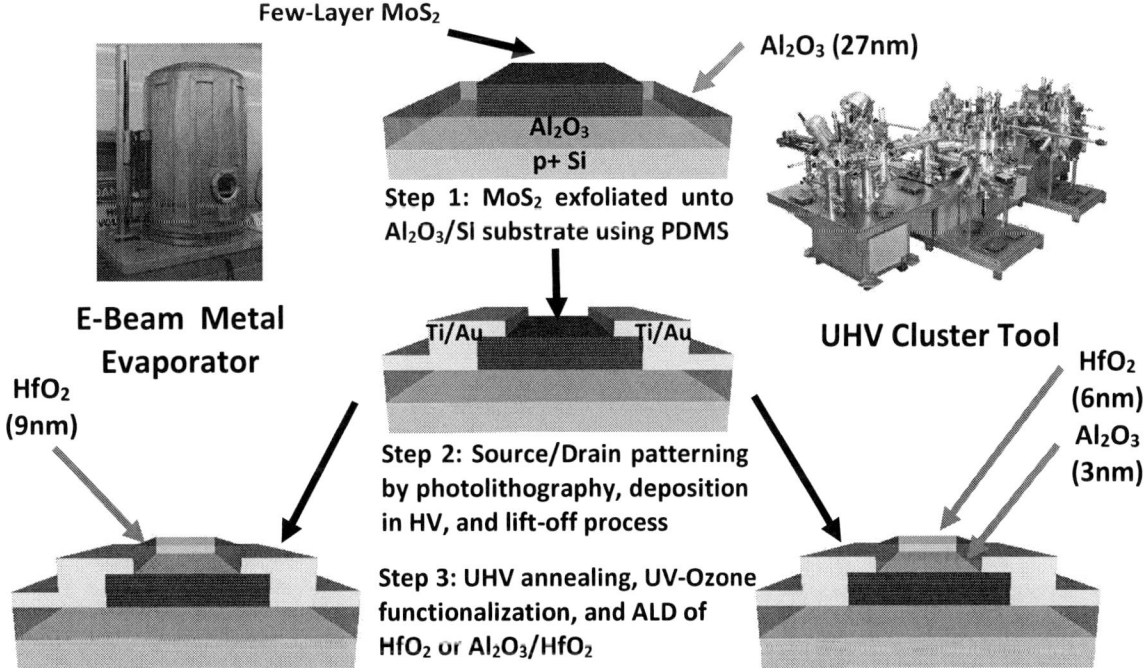

Fig. 1. Fabrication flow of high-k encapsulated MoS$_2$ FETs using photolithography with a lift-off process. MoS$_2$ functionalization and high-k deposition were done using the UHV cluster tool [10], [11].

978-1-5386-5072-1/18 $31.00 © 2018 IEEE

device fabrication. Using a PDMS Gel-Tape [16], natural MoS_2 flakes were exfoliated onto the Al_2O_3/Si substrate and then source/drain Ti/Au (10/150nm, 10^{-6} mbar) were deposited using photolithography with a lift-off process. Afterwards, the devices were subjected to a 300 °C ultra-high vacuum (UHV) anneal for 2 hours followed by a 15 minute in-situ UV-ozone treatment with subsequent high-k deposition of either HfO_2 (9nm) or Al_2O_3(3nm)/HfO_2(6nm) using ALD at 200 °C (Fig. 1). The electrical characterization was done using a Keithley 4200 semiconductor characterization system (SCS) in conjunction with a Cascade probe station at room temperature. For this study, a "medium" integration time (Normal speed mode) refers to the default speed on the Kiethley 4200 which has a delay factor of 1 and a filter factor of 1, and long integration time (Quiet speed mode) refers to the slower speed which has a delay factor of 1.3 and a filter factor of 3. The delay factor allows longer settling times for low current measurements where the higher the number, the longer the measurement time. The filter factor decreases the measurement noise by averaging multiple readings where, again, the higher the number, the longer the measurement time.

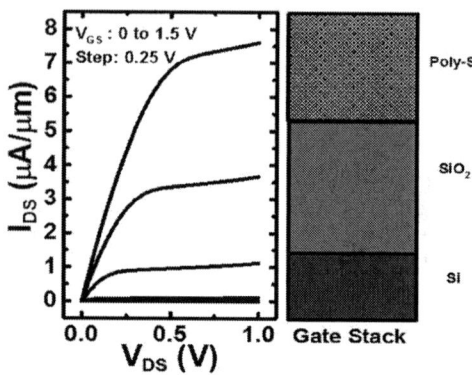

Fig. 3. The I_{DS}-V_{DS} and gate stack cross-section of the control Si transistor showing no issues with the contacts. Ohmic contacts are observed.

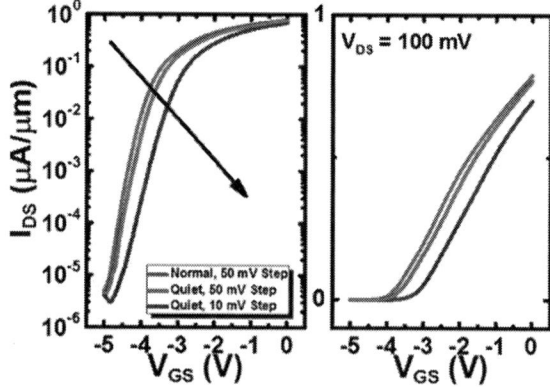

Fig. 4. The I_{DS}-V_{GS} of the Al_2O_3/MoS_2/HfO_2 demonstrating a degradation of the SS and a positive V_T shift because of increasing the gate voltage step size and measurement speed, suggesting charge trapping.

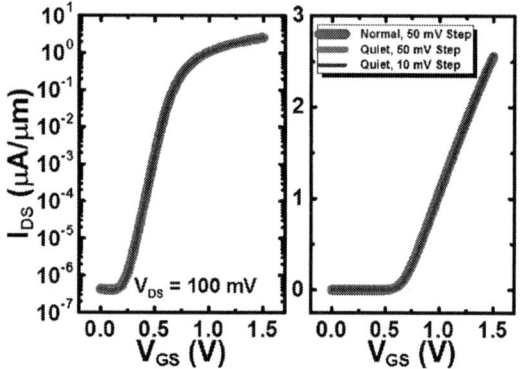

Fig. 2. The I_{DS}-V_{GS} of the control Si transistor used as a standard for calibrating the measurement setup on the Kiethley 4200.

III. Results and Discussion

A standard Si NMOS transistor, with dimensions on the same order of magnitude as the MoS_2 transistors, was measured as a reference control standard. In Fig. 2 and Fig. 3, the transfer curves and output characteristics of the control Si transistor are shown, with the 3 different measurement schemes overlaid on top of each other. The I_{DS}-V_{GS} curves in Fig. 2 do not show any major changes in the characteristic shape when changing the gate voltage step or changing the speed of the measurement, thereby showing no problems with the device or instrumentation. In Fig. 4, the I_{DS}-V_{GS} of a MoS_2 transistor with Al_2O_3/MoS_2/HfO_2 (Fig. 5) gate stack are shown to be affected by both the gate voltage step and the speed of the measurement.

The transfer curves in Fig. 4 show that changing the speed of the measurement from "Normal" to "Quiet," where Quiet mode has approximately 3x as long measurement time as Normal mode, and the gate voltage step from 50 mV to 10 mV, degrades the SS and causes a V_T shift in the positive direction. This suggests that the charge traps present as the MoS_2/HfO_2 interface are highly sensitive to measurement parameters and can negatively impact the device performance. The execution time increases from 23 sec for Normal, 50 mV step, to 69 sec for Quiet, 50 mV step, to a staggering 529 sec for Quiet, 10 mV step, corresponding to the magnitude of the positive threshold voltage shift (ΔV_T) in the transfer curve. This suggests that longer integration time enables significant charge trapping during electrical characterization. Charge relaxation/de-trapping does occur when left unbiased and then re-measured again several hours

978-1-5386-5072-1/18 $31.00 © 2018 IEEE

2018 INTERNATIONAL CONFERENCE ON MICROELECTRONIC TEST STRUCTURES, MARCH 19-22, 2018, AUSTIN, TEXAS, USA

later (observed in the original Normal, 50 mV step transfer curve in Fig. 4).

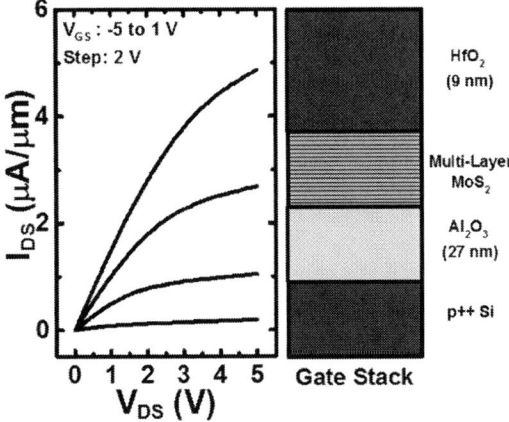

Fig. 5. The I_{DS}-V_{DS} and gate stack cross section schematic that shows the encapsulation of the MoS_2 channel in a high-k environment, isolating it from ambient. Ohmic-like contacts are observed.

Inserting an Al_2O_3 layer at the top-side interface between MoS_2 and HfO_2 results in a more positive V_T and an improved SS as shown in Fig. 6, suggesting a reduction in positively charged impurities at the top MoS_2/Al_2O_3 interface. Similar effects have been observed in InGaAs [17], [18] and Ge systems comparing HfO_2 to a bilayer dielectric (Al_2O_3/HfO_2) [19]. Furthermore, I_{DS}-V_{GS} measurements from Normal, 50 mV step scheme to Quiet, 10 mV step scheme show a steeper SS and a negative V_T shift – completely opposite to the $Al_2O_3/MoS_2/HfO_2$ gate stack. This suggests that a vastly different interface is formed at the top surface of the $Al_2O_3/MoS_2/Al_2O_3/HfO_2$ (Fig.7) gate stack when compared to the $Al_2O_3/MoS_2/HfO_2$ (Fig. 5) gate stack.

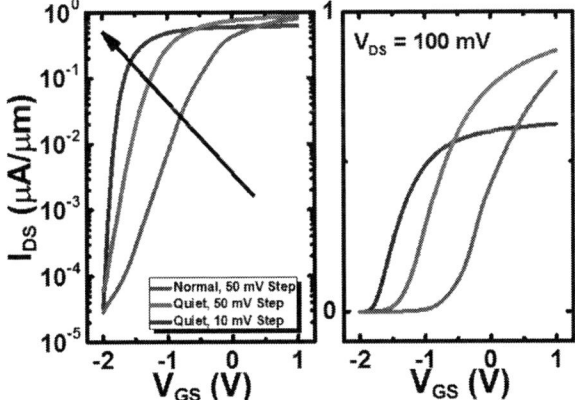

Fig. 6. The I_{DS}-V_{GS} of the $Al_2O_3/MoS_2/Al_2O_3/HfO_2$ demonstrating a steepening of the SS and a negative V_T shift because of increasing the gate voltage step size and measurement speed, suggesting a high-k screening layer is formed.

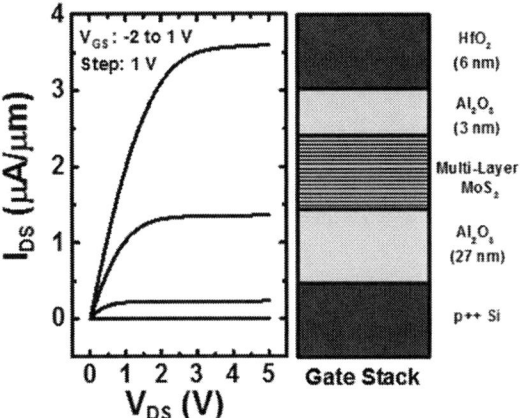

Fig. 7. The I_{DS}-V_{DS} and gate stack cross section schematic that shows the insertion of a Al_2O_3 layer between MoS_2 and HfO_2, isolating it from ambient. Ohmic-like contacts are observed.

Both are highly sensitive to the gate voltage step and test speed. The drastic change in SS from Normal, 50 mV step scheme to the Quiet, 10 mV step scheme in Fig. 8 may be attributed to a screening layer that is formed because of negative charge build-up (Fig. 9) at the top MoS_2/Al_2O_3 interface at longer execution times, allowing for a more uninhibited electron transport deeper in the MoS_2 channel. Further voltage ramp stress (VRS) and constant voltage stress (CVS) studies are needed to fully comprehend the high-k reliability with MoS_2 transistors in a high-k environment.

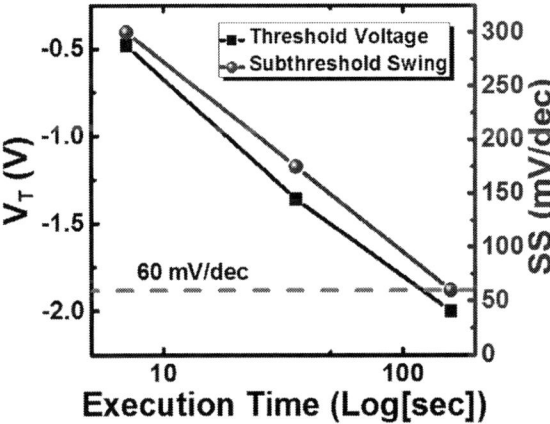

Fig. 8. The relationship between V_T and SS in terms of execution time showing logarithmic relationship as well as a near-ideal SS of ~60 mV/dec for the MoS_2 FET with the Al_2O_3/HfO_2 bilayer.

978-1-5386-5072-1/18 $31.00 © 2018 IEEE   163

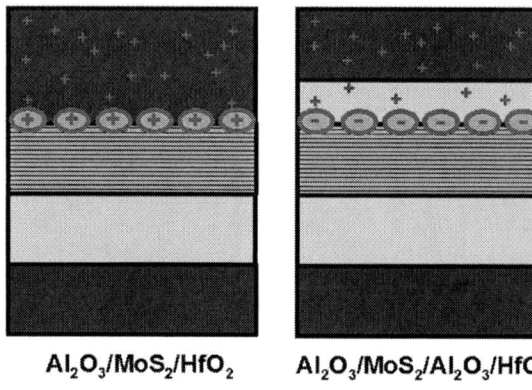

Al₂O₃/MoS₂/HfO₂
Gate Stack **Al₂O₃/MoS₂/Al₂O₃/HfO₂**
Gate Stack

Fig. 9. Cross-sectional diagrams depicting positive oxide charges present throughout the bulk of the top high-k dielectrics with positive charge at the MoS_2/HfO_2 interface or negative charge at the MoS_2/Al_2O_3 interface.

IV. Conclusion

MoS_2 FETs encapsulated in a high-k dielectric were fabricated in order to study the effects of increasing electrical measurement execution time on device characteristics such as SS and V_T by changing the gate voltage step size and measurement speed. It has been demonstrated that changes to the aforementioned parameters can either change the I-V response as a result of charge trapping and high-k screening. This work suggests that great care must be taken in the electrical characterization of 2D materials devices due to their high sensitivity to measurement parameters, and further stress studies are needed if these materials are to be truly integrated into reliable device structures.

Acknowledgment – This work is partially funded by the National Science Foundation (NSF) award 1407765.

References

[1]   C. D. English, G. Shine, V. E. Dorgan, K. C. Saraswat, and E. Pop, "Improved contacts to MoS_2 transistors by ultra-high vacuum metal deposition," *Nano Lett.*, vol. 16, no. 6, pp. 3824–3830, 2016.

[2]   C. M. Smyth, R. Addou, S. McDonnell, C. L. Hinkle, and R. M. Wallace, "Contact metal-MoS_2 interfacial reactions and potential implications on MoS_2-based device performance," *J. Phys. Chem. C*, vol. 120, no. 27, pp. 14719–14729, Jul. 2016.

[3]   Q. Qian *et al.*, "Improved Gate Dielectric Deposition and Enhanced Electrical Stability for Single-Layer MoS_2 MOSFET with an AlN Interfacial Layer," *Sci. Rep.*, vol. 6, pp. 1–4, Mar. 2016.

[4]   V. V. Afanas'ev *et al.*, "Band alignment at interfaces of few-monolayer MoS_2 with SiO_2 and HfO_2," *Microelectron. Eng.*, vol. 147, pp. 294–297, 2015.

[5]   C. D. Young *et al.*, "Evaluation of Few-Layer MoS_2 Transistors with a Top Gate and HfO_2 Dielectric,"

ECS Trans., vol. 75, no. 5, pp. 153–162, Sep. 2016.

[6]   P. Zhao *et al.*, "Probing Interface Defects in Top-Gated MoS_2 Transistors with Impedance Spectroscopy," *ACS Appl. Mater. Interfaces*, vol. 9, no. 28, pp. 24348–24356, 2017.

[7]   A. Pospischil and T. Mueller, "Optoelectronic Devices Based on Atomically Thin Transition Metal Dichalcogenides," *Appl. Sci.*, vol. 6, no. 3, p. 78, Mar. 2016.

[8]   D. Jariwala, V. K. Sangwan, L. J. Lauhon, T. J. Marks, and M. C. Hersam, "Emerging device applications for semiconducting two-dimensional transition metal dichalcogenides," *ACS Nano*, vol. 8, no. 2. pp. 1102–1120, 2014.

[9]   Q. H. Wang, K. Kalantar-Zadeh, A. Kis, J. N. Coleman, and M. S. Strano, "Electronics and optoelectronics of two-dimensional transition metal dichalcogenides," *Nat. Nanotechnol.*, vol. 7, no. 11, pp. 699–712, 2012.

[10] A. Azcatl *et al.*, "MoS_2 functionalization for ultra-thin atomic layer deposited dielectrics," *Appl. Phys. Lett.*, vol. 104, no. 11, p. 111601, 2014.

[11] A. Azcatl *et al.*, "HfO_2 on UV-O_3 exposed transition metal dichalcogenides: Interfacial reactions study," *2D Mater.*, vol. 2, no. 1, p. 14004, Jan. 2015.

[12] T. Li, B. Wan, G. Du, B. Zhang, and Z. Zeng, "Electrical performance of multilayer MoS_2 transistors on high-κ Al_2O_3 coated Si substrates," *AIP Adv.*, vol. 5, no. 5, p. 57102, 2015.

[13] P. Bolshakov, P. Zhao, A. Azcatl, P. K. Hurley, R. M. Wallace, and C. D. Young, "Electrical characterization of top-gated molybdenum disulfide field-effect-transistors with high-k dielectrics," *Microelectron. Eng.*, vol. 178, pp. 190–193, 2017.

[14] P. Bolshakov, P. Zhao, A. Azcatl, P. K. Hurley, R. M. Wallace, and C. D. Young, "Improvement in top-gate MoS_2 transistor performance due to high quality backside Al_2O_3 layer," *Appl. Phys. Lett.*, vol. 111, no. 3, p. 32110, Jul. 2017.

[15] P. Zhao *et al.*, "Effects of annealing on top-gated MoS_2 transistors with HfO_2 dielectric," *J. Vac. Sci. Technol. B, Nanotechnol. Microelectron. Mater. Process. Meas. Phenom.*, vol. 35, no. 1, p. 01A118, Jan. 2017.

[16] A. Castellanos-Gomez *et al.*, "Deterministic transfer of two-dimensional materials by all-dry viscoelastic stamping," *2D Mater.*, vol. 1, no. 1, p. 11002, 2014.

[17] P. K. Hurley *et al.*, "The characterization and passivation of fixed oxide charges and interface states in the Al_2O_3/InGaAs MOS system," *IEEE Trans. Device Mater. Reliab.*, vol. 13, no. 4, pp. 429–443, Dec. 2013.

[18] B. Shin *et al.*, "Origin and passivation of fixed charge in atomic layer deposited aluminum oxide gate insulators on chemically treated InGaAs substrates," *Appl. Phys. Lett.*, vol. 96, no. 15, p. 152908, Apr. 2010.

978-1-5386-5072-1/18 $31.00 © 2018 IEEE

[19] X.-F. Li, Y.-Q. Cao, A.-D. Li, H. Li, and D. Wu, "HfO_2/Al_2O_3/Ge Gate Stacks with Small Capacitance Equivalent Thickness and Low Interface State Density," *ECS Solid State Lett.*, vol. 1, no. 2, pp. N10–N12, Jul. 2012.

2018 INTERNATIONAL CONFERENCE ON MICROELECTRONIC TEST STRUCTURES, MARCH 19-22, 2018, AUSTIN, TEXAS, USA

Total Ionizing Dose Effects on Analog Performance of 65 nm Bulk CMOS with Enclosed-Gate and Standard Layout

Matthias Bucher[1], Aristeidis Nikolaou[1], Alexia Papadopoulou[1], Nikolaos Makris[1], Loukas Chevas[1],
Giulio Borghello[2,4], Henri D. Koch[3,4], Federico Faccio[4]

[1]School of Electrical and Computer Engineering, Technical University of Crete, 73100 Chania, Greece
Email: bucher@electronics.tuc.gr
[2]DPIA, Università degli Studi di Udine, 33100 Udine, Italy
[3]SEMi, Université de Mons, 7000 Mons, Belgium
[4]EP Dept., CERN, 1211 Geneva, Switzerland

Abstract— High doses of ionizing irradiation cause significant shifts in design parameters of standard bulk silicon CMOS. Analog performance of a commercial 65 nm CMOS technology is examined for standard and enclosed gate layouts, with Total Ionizing Dose (TID) up to 500 Mrad(SiO$_2$). The paper provides insight into geometrical and bias dependence of key design parameters such as threshold voltage, DIBL, transconductance efficiency, slope factor, and intrinsic gain. A modeling approach for an efficient representation of saturation transfer characteristics under TID from weak through moderate and strong inversion and over channel length is discussed.

Keywords— *analog parameters, enclosed layout, modeling, MOSFET, parameter extraction, radiation hardness, total ionizing dose*

I. INTRODUCTION

Radiation effects on CMOS technology have been the objective of research, with recent examples for 130 nm, 65 nm [1], and 28 nm [2] bulk CMOS. Devices with thinner oxides tend to be less sensitive to radiation [3]. The current upgrade of the High-Luminosity Large Hadron Collider (HL-LHC) experiments at CERN requires a large amount of electronics to be fabricated for high Total Ionizing Dose (TID) up to 1 Grad(SiO$_2$) levels. The 65 nm bulk CMOS node offers valuable trade-offs among performance, radiation tolerance, and cost. MOS transistors with enclosed-gate (EG) layouts have shown increased resistance to radiation effects [4], but also show improved mismatch and low frequency noise properties [5]. The impact of high TID on analog performance in 65 nm bulk CMOS [6][7] requires further investigation. The benefit of EG layout should be detailed further, and modeling approaches for all the above need to be discussed.

In the present paper, the incidence of high TID on analog parameters such as transconductance efficiency, DIBL, slope factor, and intrinsic gain will be investigated for 65 nm n- and p-MOSTs (core $T_{oxn(p)}$=26 (28) Å, standard $|V_{Tn(p)}|$=0.4 (0.48) V, $|V_{DD}|$=1.2V) with either types of layout. Devices are tested at CERN under probe cards, using an X-ray source as shown in Fig. 1, for a TID experiment up to 500 Mrad(SiO$_2$).

Fig. 1. X-ray source and 32-channel probe card mounted on a temperature-controlled prober at CERN.

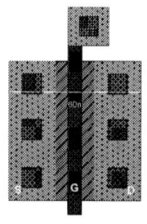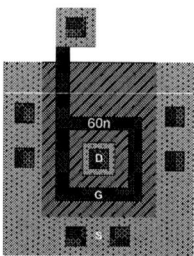

Fig. 2. Layout of MOS transistors with standard (ST) and enclosed gates (EG).

Devices under irradiation are biased at $|V_{GS}|=|V_{DS}|=|V_{DD}|$, which is considered a worst case. IV characterization is performed at the same temperature as the irradiation. In the present work, this is 25°C; annealing effects are not considered. EG transistors with a tetragonal layout, operated with drain at the center as shown in Fig. 2, have their widths varying according to their channel length [4], ranging from L = 60nm to several um. Standard (ST) transistors considered here have a fixed width of W=1um and variable length. The incidence of channel width and temperature effects will be considered elsewhere.

978-1-5386-5072-1/18 $31.00 © 2018 IEEE

2018 INTERNATIONAL CONFERENCE ON MICROELECTRONIC TEST STRUCTURES, MARCH 19-22, 2018, AUSTIN, TEXAS, USA

The overall impact of high TID on advanced CMOS technology has been investigated by several groups. Charge trapping in oxides, and in particular in lateral shallow-trench isolation (STI) regions of bulk silicon MOSTs, may lead to a severe reduction of device performance, causing notably mobility reduction, threshold voltage shifts, and increase in leakage currents. These effects have been shown to be significantly more severe for short-channel and/or narrow-width devices; thinner oxides lead to less TID sensitivity. However, analog performance parameters, such as transconductance efficiency, intrinsic gain etc. require more attention, in particular as to their geometrical scaling and layout dependence. Finally, the availability of analytical and compact models is of key importance for the circuit design community. A simple but efficient modeling approach covering all saturation characteristics of the devices at all channel lengths, will be discussed.

II. RESULTS AND DISCUSSION

The main electrical parameters under examination are threshold voltage V_T, drain induced barrier lowering DIBL $\equiv -\Delta V_T / \Delta V_{DS}$, subthreshold slope factor n, and mobility. For the extraction of threshold voltage V_T, the "adjusted constant current" (ACC) method [8] is used. This method provides an adaptation of the current criterion depending on V_{DS} conditions, an important aspect in correctly estimating DIBL, a key parameter for analog performance. Quantities such as intrinsic gain $A_V = g_m/g_{ds}$ are governed mostly by DIBL. In the following, examples are provided for selected individual transistors. Length-scaling effects are then examined in detail. Comparisons among standard and enclosed-gate layout transistors are provided.

Fig. 3 shows typical effects of TID experiments on transfer characteristics of short-channel n- and p-MOSTs with a moderate channel width. Well-known decrease of mobility, threshold voltage shifts, degradation of subthreshold slope (particularly ST n-MOSTs), and increase in leakage current with increasing TID may be observed. While p-MOSTs typically show a reduced sensitivity, EG transistors show overall significantly reduced effects with TID. Fig. 4 shows transconductance-to-current ratio g_m/I_D versus drain current. The latter is degraded in particular in weak inversion, which corresponds to a degradation of the subthreshold slope factor n.

In Fig. 5, output characteristics are shown. Drain current degradation of ST n-MOSTs may exceed 50% for a TID of 500 Mrad, while EG transistors show significantly less losses. The output conductance g_{ds} is also degraded; the impact of high TID is less severe in saturation as compared to conduction.

Fig. 6 shows threshold voltage V_{TO} in saturation and linear modes versus channel length L, and the impact of TID according to the type and layout of transistors. Clearly, the linear mode threshold voltage exhibits a strongly increased RISCE with TID [1] in n-MOSTs, while saturated transistors are much less affected.

In Fig. 7, the resulting DIBL effect derived from the data in Fig. 6 is shown. DIBL effect is most sensitive to TID in ST n-MOSTs. Significantly lower DIBL is observed in EG vs. ST layout devices. For short channel n-MOSTs, DIBL worsens

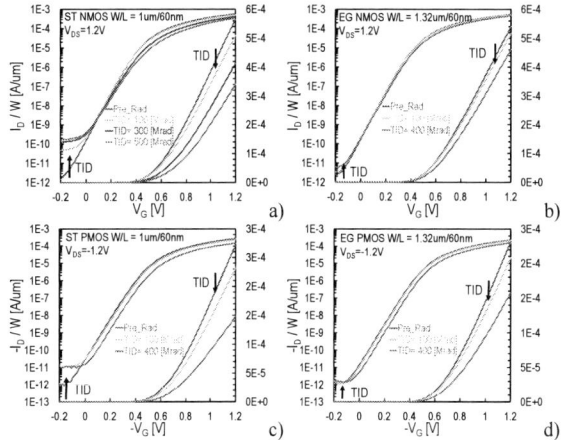

Fig. 3.   Transfer characteristics with increasing TID for short-channel ST and EG n- and p-MOSTs in saturation.

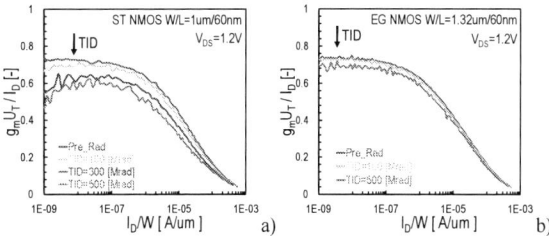

Fig. 4.   Transconductance-to-current ratio vs. drain current with increasing TID for short-channel ST and EG n-MOSTs in saturation.

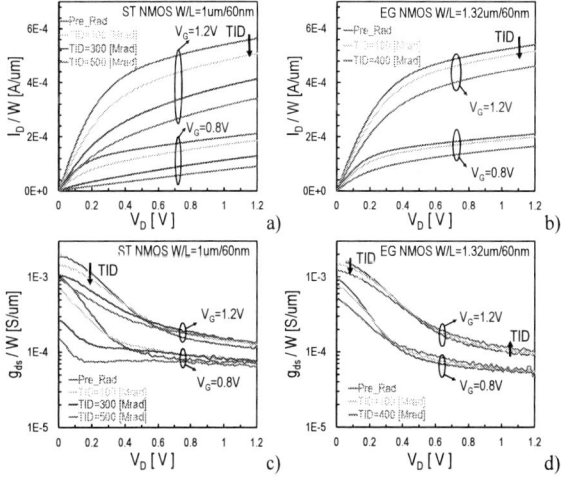

Fig. 5.   Output characteristics at different TID for short-channel ST and EG n-MOSTs at different gate voltages.

with TID, and shows an $\sim L^{-1.2}$ dependence. EG p-MOSTs are insensitive to TID at all channel lengths. In Fig. 8, the subthreshold slope factor n is examined. n is obtained from the inverse of the maximum transconductance efficiency $g_m U_T/I_D$ shown in Fig. 4. The degradation of slope factor n for ST n-MOSTs with increasing TID is prominent at all channel lengths. The same phenomenon is significantly less pronounced for ST p-MOSTs. For both channel types, the EG

978-1-5386-5072-1/18 $31.00 © 2018 IEEE   167

2018 INTERNATIONAL CONFERENCE ON MICROELECTRONIC TEST STRUCTURES, MARCH 19-22, 2018, AUSTIN, TEXAS, USA

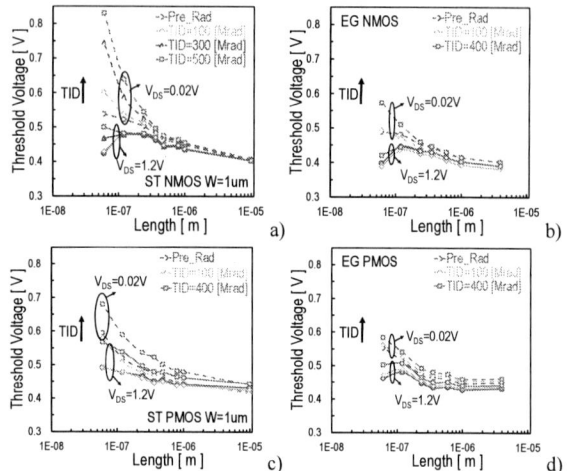

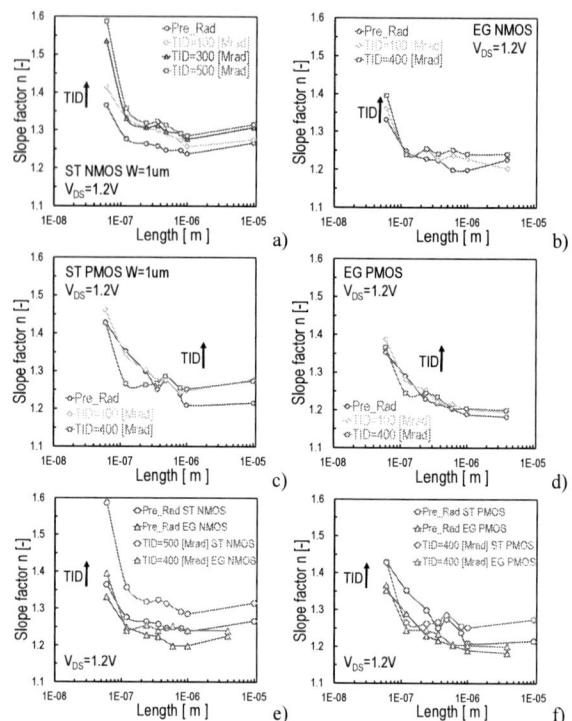

Fig. 6. Threshold voltage vs. channel length for varying TID for ST and EG n- and p-MOSTs in linear and saturation modes.

Fig. 8. Slope factor vs. channel length at increasing TID for ST and EG n- and p-MOSTs in saturation.

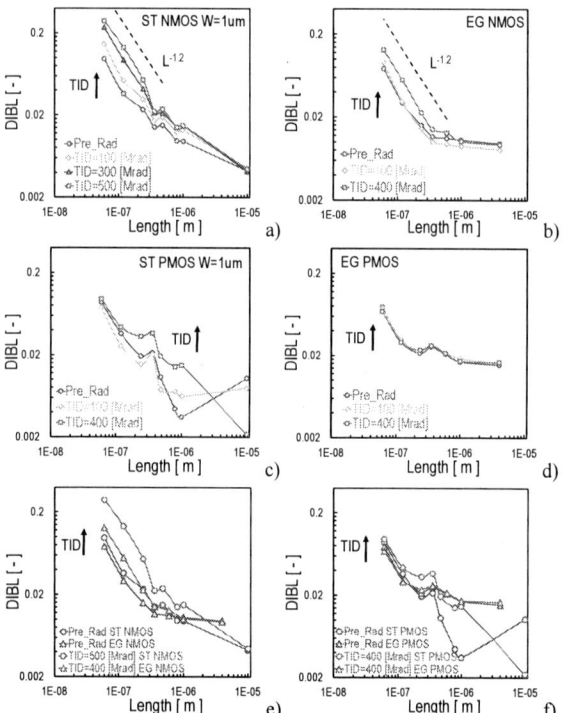

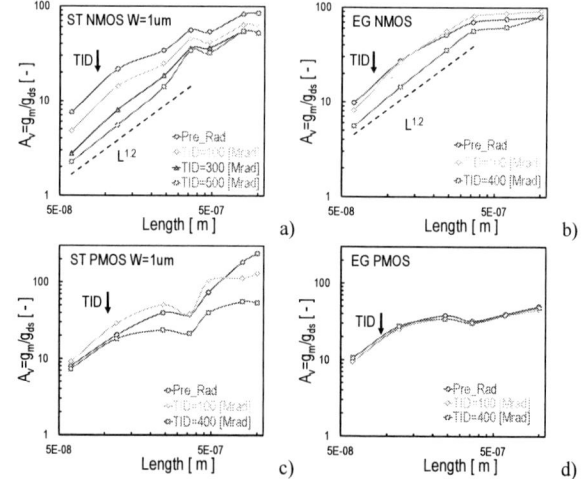

Fig. 7. DIBL vs. channel length for varying TID for ST and EG n- and p-MOSTs.

Fig. 9. Intrinsic gain (g_m/g_{ds}) @ $V_G \approx V_T$ vs. channel length at increasing TID for ST and EG n- and p-MOSTs in saturation.

layout transistors show a definitely improved slope factor, at pre-rad as well as at high TID conditions.

The combination of lower DIBL (and better slope factor) should lead to improved intrinsic gain $A_v = g_m/g_{ds}$, as is indeed confirmed in Fig. 9. Note that here, the intrinsic gain is evaluated around-threshold, with an inversion coefficient of about unity [9][10]. Overall, the intrinsic gain is mainly dictated by the behavior of DIBL. Short-channel ST n-MOSTs show a clear sensitivity to high TID. EG n-MOSTs show an

improved intrinsic gain, combined with reduced sensitivity to TID. A scaling tendency of roughly $A_v \sim L^{1.2}$ can be observed for short-channel devices. A_v scaling is seen to level off at longer channel lengths, which is attributed to an adverse impact of pocket-implants at long channel.

Fig. 10 provides an evaluation of on-state ($|V_G|=|V_D|=1.2V$) and off-state ($|V_G|=0V$, $|V_D|=1.2V$) currents versus channel length. On-state current degradation at shorter channel length is also impacted by mobility and velocity saturation effects,

978-1-5386-5072-1/18 $31.00 © 2018 IEEE 168

2018 INTERNATIONAL CONFERENCE ON MICROELECTRONIC TEST STRUCTURES, MARCH 19-22, 2018, AUSTIN, TEXAS, USA

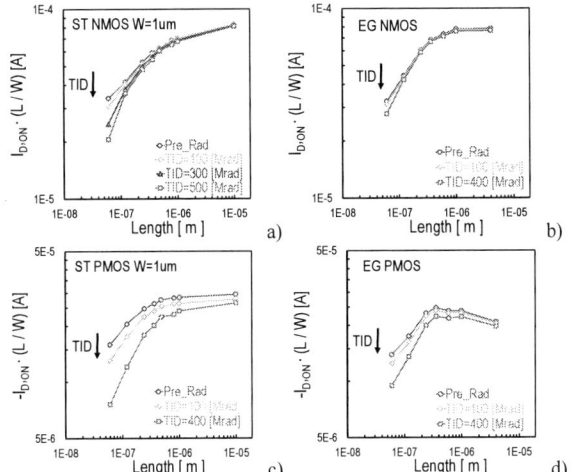

Fig. 10.   Normalized on-state current vs. channel length at increasing TID for n-and p-type transistors with ST and EG layout.

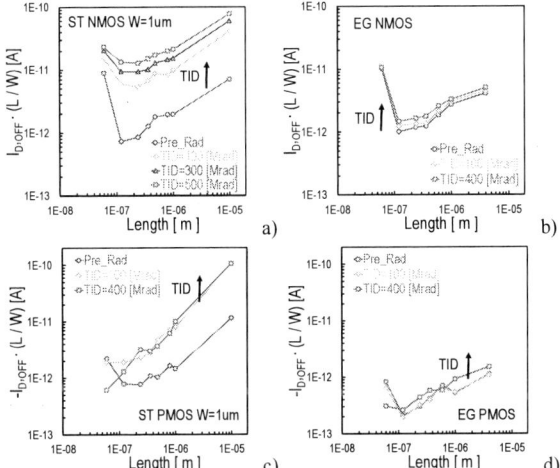

Fig. 11.   Normalized off-state current vs. channel length at increasing TID for n-and p-type transistors with ST and EG layout.

which are more pronounced for n-MOSTs as compared to p-MOSTs. Fig. 11 provides equivalent off-state ($|V_G|$=0V, $|V_D|$=1.2V) current. The observed I_{off} scaling is expected and coherent with the behavior of threshold voltage.

III. MODELING APPROACH

In the present section, a simplified EKV-type model proposed in [2] will be used to capture the effects of increasing TID in both ST and EG MOSTs. Drain current (in saturation) is normalized according to,

$$IC = I_{Dsat}/I_{SPEC} \qquad (1)$$

where $I_{SPEC} = I_0\,(W/L)$ is the specific current, which in turn depends on the so-called technology current $I_0=2nU_T^2\mu C_{ox}$. In the latter, $U_T=kT/q$ is the thermal voltage, n is the slope factor, μ is the carrier mobility and C_{ox} is the gate oxide capacitance

per unit area. The basic voltage-inversion charge equation is given by,

$$v_p-v_s \approx \frac{V_G-V_{TO}-nV_S}{nU_T} = 2q_s+ln(q_s), \qquad (2)$$

where v_p is the normalized pinch-off voltage and q_s the normalized inversion charge density at the source end of the channel. The impact of velocity saturation can be described with parameter λ_c and in this case IC can be expressed as [11],

$$IC = \frac{4(q_s^2+q_s)}{2+\lambda_c+\sqrt{4(1+\lambda_c)+\lambda_c^2(1+2q_s)^2}} \qquad (3)$$

Note that (3) can be inverted,

$$2q_s = \sqrt{(\lambda_c IC+1)^2+4IC} \; - 1. \qquad (4)$$

Neglecting the velocity saturation effect (λ_c=0) in (3), the basic EKV current-inversion charge relation can be obtained,

$$IC = q_s^2+q_s . \qquad (5)$$

Equations (1), (2) and (3) describe an EKV-type model that can be used to extract a set of parameters (I_0, V_{TO}, n and λ_c) that can predict accurately the effect of high TID in saturation. I_0 may be extracted from the longest- and widest-channel device available and scaled for other devices according to their geometry [8]. Threshold voltage parameter V_{TO} is extracted from the I_D vs. V_G characteristics, slope factor n from the maximum g_mU_T/I_D vs. I_D plateau in weak inversion and λ_c by fitting the model in strong inversion. In each case, leakage current is adapted to the observed leakage level.

In Fig. 12, the above model is adapted to the transfer characteristics of short-channel ST and EG n-MOSTs at different levels of TID. The parameter λ_c is related to the intersection of the weak and strong inversion asymptotes of the normalized transconductance to current ratio. Other parameters, namely I_0, n, V_{TO} are dependent on channel length L as well as on the level of TID, as analyzed in Section II.

Fig. 13 finally shows transfer characteristics resulting for high TID for all available channel lengths and both types of transistors. The λ_c parameter, also shown in the same figure, is dependent, essentially, on channel length, and only marginally on TID. The channel length dependence $\lambda_c = L_{sat}/L$ [2] is somewhat coarse. Here, a different model for the dependence of λ_c on channel length is proposed,

$$\lambda_c(L) = \lambda_0\left(1+\left(\frac{L_b}{L}\right)^a\right), \qquad (6)$$

for which the parameters λ_0, L_b and a are listed in Table I. The same model parameters apply for both ST and EG n-MOSTs. Interestingly, the velocity saturation effect is, finally, basically unaffected by TID, and EG transistors show the same length-dependence as ST transistors.

978-1-5386-5072-1/18 $31.00 © 2018 IEEE   169

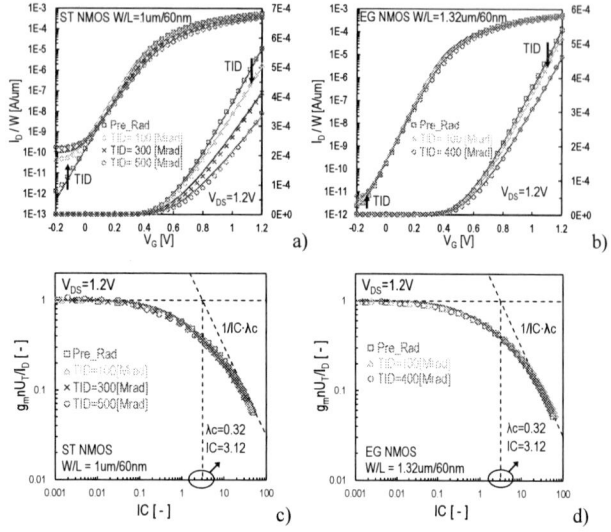

Fig. 12. Transfer characteristics and normalized transconductance efficiency vs. inversion coefficient IC for different TID levels, for short-channel ST and EG nMOSTs in saturation. Markers: measurements, lines: model. Asymptotes (dashed) and extracted parameter λ_c are shown.

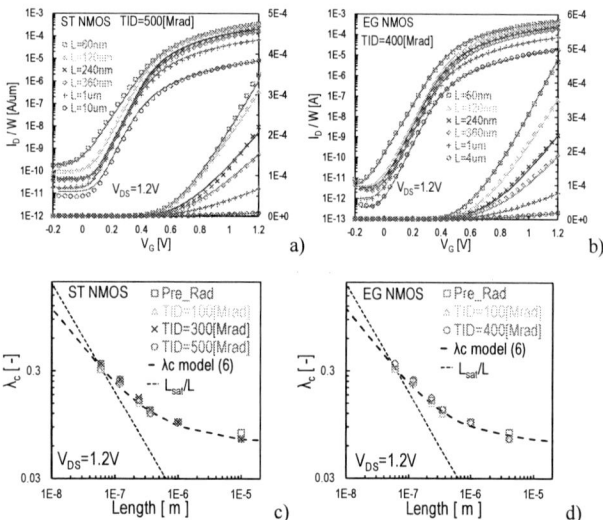

Fig. 13. Transfer characteristics for ST and EG n-MOSTs at high TID in saturation, for different channel lengths. Parameter λ_c vs. channel length at various TID. Markers: measurements, lines: model (1)-(3), dashed line: model (6), dotted line: L_{sat}/L.

TABLE I.   LIST OF λ_C MODEL PARAMETERS

Parameters	Units	ST & EG NMOS
λ_0	-	0.065
L_b	m	330n
α	-	0.8
L_{sat}	m	19.2n

IV. CONCLUSIONS

In conclusion, a 65 nm bulk CMOS process has been investigated for analog performance with TID experiments up to 500 Mrad(SiO$_2$). The present work provides detailed insight into TID sensitivity of key analog performance parameters of devices, as well as their basic length-dependent scaling. Enclosed gate layout offers significant advantages over standard layout, and is particularly effective for EG n-MOSTs in suppressing effects related to high TID. EG layout devices are shown to have better DIBL, slope factor, and intrinsic gain, even for pre-radiation conditions, as compared to devices with standard layout. An EKV-type modeling approach has been shown to be highly effective in describing saturated drain current and transconductance throughout all inversion conditions. The model is based on four TID- and length-dependent parameters, namely threshold voltage, slope factor, technology current, and leakage current. The fifth parameter λ_c, related to saturation behavior, is shown to be rather insensitive to TID, for both enclosed gate and standard layout devices. Velocity saturation behavior is only marginally affected by TID and layout.

REFERENCES

[1] F. Faccio, S. Michelis, D. Cornale, A. Paccagnella, S. Gerardin, "Radiation-induced short channel (RISCE) and narrow channel (RINCE) effects in 65 and 130 nm MOSFETs," IEEE Trans. Nucl. Sci., vol. 62, nr. 6, pp. 2933-2940, Dec. 2015.

[2] C.-H. Zhang, F. Jazaeri, A. Pezzotta, C. Bruschini, G. Borghello, S. Mattiazzo, A. Baschirotto, C. Enz, "Total ionizing dose effects on analog performance of 28 nm bulk MOSFETs," ESSDERC, pp. 1-4, Sept. 11-14, 2017.

[3] N. S. Saks, M. G. Ancona, J. A. Modolo, "Generation of interface states by ionizing radiation in very thin MOS oxides," IEEE Trans. Nucl. Sci., vol. 33, nr. 6, pp. 1185-1190, Dec. 1986.

[4] W. Snoeys, F. Faccio, et al., "Layout techniques to enhance the radiation tolerance of standard CMOS technologies demonstrated on a pixel detector readout chip," Nucl. Instr. Meth. Phys. Res. A, vol. 439, pp. 349-360, 2000.

[5] M. Bucher, A. Nikolaou, N. Mavredakis, N. Makris, M. Coustans, J. Lolivier, P. Habas, A. Acovic, R. Meyer, "Variability of low frequency noise and mismatch in enclosed-gate and standard nMOSFETs," IEEE ICMTS, pp. 177-180, Mar. 28-30, 2017.

[6] M. Menouni, M. Barbero, F. Bompard, S. Bonacini, D. Fougeron, R. Gaglione, A. Rozanov, P. Valerio, A. Wang, "1-Grad total dose evaluation of 65nm CMOS technology for the HL-LHC upgrades," J. of Instr., vol. 10, C05009, 2015.

[7] M. Krohn, B. Bentele, D. C. Christian, J. P. Cumalat, G. Deptuch, F. Fahim, J. Hoff, A. Shenai, S. R. Wagner, "Radiation tolerance of 65nm CMOS transistors," J. of Instr., vol. 10, C05009, 2015.

[8] A. Bazigos, M. Bucher, J. Assenmacher, S. Decker, W. Grabinski, Y. Papananos, "An adjusted constant-current method to determine saturated and linear mode threshold voltage of MOSFETs," IEEE Trans. Electron Devices, vol. 58, nr. 11, pp. 3751-3758, Nov. 2011.

[9] M. Bucher, D. Kazazis, F. Krummenacher, D. Binkley, D. Foty, Y. Papananos, "Analysis of transconductances at all levels of inversion in deep submicron CMOS," IEEE ICECS, vol. III, pp. 1183-1186, Sep. 15-18, 2002.

[10] M. Bucher, G. Diles, N. Makris, "Analog performance of advanced CMOS in weak, moderate, and strong inversion," MIXDES, pp. 54-57, June 24-26, 2010.

[11] A. Mangla, C. C. Enz, J.-M. Sallese, "Figure-of-merit for optimizing the current-efficiency of low-power RF circuits," MIXDES, pp. 85-89, June 16-18, 2011.

Session 9

MEMS

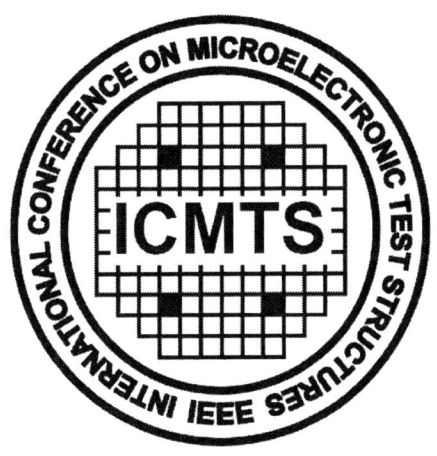

March 22, 2018

09:00-10:20

Co-Chairs:

Carlo CAGLI, CEA, France

Kiyoshi TAKEUCHI, University of Tokyo, Japan

978-1-5386-5072-1/18 $31.00 © 2018 IEEE

An On-chip Test Structure for Studying the Frictional Behavior of Deep-RIE MEMS Sidewall Surfaces

R Ranga Reddy[1], Yuki Okamoto[1], Yoshio Mita[1,]
[1]Electrical Engineering and Information Systems,
The University of Tokyo,
7-3-1, Hongo, Bunkyo-ku, Tokyo,113-8656 Japan.
E-mail: (ranga, mems) @if.t.u-tokyo.ac.jp

Abstract—In this paper, an on-chip micro-mechanical test structure has been developed to investigate the frictional behavior of Deep-RIE sidewall contacting surfaces of single crystal silicon which is most widely used in micromechanical systems (MEMS). The test structure is fabricated on Silicon on Insulator (SOI) wafer using standard MEMS process. Two orthogonally placed electrostatic comb-drive actuators are adopted, one comb drive is used to align a contact with the friction surfaces under a certain normal load and another one is used to generate the tangential motion on contacted sidewall surfaces. To assess the frictional behavior, both static and dynamic friction coefficients were observed on the contacted surfaces during the experiment with different DRIE process parameters. Through experiments, it was found that with the increment of normal forces, the static friction coefficient is no longer a constant value and it has less effect on dynamic friction coefficient. DRIE process parameters greatly influence the frictional properties on both static and dynamic friction coefficients.

Keywords— Friction, static friction coefficient, dynamic friction coefficient, roughness, Deep Reactive ion etching (DRIE).

I. INTRODUCTION

Owing to the advancement of Microelectromechanical systems (MEMS) technology, devices on a micrometer scale integrated with mechanical elements, sensors, actuators, electronics on a common substrate were employed in many commercial fields such as entertainment, communication, automation and as well as strategic fields such as defense and space [1-2]. However, due to the insufficient knowledge of tribological issues at contact sidewall surfaces on a micro scale the commercialization of MEMS devices that consist of components with sliding contact surfaces during their operation such as stepper motors, micrometers, gas bearings are still limited to laboratory studies. Friction and wear consider as key parameters that decide the reliability and lifespan of MEMS devices that contain contacting surfaces. Due to the large surface-to-volume ratio at the micrometer-sized structure, the mechanical characteristic of MEMS devices plays strong size effect which makes surface effect greatly enhanced [3]. As for movable MEMS devices, because the gaps of contacting surfaces are in the nanometer range, even zero, the influence of the frictional force of these contacting surfaces increases greatly [4]. On the other hand, well-developed theories and frictional laws are not feasible to MEMS devices, it is vital to rebuild a micro-tribology theory in which the size effect must be concerned. Therefore, a specially well-designed MEMS

devices for friction measurements at the microscale are necessary.

MEMS devices for measurement of frictional characteristics on surface micromachined sidewall structures were reported in the previous research [4-8]. Despite the significant efforts devoted to the study of the friction behavior at different contacting surfaces, the friction transitions between the static and dynamic on the sidewall surfaces have not been fully characterized. And also, the frictional properties of deep reactive ion etching (DRIE) silicon sidewall surfaces have not been fully investigated, even though the DRIE process is widely used for MEMS device fabrication.

This paper describes the design and fabrication of an on-chip micromechanical test structure to investigate the static and dynamic friction coefficients on the sidewall surfaces with consideration of different DRIE process parameters i.e., with different asperities (scalloping) size on sidewall surfaces by varying etching cycle rate. And the influence of normal load on friction coefficients was investigated for the above conditions. The main aim is to obtain the fundamental information of tribology properties on the sidewall surface with DRIE process parameters and also the influence of normal force on the contact pair.

II. DESIGN AND OPERATION PRINCIPLE

The test structure used to investigate the friction properties on sidewall surfaces of Deep-RIE MEMS devices reported in this paper is inspired from the microinstrument developed by N Ansari and W R Ashurst [5]. The schematic diagram of the proposed test structure device is shown in Fig.1 and its operation measurement schematic is shown in Fig.2. The device consists of simple beam springs and two electrostatic comb drive actuators integrated in such a way that they can produce motion in orthogonal directions. One comb actuator (normal arm) is used to make a contact with the friction sidewall surface under certain normal load and another comb actuator (tangential arm) is used to generate the tangential motion on the contacted sidewall surface. Each shuttle has extended beams in its movable part, supported by two pairs of suspended beams identically. The extended beams of these shuttles are 375μm long and 4μm (nominal) wide are connected perpendicularly and supporting beams are 550μm long and 4μm (nominal) wide.

978-1-5386-5072-1/18 $31.00 © 2018 IEEE

2018 INTERNATIONAL CONFERENCE ON MICROELECTRONIC TEST STRUCTURES, MARCH 19-22, 2018, AUSTIN, TEXAS, USA

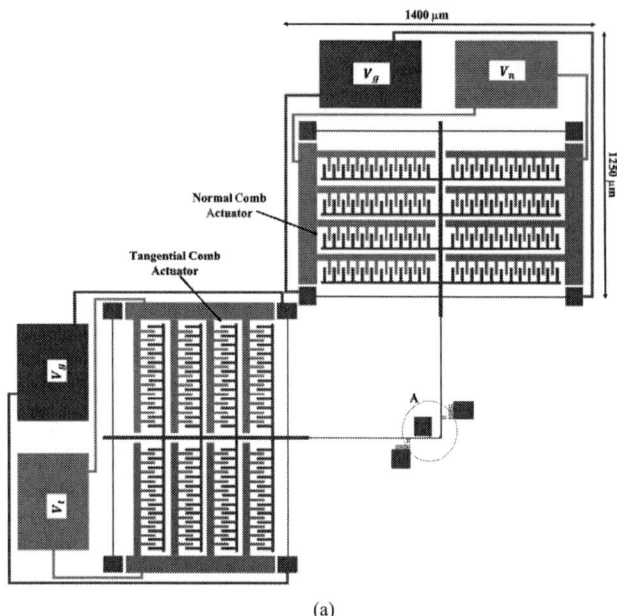

(a)

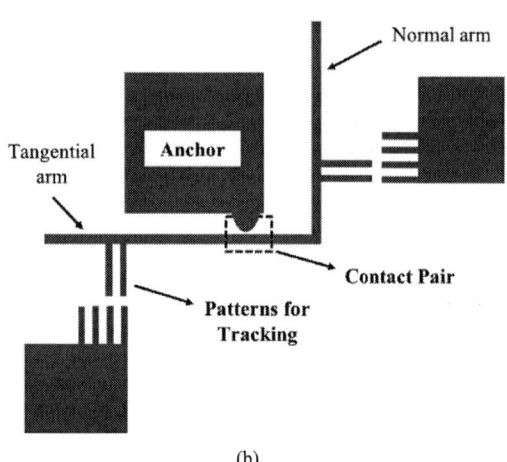

(b)

Fig.1. (a) A schematic diagram of the proposed test structure used to study the frictional behavior on sidewall surfaces (b) enlargement of segment **A**

Initial gap between the contacting sidewall surfaces is 2µm and those surfaces can be brought into contact by applying a DC voltage on the normal arm (V_n). When the voltage on normal arm $V_n = V_a$ the frictional pair is brought into contact then the voltage on the tangential arm (V_t) is increased step by step to generate tangential motion on contacted sidewall surfaces. At voltage $V_t = V_{cri}$ the electrostatic force on tangential arm overcome the restoring force caused by the deflection of suspended beams of both tangential actuator and normal arm and also the frictional force between the contact pair. During the course of an experiment, the image patterns were tracked by varying the voltage V_t on tangential arm under different normal load conditions to study the frictional behavior of contract pairs.

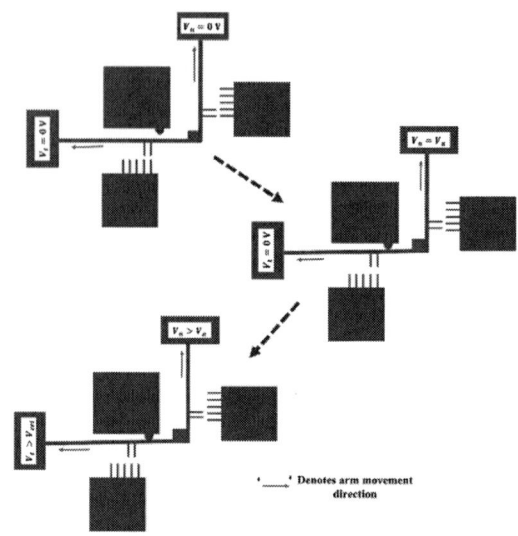

Fig.2. A Schematic diagram of operation principle of the test structure with its notations are V_n =Voltage on Normal arm, V_t =Voltage on Tangential arm, V_n = V_a=Voltage required to bring the surfaces into contact, V_t = V_{cri} = Voltage at which contact slip starts.

In addition to that, two different DRIE process parameters (i.e., by varying the asperities size on sidewall surfaces) were adopted to observe its influence on frictional characteristics for the above design conditions.

III. FABRICATION

The test structures were fabricated on Silicon on Insulator (SOI) substrate by using standard surface micromachining process as shown in Fig.3. The SOI wafer, which was used to fabricate the test structure, consists of a 20 µm silicon device layer and a 2 µm silicon dioxide sacrificial layer on the silicon substrate of 625 µm. First, the designs were patterned with high-speed EB lithography and then using the patterned photoresist as an etching mask, DRIE process was performed to etch the silicon device layer. Here two different DRIE processes were adopted to etch silicon (Si) layer and the process parameters of DRIE-I and DRIE-II are shown in Table 1, to vary the asperities size on the contacting sidewall surfaces. The DRIE-I process used 5 sec SF_6 for Si etching followed by 2 sec C_4F_8 for passivation and DRIE-II uses 1.5 sec SF_6 for Si etching followed by 1 sec C_4F_8 for passivation on different test devices. Finally, devices were released by Vapor phase HF etching. The fabricated test structure is shown in Fig.4.

TABLE 1. Process parameters of DRIE process used in experiment

Process	Step	SF₆ flow rate (sscm)	C₄F₈ flow rate (sscm)	Process gas pressure (Pa)	Coil power (W)	Bias power (W)	Step time (sec)
DRIE-I	Passivation	–	150	3.5	1800	100	2
	Etching	300	–	4.5	1800	100	5
DRIE-II	Passivation	–	150	3.5	1800	100	1
	Etching	300	–	4.5	1800	100	1.5

978-1-5386-5072-1/18 $31.00 © 2018 IEEE

2018 INTERNATIONAL CONFERENCE ON MICROELECTRONIC TEST STRUCTURES, MARCH 19-22, 2018, AUSTIN, TEXAS, USA

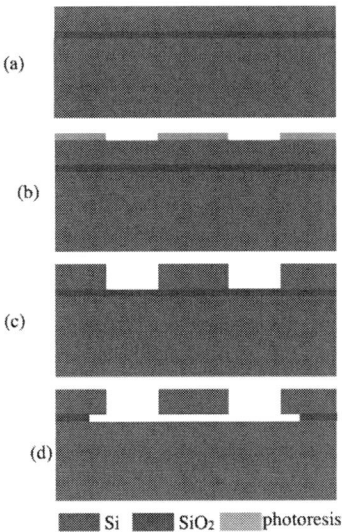

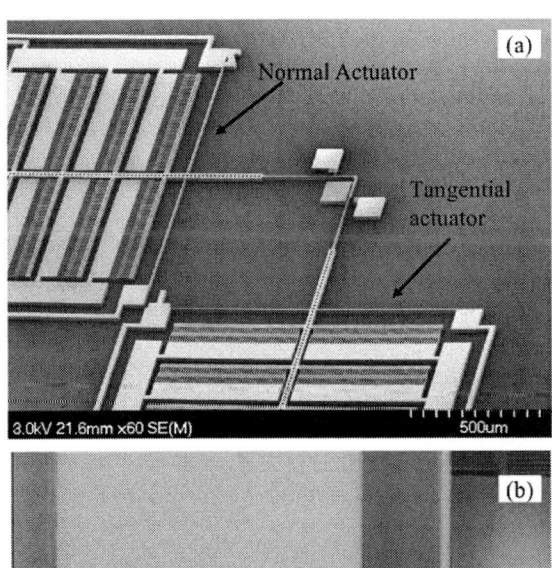

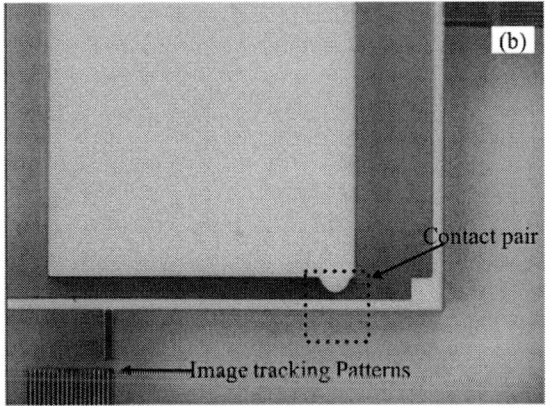

Fig.3. Fabrication process (a) Picking of SOI substrate (b) photoresist patterning on device layer using EB lithography (c) DRIE etching of silicon layer and then stripping of photo resist (d) removal of SiO₂ sacrificial layer using vapor phase HF etching.

Fig.4. (a) SEM image of fabricated test structure (b) shows the optical microscope image of contact pair with tracking patterns used for measurements.

IV. EXPERIMENTAL CALCULATIONS

Theoretical calculations were carried out during the design of the test structure. Relation between electrostatic force and driving voltages of shuttles, stiffness of test structure with frictional forces are all described. During the operation, three electric potential potentials V_n, V_t, V_g, were applied. The ground potential, V_g, was applied to all the grounded parts of the components, a dc bias V_n, was applied to the normal comb actuator and V_t, was applied to the tangential come actuator.

The electrostatic force (F) generated by a comb actuator in the direction parallel to the comb fingers is given by (neglecting fringing effects)

$$F = \frac{n\varepsilon t}{g} * V^2 \qquad - (1)$$

where n is the number of comb fingers, ε is the permittivity of air, t is the thickness of comb fingers, g is the gap between the comb fingers, V is the driving voltage applied on the shuttle. For simplicity, the constants are lumped into geometry parameter c. Therefore

$$c = \frac{n\varepsilon t}{g} \qquad - (2)$$

Accordingly,

$$F = cV^2 \qquad - (3)$$

In the experiment, two orthogonal comb actuators were used to initiate normal and frictional force, respectively and to distinguish between them subscripts n and t will be used for the normal and tangential arms, respectively. Accordingly,

$$c_t = \frac{n\varepsilon t}{g_t} \qquad - (4a)$$

$$c_n = \frac{n\varepsilon t}{g_n} \qquad - (4b)$$

since the normal and tangential comb drive actuators of test structure are exactly identical to each other, therefore

$$c_t = c_n = c \qquad - (5)$$

combining (3) and (5) gives

$$F_n = cV_n^2 \qquad - (6a)$$

$$F_t = cV_t^2 \qquad - (6b)$$

A certain driving voltage is applied on normal arm to bring the loading beam in contact with anchor post and the amount of force required is given by

978-1-5386-5072-1/18 $31.00 © 2018 IEEE

$$F_a = \delta_a(k_n + k_{arm1}) = cV_a^2 \qquad \text{- (7)}$$

where V_a is the voltage that is required to be applied on the normal comb actuator to bring the frictional pairs in to contact, k_n is stiffness of suspended beams of normal comb actuator, k_{arm1} is the stiffness of extended beam (tangential arm) that connected to normal arm as shown in Fig.1, δ_a is the displacement of normal arm to bring the frictional pair in to contact.

In a separate measurement when the beam is not in contact with the anchor, electrostatic force initiated by the tangential comb actuator must overcome the restoring force caused by the deflection of suspended beams of tangential actuator and extended beam (normal arm) that connected to the tangential arm and it is given by

$$F_c = \delta_c(k_t + k_{arm2}) = cV_c^2 \qquad \text{- (8)}$$

where V_c is the voltage applied on the tangential comb actuator when friction pair is not in contact, k_t is stiffness of suspended beams of tangential comb actuator, k_{arm2} is the stiffness of extended beam (normal arm) that connected to tangential arm as shown in Fig.1, δ_c is the displacement of a tangential arm.

The coefficient of static friction of the contacting surfaces can be determined using

$$\mu_s = \frac{F_t - F_c}{F_n - F_a} \qquad \text{- (9)}$$

where F_n and F_t are the total tangential and normal forces generated, at the instant of slip between the frictional pair using the normal and tangential comb actuators, respectively. Substituting (6), (7) and (8) in the above equation gives

$$\mu_s = \frac{V_t^2 - V_c^2}{V_n^2 - V_a^2} \qquad \text{- (10)}$$

Once the initiation of slip starts at the sidewall surface, the loading beam continues to slide with the anchor post until the sum of restoring force generated by the suspended beams of a tangential arm and extended beams connects to normal arm and the dynamic friction force equal to the total force generated by the tangential comb actuator. Therefore, the coefficient of dynamic friction of contacting sidewall surface can be determined using

$$\mu_k = \frac{F_t - F_c - \delta_s(k_t + k_{arm2})}{F_n - F_a} \qquad \text{- (11)}$$

where δ_s is the distance which the loading beam slides on the contacting surface after its initiation. Substituting (6), (7) and (8) into the equation gives

$$\mu_k = \frac{V_t^2 - V_c^2 - \delta_s(\frac{k_t + k_{arm2}}{c})}{V_n^2 - V_a^2} \qquad \text{- (12)}$$

V. RESULTS AND DISCUSSION

The fabricated MEMS test structures were tested in a clean room (Class 100, 20 °C, 40% humidity). The testing equipment composed of power supply system, microscope, processing circuit board, a high-speed camera (plexlogger) for image acquisition and a computer. Fig.5 shows Scanning Electron Microscope (SEM) images of sidewall surfaces of fabricated test structures with DRIE-I and DRIE-II process and whose measured scallop dimensions are 600 nm and 180 nm respectively. Fig.6 shows the optical image of contact pair which brought into contact by applying DC voltage ($V_n = V_a = 20V$) on a normal actuator to study the frictional behavior on the sidewall surfaces. During the course of an experiment, image patterns of test structure were captured through high speed camera (50fps) in corresponding with the applied voltages V_t and V_n. Fig.7 shows the relative displacement of contacting surface pair with the driving voltage (V_t) applying on tangential arm under different normal holding voltage (V_n) on a normal arm. At the driving voltage $V_t = V_{cri}$, the contact beam starts slipping where the electrostatic force on tangential comb actuator overcomes the restoring force caused by suspended beams and extended beam that connected to the tangential arm and also the maximum static frictional force between contact pair.

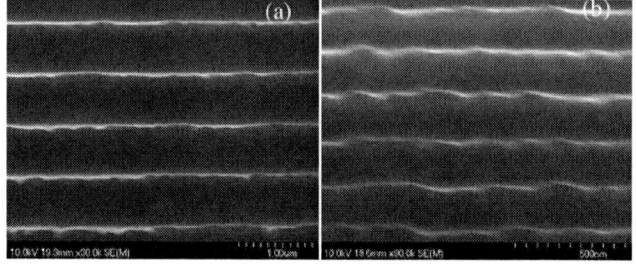

Fig.5. (a) SEM images of sidewall surfaces of fabricated test structures with DRIE-I process (scalloping size ≈ 600 nm) and (b) with DRIE-II process (scalloping size ≈ 180 nm).

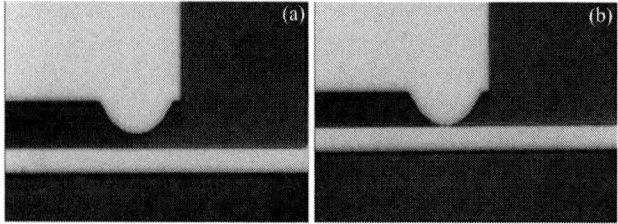

Fig.6. Shows the optical microscope image of contact pair before (a) and after (b) applying voltage ($V_n = V_a = 20V$) on normal comb actuator to bring the contact pair surfaces into contact.

2018 INTERNATIONAL CONFERENCE ON MICROELECTRONIC TEST STRUCTURES, MARCH 19-22, 2018, AUSTIN, TEXAS, USA

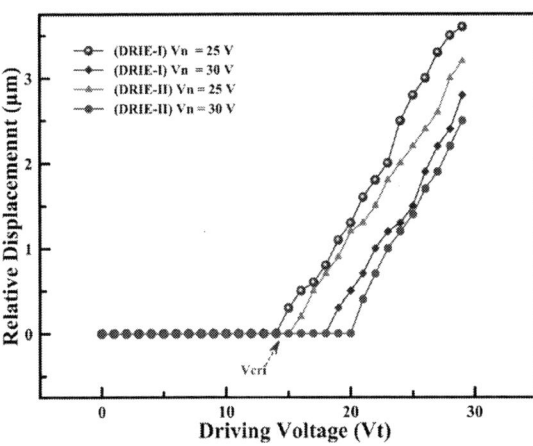

Fig.7. Shows the relative displacement of contacting surface pair with the increment of driving voltage (V_t) on tangential comb actuator under different normal holding voltage (V_n) on normal comb actuator of two different DRIE process parameter test structures and V_{cri} is the point at which motion starts of corresponding V_n.

From Fig.7, it shows that with the increment of V_n, V_{cri} increased correspondingly. For each test condition, the experiment was repeated five times to obtain more accurate data. The error bar of each point indicates the one standard deviation above and below the corresponding average value. Table 2. shows the measured data of static friction coefficient of DRIE-I and DRIE-II at two different normal loads. Fig.8 shows the influence of static friction coefficient (μ_s) with increasing normal load (μN) on the surface of contact pair. At low normal load, higher μ_s is observed where the Vander walls and capillary forces become significant and with the increase in normal load, these Vander walls and capillary forces become secondary compared to the external force. DRIE-II process chip experience slightly higher static friction coefficient (μ_s) than the DRIE-I process chip, since the real contact area between

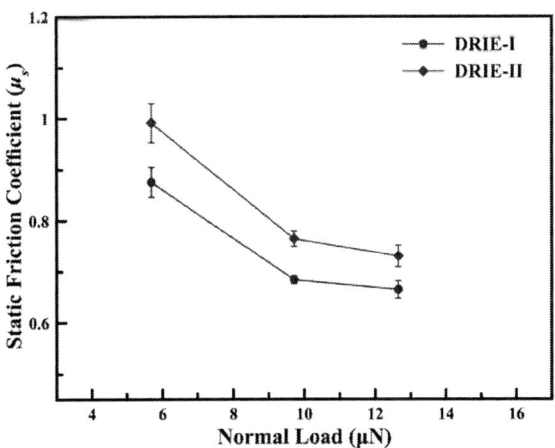

Fig.8. Shows the static friction coefficient (μ_s) versus normal load on contacting surfaces for the two DRIE test structures.

TABLE.2 Static friction coefficient data of DRIE-I and DRIE-II at two normal loads. All the values in the reported in the table are average of several measurements.

Process	Normal Load (μN)	Static friction coefficient $\mu_s \pm$ std. dev.
DRIE-I	5.7	0.87 ± 0.03
	12.6	0.67 ± 0.02
DRIE-II	5.7	0.99 ± 0.04
	12.6	0.73 ± 0.02

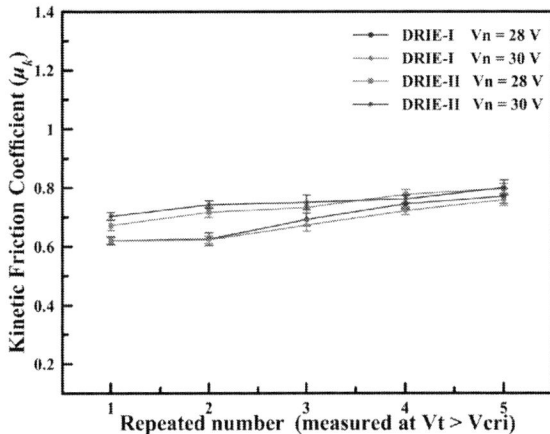

Fig.9. Shows the dynamic friction coefficient (μ_k) measured at different driving voltages ($V_t > V_{cri}$) on sidewall surface of two test structures at different normal holding voltage (V_n).

the frictional pair of DRIE-II is higher than the DRIE-I for corresponding normal loads due to their asperity size difference on sidewall surfaces.

Fig.9 shows the dynamic friction coefficient (μ_k) on sidewall surfaces which are measured at different driving voltages ($V_t > V_{cri}$) once the contact surface pair starts slipping under certain holding voltages (V_n) on the normal arm. It shows that normal forces may have less effect on dynamic friction coefficient (μ_k) than the static friction coefficient (μ_s) on both DRIE process, however DRIE-II process experiences slightly higher dynamic friction coefficient than the DRIE-I due to their asperity size difference. Average values of the dynamic friction coefficient of DRIE-I and DRIE -II are 0.64±0.03, 0.72±0.04 respectively. Through experiments, it was found that with the increment of normal forces, the static friction coefficient is no longer a constant value and it has less effect on dynamic friction coefficient. DRIE process parameters (i.e. by varying scalloping size) greatly influence the frictional properties on both static and dynamic friction coefficients.

978-1-5386-5072-1/18 $31.00 © 2018 IEEE

VI. Conclusion

An on-chip test structure with two orthogonally placed comb-drive actuators was developed to study the frictional behavior of DRIE etched sidewall surfaces of silicon micromachined structures. The static and dynamic friction coefficients of DRIE sidewall surfaces were studied under the influence of normal load with different DRIE process parameters. Experiment results indicate that with the increment of normal load on the contacting pair, static friction is no longer a constant value and it decreases with increasing load and it has less influence on dynamic friction coefficient on sidewall surfaces. The DRIE process parameters (i.e. scalloping size) has great influence on friction coefficients, variation in asperity size changes the real contact area between the contact pairs during their operation.

ACKNOWLEDGEMENTS

The part of the work is supported by Japan Science and Technology (JST), and Department of Science and Technology (DST), India and Japan Society of Promotion of science (JSPS) KAKENHI grant-in-aid 16H04345 and 17J04382. Chip design and fabrication are supported by VLSI Design and Education Center (VDEC) at The University of Tokyo in collaboration with Cadence Corporation and Mentor Graphics Corporation. The process was carried out using the open facilities maintained by MEXT's Nanotechnology Platform Program.

REFERENCES

[1] Bhushan, B. (Ed.) Handbook of Micro/Nano Tribology, 2nd ed.; CRC: Columbus, OH, USA, 1999.)

[2] D. V. Dao K. Nakamura T. T. Bui S. Sugiyama "Micro/nano-mechanical sensors and actuators based on SOI-MEMS technology," Adv. Natural Sci Nanosci. Nanotechnol., vol. 1 no. 1 pp. 013001 2010.

[3] N. R. Tas, C. Gui, M. Elwenspoek, "Static friction in elastic adhesion contacts in MEMS", J. Adhesive Sci. Technol., vol. 17, pp. 547, 2003.

[4] J. Wu, S. Wang, J. Miao "A MEMS device for studying the friction behavior of micromachined sidewall surfaces," Journal of Microelectromechanical Systems, 17 (2008), pp. 921-933

[5] N. Ansari, W. R. Ashurst, "Single-crystal-silicon-based microinstrument to study friction and wear at MEMS sidewall interfaces", J. Micromech. Microeng., vol. 22, no. 2, pp. 025008, 2012.

[6] Guo, Z., Meng, Y., Su, C., Wu, H.: An on-chip micro-friction tester for tribology research of silicon based MEMS devices. Microsyst. Technol. 14, 109–118 (2007)

[7] Wu J, Wang S and Miao J M 2009 Friction characteristics of the curved sidewall surfaces of a rotary MEMS device in oscillating motion J. Micromech. Microeng. 19 65020

[8] I. H. Hwang, Y. G. Lee, J. H. Lee, "A micromachined friction meter for silicon sidewalls with consideration of contact surface shape", J. Micromech. Microeng., vol. 16, no. 11, pp. 2475-2481, Nov. 2006.

Wafer Level Characterisation of Microelectrodes for Electrochemical Sensing Applications

E.O. Blair*, L. Parga Basanta[†], I. Schmueser[‡], J.R.K. Marland*, A. Buchoux[§], A. Tsiamis[†]
C. Dunare*, M. Normand*, A.A. Stokes*, A.J. Walton* and S. Smith[†]

*School of Engineering, Institute for Integrated Micro and Nanosystems
[†]School of Engineering, Institute for Bioengineering
[‡]School of Chemistry
[§]School of Engineering, Institute for Multiscale Thermofluids

The University of Edinburgh, Alexander Crum Brown Road,
Edinburgh, EH9 3FF, UK
Email: Stewart.Smith@ed.ac.uk
Tel: +44 (131) 650 7471

Abstract—This work presents a system for the in-line wafer-level characterisation of electrochemical sensors. Typically, such sensors are first diced and packaged before being electrochemically tested. By integrating their characterisation into the manufacturing process, the production of electrochemical sensors becomes more efficient and less expensive as they can be parametrically tested midway through the fabrication process, without the need to package them. This enables malfunctioning or failed devices to be identified before dicing and reduces costs as only functional devices are packaged (in many cases this can be more expensive than the sensor fabrication). This study describes wafer-level characterisation of a simple electrochemical sensor design using a photoresist hydrophobic corralling film for the electrolyte and a probe station for contacting to individual dies.

I. INTRODUCTION

Electrochemical sensors are a popular tool in a wide variety of applications, ranging from industrial monitoring to fundamental science [1]–[4]. One of the most common methods of manufacturing electrochemical sensors is using microfabrication, which allows for controlled and reproducible production of nominally identical sensors. Such sensors are also easily miniaturised to the micron scale and are compatible with CMOS processes [5]–[7]. This presents the opportunity to adapt metrology techniques typically used in semiconductor manufacturing, specifically wafer-level measurements of devices, for characterising electrochemical systems before packaging [8]. Typically, electrodes are characterised after dicing and packaging, which is an expensive and time-consuming element of sensor production. This work demonstrates a system for characterising electrochemical sensors at wafer level. The benefits of this include the ability to rapidly characterise many different electrode designs without the need to package devices. This will help expedite the development of electrode technologies through systematic assessment of adjustments to the electrode design, by monitoring electrochemical test structures. Additionally, this approach enables the identification of

malfunctioning devices, as well as the verification and process control of microfabrication steps.

To demonstrate the validity of the proposed system, quantifiable electrochemical measurements are performed in a manual probe station using electrochemical test structures. These are compared with measurements made in a traditional electrochemical measurement setup. Following this, the capability of the probe station setup to spatially map electrochemical responses across the wafer is demonstrated. The requirements for a completely automatic system are also discussed.

II. TEST STRUCTURE DESIGN AND FABRICATION

A. Test Structure Design

The test structures consist of a three-electrode cell on a 3 mm square die, pictured in Fig. 1. The central working electrode (WE) is where the potential and current of interest are applied and measured. The inner ring electrode around the WE is a pseudo-reference electrode (RE), which provides a reference potential against which potentials at the WE are applied and measured. The outer ring electrode forms the counter electrode (CE), which supplies the current required for the chemical reactions taking place at the WE. The WE designs in this work are single square and disc microelectrodes with a range of edge lengths/diameters (20, 50, and 100 μm), along with Cartesian arrays of discs of various sizes and spacings (20 and 50 μm diameter with pitches of 100 and 150 μm). A layer of patterned positive photoresist was used as a corral to help contain the liquid over the electrodes, and this is visible as a square pattern around the electrodes in Fig. 1.

B. Test Structure Fabrication

The test structure design consisted of a metal layer sandwiched between two insulation layers. Firstly, a 500 nm thick thermal silicon dioxide (SiO_2) insulator was grown on a 100 mm silicon (Si) substrate. 50 nm of platinum (Pt) with a

2018 INTERNATIONAL CONFERENCE ON MICROELECTRONIC TEST STRUCTURES, MARCH 19-22, 2018, AUSTIN, TEXAS, USA

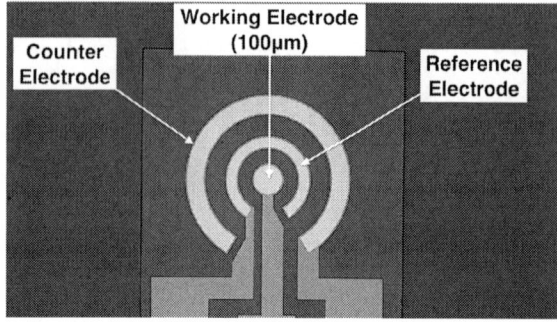

TABLE I
THE AVERAGE MEASURED CONTACT ANGLES OF THE SIN, PT, AND
PHOTORESIST SURFACES, $N = 6$ $(\pm 3\sigma)$

Material	Contact Angle
Silicon Nitride	$36° \pm 7°$
SPR350-3	$70° \pm 8°$

and SiN made using de-ionised (DI) water. The difference in mean contact angle of $34°$ is sufficient to help confine the liquid to the electrode area. Fig. 3(a) and (b) show a photograph of the liquid electrolyte contained within the more hydrophilic pattern in the photoresist. If a larger difference in hydrophobicity is desired, more hydrophobic materials such as polytetrafluoroethylene (PTFE) or Parylene-C^{TM} could be employed [9], [10]. However, this requires considerably more processing, which is not desirable in the middle of a pre-existing process. Although photoresist is not hydrophobic, its benefit lies in its simplicity to apply and then remove once testing is finished.

III. EXPERIMENTAL METHODS

A. Experimental Setup

The electrochemical measurements were made using an Everbeing EB8 manual probe station inside a shielded dark box. The probe needles were placed on the die contact pads and connected to an Autolab Micro III potentiostat (Metrohm). The solution was dispensed through a micro-pipette tip which was attached to another micromanipulator arm and placed over the electrode area. Silicone tubing connected the pipette tip to a syringe, enabling control over the flow of liquid. The full experimental set up on the manual probe station is presented in Fig.4.

Results from this setup were compared against benchmark measurements made in a traditional electrochemical setup. This consisted of a glass beaker in a Faraday cage, with separate Pt CE and pseudo RE. These electrodes were shards of Si, coated in 500 nm SiN and then 50 nm of Pt on a 10 nm Ti adhesion layer. The WE was a 20 μm diameter Pt disc.

B. Measurements

A standard electrochemical measurement technique called cyclic voltammetry was used to characterise the electrodes in the probe station setup. This involves sweeping the potential at the WE (vs. the RE) between two defined values and back while monitoring the current. This measurement is typically performed in an electrolyte containing a substance which undergoes electrochemical reduction and oxidation in the chosen potential range, a redox couple. Ferri/ferrocyanide ($[Fe(CN)_6]^{3-}$/$[Fe(CN)_6]^{4-}$) is a commonly used and well-understood redox couple [11], [12], and was hence used for the experiments in this paper. The solution used throughout this work was 1 mM of both potassium ferricyanide and potassium ferrocyanide in a background electrolyte of 100 mM potassium chloride.

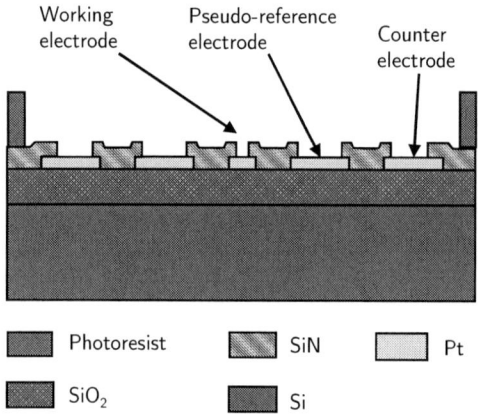

Fig. 2. A cross section of the test structure design showing the layers, with the electrodes labelled. The thickness of the layers have been exaggerated for clarity.

10 nm titanium (Ti) adhesion layer was then deposited and patterned to create electrodes, contact pads, and interconnects. 500 nm of silicon nitride (SiN) was then deposited using plasma enhanced chemical vapour deposition to insulate the metal from the solution. Windows were then etched into this insulation to provide access to the contact pads and electrodes. A cross section (left to right in Fig. 1) of the die through the WE is presented in Fig. 2.

C. Hydrophobic Liquid Corral

To help corral the liquid electrolyte over the electrodes, a layer of SPR 350-1.2 (Rohm and Haas) photoresist was spin coated onto the wafer and patterned to expose the contact pads (for electrical connections) and a 900 μm edge length square area over the electrodes. As the photoresist is less hydrophilic than the SiN insulator, it helped prevent liquid wicking up the to the contact pads. To demonstrate this, Table I presents contact angle measurements of the photoresist

978-1-5386-5072-1/18 $31.00 © 2018 IEEE

2018 INTERNATIONAL CONFERENCE ON MICROELECTRONIC TEST STRUCTURES, MARCH 19-22, 2018, AUSTIN, TEXAS, USA

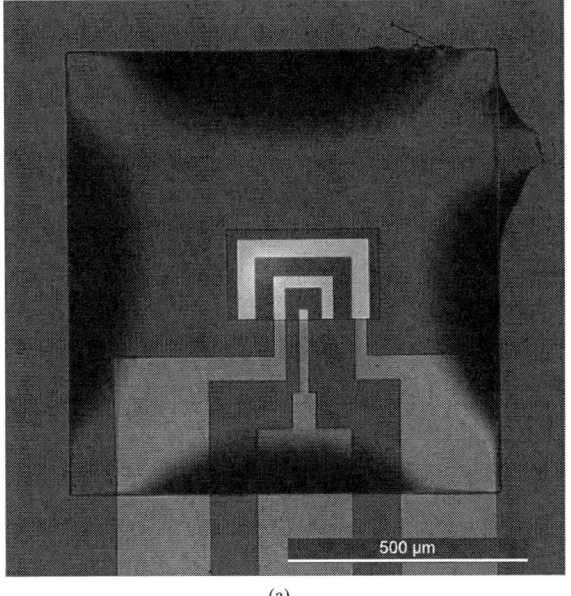

(a)

(b)

Fig. 3. Photographs showing the photoresist corralling DI water over an electrode (a) from the top down and (b) at an angle.

Despite the shielding of the probe station, it was observed that the measurements recorded on the probe station setup suffered from electrical noise superimposed on the electrochemical response. Therefore, a Fast Fourier Transform smoothing filter over an 8 point window was used to filter the data using Origin 2016 graphing and analysis software (Originlab).

IV. MEASUREMENT RESULTS

A. Comparison of On-Wafer and In-Beaker Results

Firstly, the capability of making electrochemical measurements at wafer level was assessed by comparing those made using in the probe station set up, with those made in a traditional electrochemical system. A cyclic voltammogram (CV) of the redox reaction of ferri/ferrocyanide recorded in (a) a beaker and (b) the probe station setup is presented in Fig. 5.

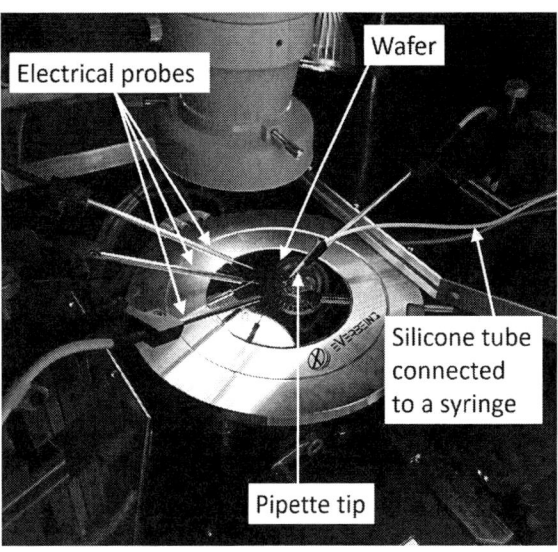

Fig. 4. Photographs of the experimental set up showing the probe station, electrical probes, and pipette tip.

TABLE II
THE AVERAGE MEASURED LIMITING CURRENT OF THE TWO SETUPS,
$N = 5(\pm3\sigma)$, COMPARED AGAINST THE PREDICTED VALUE.

Experiment	Current (nA)
Predicted	2.57
Wafer Level	2.77 $\pm$ 0.45
Beaker	2.48 $\pm$ 0.29

The shape of the CV is the expected wave in both set ups, characteristic of the electrochemical response of a microelectrode. As the potential is swept positive, ferrocyanide is oxidised to ferricyanide generating a positive current, while during the negative potential sweep, a negative current is generated by the reduction of ferricyanide to ferrocyanide. Under the presented conditions, the current is mass transport controlled by the diffusion of analyte to the electrode and reaches a steady state current (i_L). This current can be quantified using [13], [14],

$$i_L = BnFDcr \qquad (1)$$

where n is the number of electrons transferred in the reaction, F is the Faraday constant, D and c are the diffusion coefficient and concentration of the reactant respectively, and r is the radius of a disc electrode (or l, the edge length for a square). The B term is a dimensionless coefficient dependant on the geometry of the electrode, and is equal to 4 for an inlaid disc and 2.3 for an inlaid square. This enabled the comparison of predicted and measured currents. Using a literature value for D of 6.67×10^{-10} m^{2}s^{-1} [15], the predicted limiting current for the reduction reaction was calculated and is presented in Table II alongside the measured currents from both systems.

The three values match each other within error, lending further confidence that the probe station set up can be used to record quantifiable electrochemistry. Furthermore, the wave

978-1-5386-5072-1/18 $31.00 © 2018 IEEE

2018 INTERNATIONAL CONFERENCE ON MICROELECTRONIC TEST STRUCTURES, MARCH 19-22, 2018, AUSTIN, TEXAS, USA

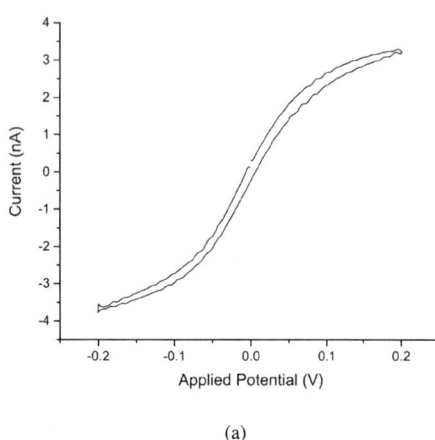

(a)

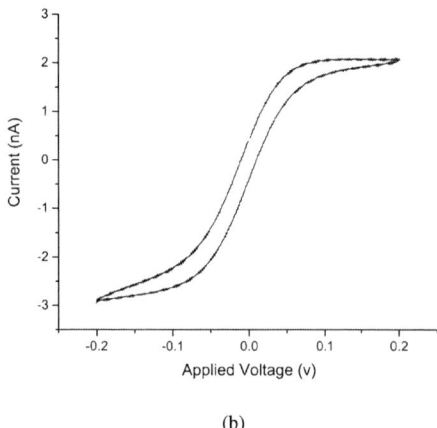

(b)

Fig. 5. (a) CV recorded on a 20 μm diameter disc electrode at $100\ \mathrm{mVs}^{-1}$ in the probe station and (b) the same measurement in the beaker setup.

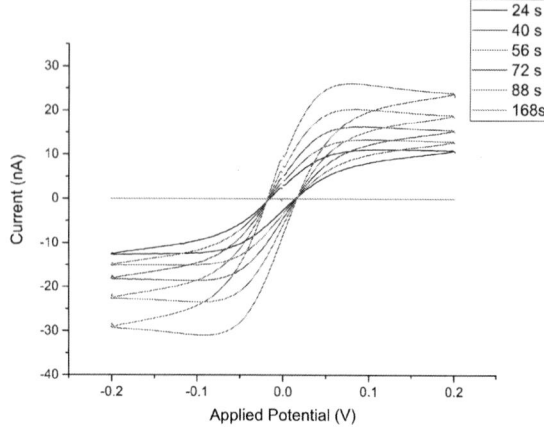

Fig. 6. CVs recorded over time on a 50 μm diameter disc electrode at 100 mVs^{-1}.

TABLE III
DETAILS OF ELECTRODE DESIGNS.

Electrode (Fig. 7(b))	Geometry	Size (μm)	Pitch (μm) /no. of electrodes
(i)	Disc Array	20	100 / 12
(ii)	Disc Array	50	100 / 9
(iii)	Disc Array	50	150 / 9
(iv)	Disc	20	- / 1
(v)	Disc	50	- / 1
(vi)	Disc	100	- / 1
(vii)	Square	20	- / 1
(viii)	Square	50	- / 1
(ix)	Square	100	- / 1

shape of the CVs are typical of an electrochemical reaction happening at a micro-scale electrode (where diffusion is the rate limiting step).

It is noteworthy that the variation in the wafer-level value is higher than the beaker. This could be due to the noisier measurements from the probe station, potentially arising from the leads to the potentiostat, which was not in the shielded box around the probe station. Another source of variation is the difference in ambient temperature between the two setups or, more likely, changes in concentration of the redox couple over time. This was seemingly caused by evaporation of water from the droplet, resulting in the relative concentration of redox couple increasing. This effect can be seen in the CVs presented in Fig. 6, where the currents associated with the reduction and oxidation reactions increase over time. Eventually the current suddenly decreases to near zero indicating complete evaporation of the liquid and thus loss of electrical contact between the electrodes. Another difference between the two setups is the difficulty in electrochemically cleaning electrodes on the wafer, arising from the requirement of changing be-

tween a cleaning solution and one containing the analyte for measuring. As a consequence, this likely contributed to the slight differences in shape between the CVs as well as the increased error for the probe station measurement.

B. Spatial Electrode Measurements

Wafer-level mapping of devices can be a powerful tool for understanding fabrication processes, as well as quantifying parameters such as yield and uniformity [16]. To demonstrate that capability can be expanded to electrochemical sensors, the same cyclic voltammetry measurements were performed across nine neighbouring electrodes shown in Fig. 7(a). The recorded CVs are shown in Fig. 7(b), and Table III lists the details of the designs of the electrodes. All the electrodes gave the expected electrochemical response, with the larger 100 μm electrodes shown in Fig. 7(b) (vi) and (ix) showing peaks characteristic of electrochemical reactions occurring at the surface of larger, macro-scale electrodes. This transition between macro- and microelectrode behaviour typically occurs between sizes of 20 50 μm and can be observed in Fig. 7(b) (v) and (vi). The arrays (Fig. 7(b) (i), (ii) and (iii)) show larger currents, proportional to the number of electrodes, while retaining the wave-like CV response.

2018 INTERNATIONAL CONFERENCE ON MICROELECTRONIC TEST STRUCTURES, MARCH 19-22, 2018, AUSTIN, TEXAS, USA

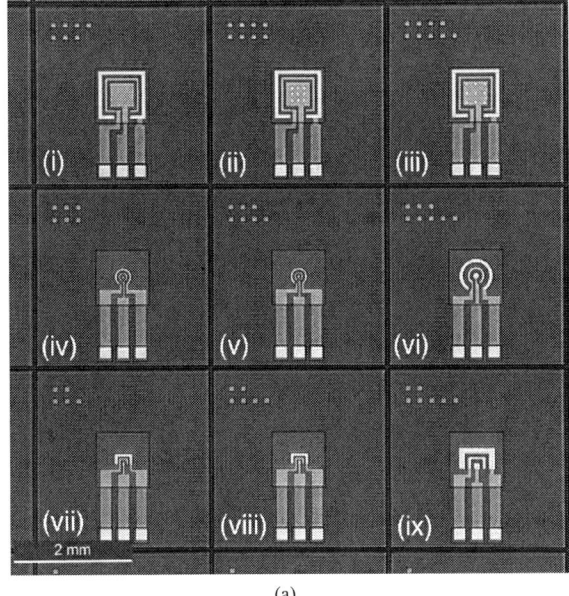

(a)

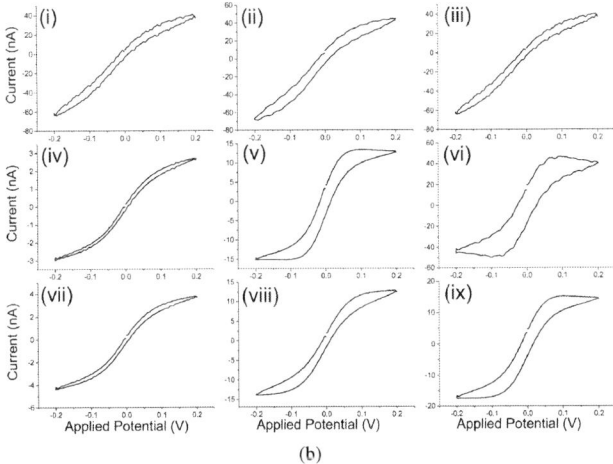

(b)

Fig. 7. (a) The nine electrodes on the wafer surface which were used to record the (b) CVs.

V. DISCUSSION

The results in Table II and Fig. 5 confirm that quantifiable electrochemical measurements can be made at wafer level using a standard probe station setup. However, the evaporation of the droplet presents a complication, as it both imposes a maximum measurement time and a change in analyte concentration over time. This makes quantification of the current, which typically requires a known concentration, beyond 30 to 60 seconds challenging. This also limited the time of the experiment as the droplets, which were typically around 1 mm across, evaporated completely in under 5 minutes in ambient conditions. While this presents a challenge for some measurements, it enables others, such as electrochemical monitoring of droplet evaporation. There are also methods of removing the evaporation factor. For example, a solution could be to employ

a seal such as that used by [17]. which would prevent the liquid evaporating or to have a constant new supply of solution to compensate. Although salt residue is left on the surface of the wafer once the droplet is dried, this is easily washed off with DI water. Electrochemical cleaning also presents a difficulty as changing between cleaning and measurement solutions would expose the electrode air, allowing reformation of surface oxides. This could be resolved by combining the redox and cleaning solution, although this is not possible in all cases.

Even with evaporation, the set up can be used to confirm operation of electrochemical sensors and identify faulty devices, as the value of limiting current is dependent on radius (or edge length of a square) through equation 1. Should the current deviate from the predicted value, then this could be indicative of pinholes or other defects in the top insulator. This process is much faster than making measurements in a beaker setup, especially in the case miniaturised sensors like those presented in this paper, which typically require packaging to insulate nearby wire bonds and contact pads. Being able to perform wafer-level electrochemical testing enables the fast collection of a large number of measurements. For example, setting up and then recording 10 CVs at each of the nine different electrodes presented in Fig. 7(b) took around 10 minutes. This process could be further expedited with the incorporation of both automatic liquid dispenser and wafer stage, enabling a fully automatic electrochemical measurement setup.

Future work is to develop the set up along the lines described above. An example of this is the bespoke microfluidic system, under development to replace the pipette tip shown in Fig. 8. The component is made from PDMS and was manufactured using a 3D printed scaffold made of Acrylonitrile Butadiene Styrene. PDMS was then cast around this and the scaffold dissolved in acetone [18]. It consists of two inlets at the top, one for solution and one for a non-pseudo reference electrode which will expand the range of possible electrochemical measurements. The outlet at the bottom is designed to seal around the electrode area, mitigating evaporation of the solution.

VI. CONCLUSION

A method of making electrochemical measurements at wafer-level has been demonstrated. This consists of a fluidic system for dispensing droplets in a standard probe station and was validated with wafer-level electrochemical test structures. Electrochemical measurements were performed on a wafer and, when quantified, matched the performance of those made in a conventional electrochemical cell. Electrodes of different designs were then measured and gave the expected response, demonstrating the rapid and simple characterisation of multiple electrode designs across a wafer.

ACKNOWLEDGMENT

This work was funded by the Engineering and Physical Sciences Research Council, through the following schemes:

- IMPACT programme grant (EP/K034510/1)

Fig. 8. Photograph of a microfluidic component, designed to replace the pipette tip.

- PACIFIC programme grant (EP/L018616/1)
- EPSRC CDT in Intelligent Sensing and Measurement (EP/L016753/1)
- New Engineering Concepts from Phase Transitions: A Leidenfrost Engine (EP/L018616/1)

All data presented in this paper can be accessed at http://dx.doi.org/10.7488/ds/2295.

The authors would like to acknowledge the support of Everbeing for the donation of the manual probe station used in this work.

REFERENCES

[1] D. K. Corrigan, J. P. Elliott, E. O. Blair, S. J. Reeves, I. Schmüser, A. J. Walton, and A. R. Mount, "Advances in electroanalysis, sensing and monitoring in molten salts," *Faraday Discussions*, vol. 190, pp. 351–366, 2016.

[2] E. González-Fernández, N. Avlonitis, A. F. Murray, A. R. Mount, and M. Bradley, "Methylene blue not ferrocene: Optimal reporters for electrochemical detection of protease activity," *Biosensors and Bioelectronics*, vol. 84, pp. 82–88, Oct. 2016.

[3] M. A. Johnson, "In vivo electrochemical measurements: past, present and future," *Bioanalysis*, pp. 119–122, 2013.

[4] R. J. Forster, "Microelectrodes: new dimensions in electrochemistry," *Chemical Society Reviews*, vol. 23, no. 4, pp. 289–297, 1994.

[5] J. G. Terry, I. Schmüser, I. Underwood, D. K. Corrigan, N. J. Freeman, A. S. Bunting, A. R. Mount, and A. J. Walton, "Nanoscale electrode arrays produced with microscale lithographic techniques for use in biomedical sensing applications," *IET nanobiotechnology*, vol. 7, no. 4, pp. 125–134, Dec. 2013.

[6] N. Jamil, S. Smith, Y. Yang, J. Jia, P. Bagnaninchi, and E. González-Fernández, "Design and fabrication of microelectrodes for electrical impedance tomography of cell spheroids," in *IECBES 2016 - IEEE-EMBS Conference on Biomedical Engineering and Sciences*, Feb. 2017, pp. 426–431.

[7] E. O. Blair, D. K. Corrigan, J. G. Terry, A. R. Mount, and A. J. Walton, "Development and Optimization of Durable Microelectrodes for Quantitative Electroanalysis in Molten Salt," *Journal Of Microelectromechanical Systems*, vol. 24, no. 5, pp. 1346–1354, 2015.

[8] S. Bahukudumbi and K. Chakrabarty, *Wafer-Level Testing and Test During Burn-In for Integrated Circuits.* Artech House, 2010.

[9] H. Zou, Y. Li, S. Smith, A. Bunting, A. J. Walton, and J. Terry, "Modification and characterisation of material hydrophobicity for surface acoustic wave driven microfluidics," in *Microelectronic Test Structures (ICMTS), 2012 IEEE International Conference on*, 2012, pp. 61–65.

[10] Y. Li, E. O. McKenna, W. Parkes, P. A. R., and A. J. Walton, "The application of fixed hydrophobic patterns for confinement of aqueous solutions in proteomic microarrays," *Appl. Phys. Lett.*, vol. 99, pp. 073 703–1 – 073 703–3, 2011.

[11] P. H. Daum and C. G. Enke, "Electrochemical kinetics of the ferri-ferrocyanide couple on platinum," *Analytical Chemistry*, vol. 41, no. 4, pp. 653–656, May 2002.

[12] J. J. Van Benschoten, J. Y. Lewis, W. R. Heinemann, D. A. Roston, and P. T. Kissinger, "Cyclic voltammetry experiment," *Journal of CHemical Education*, vol. 60, no. 9, pp. 772 – 776, 1983.

[13] Y. Saito, "A theoretical study on the diffusion current at the stationary electrodes of circular and narrow band types," *Review of Polarography (Japan)*, vol. 15, no. 6, pp. 177 – 187, 1968.

[14] I. Schmueser, A. J. Walton, J. G. Terry, H. L. Woodvine, N. Freeman, and A. R. Mount, "A systematic study of the influence of nanoelectrode dimensions on electrode performance and the implications for electroanalysis and sensing," *Faraday Discussions*, vol. 164, pp. 295 – 314, 2013.

[15] S. J. Konopka and B. McDuffle, "Diffusion coefficients of ferri- and ferrocyanide ions in aqueous media, using twin-electrode thin-layer electrochemistry," *Analytical Chemistry*, vol. 42, no. 14, pp. 1741 – 1746, 1970.

[16] S.-C. Hsu and C.-F. Chien, "Hybrid data mining approach for pattern extraction from wafer bin map to improve yield in semiconductor manufacturing," *International Journal of Production Economics*, vol. 107, pp. 88 – 103, 2007.

[17] A. Poghossian, K. Schumacher, J. Kloock, C. Rosenkranz, J. Schultze, M. Müller-Veggian, and M. Schöning, "Functional Testing and Characterisation of ISFETs on Wafer Level by Means of a Micro-droplet Cell," *Sensors*, vol. 6, no. 4, pp. 397–404, Apr. 2006.

[18] A. Buchoux, P. Valluri, S. Smith, A. A. Stokes, P. R. Hoskins, and V. Sboros, "Manufacturing of microcirculation phantoms using rapid prototyping technologies," *2015 37th Annual International Conference of the IEEE Engineering in Medicine and Biology Society (EMBC)*, pp. 5908–5911, Aug. 2015.

2018 INTERNATIONAL CONFERENCE ON MICROELECTRONIC TEST STRUCTURES, MARCH 19-22, 2018, AUSTIN, TEXAS, USA

Test Structure for Electrical Assessment of UV Laser Direct Fine Patterned Material

Naoto Usami[1], Akio Higo[2] Ayako Mizushima[2], Yuki Okamoto[1], and Yoshio Mita[1,3]

[1]Dept. of Electrical Engineering and Information Systems, The University of Tokyo, 113-8656 Japan

Email: {usami,mems}@if.t.u-tokyo.ac.jp

[2]VLSI Design and Education Center, The University of Tokyo, 2-11-16, Yayoi, Bunkyo-ku, Tokyo 113-0032, Japan

[3]Dept. of Spacecraft Engineering, Institute of Space and Astronautical Science, Japan Aerospace Exploration Agency, Japan

Abstract—We propose a test structure to electrically assess direct laser fine patterning, which is entering a microelectronic era (below $10\mu m$). Indium-Tin-Oxide (ITO) was used as a material example. High speed ITO patterning with laser ablation can contribute short turn-around-time development of opto-electrical devices, such as organic light emitting diode. However, not only machine-induced line-edge fluctuation but also the process (e.g. heat) induced material degradation may affect electrical linewidth. The aim of our test structure is to assess such critical dimension change through measurement of electrical property (i.e. conductivity). It consists of Kelvin-connection straight lines and Greek crosses with various widths. Ultraviolet (UV) laser process as well as lithography and plasma etching were applied with the same test structure. The measurement revealed that the applied direct patterning condition induced small damage, showing applicability of direct patterning in microelectronics R&D.

I. INTRODUCTION

Direct laser patterning process is an attractive alternative for short-turn-around-time microelectronic devices fabrication. Laser patterning may supersede classical lithography-and-etching and can drastically reduce process steps as well as chemical wastes. Due to the continuous evolution on laser source and optical scanning methods, a number of machines with reasonable price are being introduced. One of the most attractive applications is an optical device: for example, organic and quantum light emitting diodes (OLED [1] and QLED [2], respectively) and liquid crystal display (LCD [3]). The main target material of direct laser patterning for optical devices is indium-tin oxide (ITO) thin film, as reported by Li *et al.* [4]. There are a couple of reports on degradation issues such as high temperature sensitivity [5], thermal damage on surrounding areas [6], and electrical characteristic change [7]. To obtain fine (referring to as fine as $5\,\mu m$) patterns, such damages as well as critical dimension (CD) dependence on cutting condition and location must be assessed carefully.

Laser direct patterning is essentially a process to remove material as shown in Fig. 1(a). In a standard procedure, the contour of the pattern is firstly removed and then remaining large "rubout" area is removed. Fig. 1(b) show four potential issues affecting CD loss. Two issues may result in systematic CD discrepancy are: (1) material removal radius r may differ from presumed one r_p, (2) long-time exposure may introduce heat damage to the material. The other two issues may introduce random CD fluctuation: (3) the fluctuation of laser

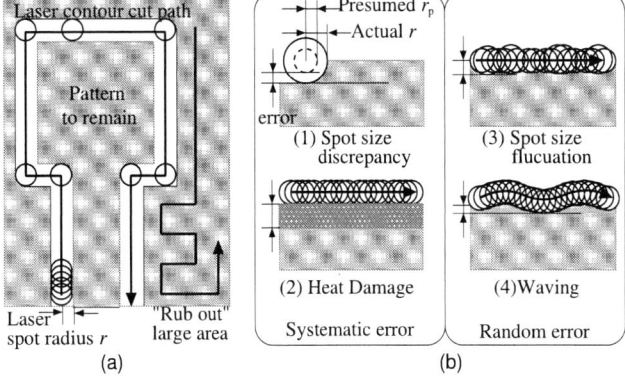

Fig. 1. (a) Laser direct patterning . (b) Issues of laser patterning all affecting electrical critical dimension width.

spot size will directly appear as an edge roughness, (4) laser spot location may wave. Engineers may optically observe for first aid but the issues should be evaluated in electrical manner in the end.

In this paper, we propose a test method to assess the laser direct patterning process. The aim of our test structure is to assess such critical dimension change through measurement of electrical characteristics. We fabricated laser-patterned and classically-processed samples and examine dimension and / or heating influences through measurements of them.

II. DESIGN AND EXPERIMENTS

Our test structure consists of each four-taps straight wires with various widths and Greek crosses with various arm widths as shown in Fig. 2. The influence of the laser process can be assessed by conductivity measurements of the structures. Fig. 3(a) presents the entire layout of our test structure. It is composed of various widths (w =2, 4, 8, 16, 32, 64, 128, $256\mu m$) and lengths (l =1, 2, 4 mm) bridge wires with four taps. We also prepared Greek cross test structures for sheet resistance measurements with various arm width (w =2, 4, 8, 16, 32, 64 μm) according to Ref. [8].

In this experiment, we used an UV laser processing machine (λ = 355 nm) for direct patterning. The processing condition was optimized for smallest spot size: Laser pulse frequency was 140kHz, the power was 2 W and the scanning speed

978-1-5386-5072-1/18 $31.00 © 2018 IEEE   185

2018 INTERNATIONAL CONFERENCE ON MICROELECTRONIC TEST STRUCTURES, MARCH 19-22, 2018, AUSTIN, TEXAS, USA

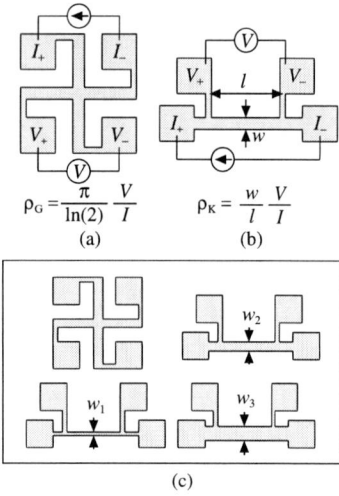

$$\rho_G = \frac{\pi}{\ln(2)}\frac{V}{I} \qquad \rho_K = \frac{w}{l}\frac{V}{I}$$

(a)   (b)

(c)

Fig. 2. (a)Greek cross for sheet resistance measurement, (b) Line width extraction structure. (c) Test structure consists of Greek cross and line width extraction structure with a couple of different line widths.

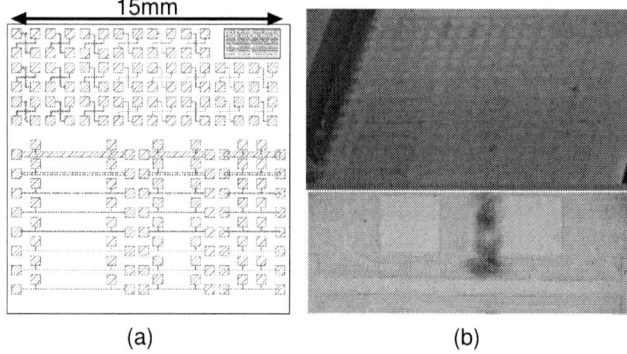

(a)   (b)

Fig. 3. The test structure for the assessment under the laser ablation process (a) layout (b) actual fabricated sample and the four-taps wire.

was 500 mm/s. The processed sample is shown in Fig. 3(b). To compare direct laser patterning process with classical photolithography, we also prepared plasma-etched sample. A 5-inch Cr/CrO_2 photomask (type 5009) was made with EB lithography and wet etching. The pattern was transfered to ITO/glass substrates by Süss MA6 mask aligner. One sample was etched by $Cl_2 + BCl_3 + Ar$ reactive ion etching (RIE, ULVAC NE-550), and oxide plasma ashing was also applied after etching. Another sample was etched by mix-acid ITO etchant (ITO-02, Kanto Chemical Co., Inc.)

Fig. 4 presents a test strategy with our structure. As seen Fig. 1(b), there may be spot size discrepancy that produce systematic size difference, and laser fluctuation and waving may make random variation. They end up with resistance variation of four-taps-bridge wires through dimension error. To calculate sheet resistance from four-taps-bridge wires, measured resistance R_K is divided by designed length per width:

$$\rho'_K = \frac{V}{I}\cdot\frac{w}{l} = R_K\frac{w}{l}. \qquad (1)$$

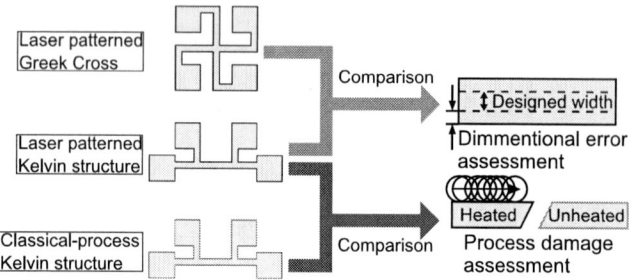

Fig. 4. Assessment procedure. Line width structures provide dimension error and can be doubly checked with Greek cross. Greek cross comparison with other patterning methods provides heat influence.

However the value given by equation (1) contains dimension error Δw. Therefore we call it *apparent* sheet resistance ρ'_K in this article. True (or intrinsic) sheet resistance ρ_K is given by following equation:

$$\rho_K = R_K\frac{w+\Delta w}{l} = R_K\frac{w}{l}\cdot\frac{w+\Delta w}{w} = \rho'_K\frac{w+\Delta w}{w}. \qquad (2)$$

On the contrary, measured sheet resistance ρ_G of Greek cross structure is not influenced by absolute width fluctuation. The only possible cause of ρ_G variation is heat damage. Classical dry and wet etched sample may not be affected by heat damage in fig. 1(b). Therefore a comparison between laser and classically etched sample is helpful for heat damage assessment.

III. MEASUREMENTS AND ANALYSES

The measurements were made for all test patterns using probe station (Süss PM8) and semiconductor parameter analyzer (Keysight B1500). Line width structures were measured with constant voltage $V = 0.5$V, and Greek crosses were measured with constant current $I = 1$mA. Figure 5(a) shows the resistances of the wires on the laser-processed sample. These values were then converted to apparent sheet resistances ρ'_K by taking designed wire widths and lengths, as shown in Fig. 5.(b). As seen clearly in the figure, apparent sheet resistances ρ'_K decreased especially for smaller w. It is due to the discrepancy between preset spot radius r_p and actual effective laser spot radius r. As shown in Fig 6, the machine defines laser path r_p apart from the edge of the pattern to be remained. The spot size was kept constant regardless of r_p; In reality, because the laser beam is Gaussian, the effective laser spot size r becomes smaller (or larger) than that of presumed one r_p, depending on the power condition and absorption coefficient of the material. The difference of real and preset (expected) radii causes the width difference: $\Delta w = 2(r_p - r)$. At the process development an engineer must presume some value for laser spot radius r_p; in the experiments of Fig. 5 was 10 μm, derived from the standard value of our machine. The measured ρ'_K could be least square fit with equation (2). The fitting resulted in $\rho_K = 10\,\Omega$ and $\Delta w = 8.26\,\mu$m for $L = 1$ mm and $\Delta w = 8.98\,\mu$m for $L = 2$ and 4 mm. It means that the actual size was larger than designed one, indicating smaller effective

978-1-5386-5072-1/18 $31.00 © 2018 IEEE   186

2018 INTERNATIONAL CONFERENCE ON MICROELECTRONIC TEST STRUCTURES, MARCH 19-22, 2018, AUSTIN, TEXAS, USA

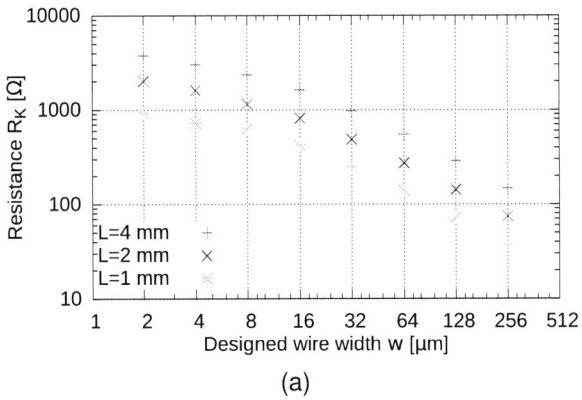

(a)

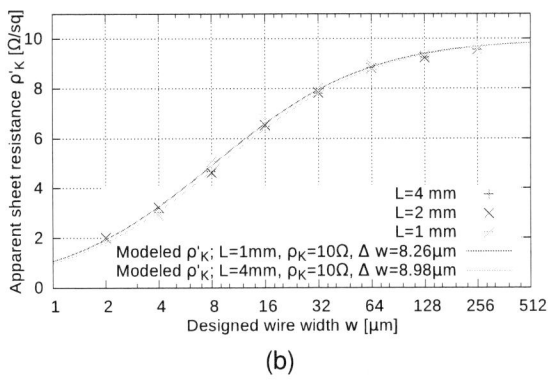

(b)

Fig. 5. (a) Resistance R_K and (b) sheet resistance ρ_K vs. wire width characteristics from line width test structure with laser patterning.

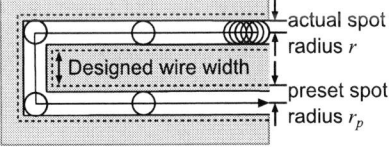

Fig. 6. Actual laser spot becomes smaller than supposed one derived from machine spec.

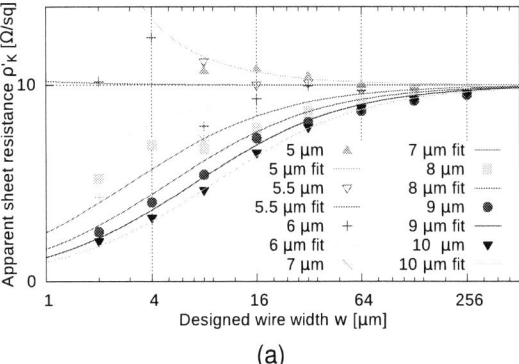

(a)

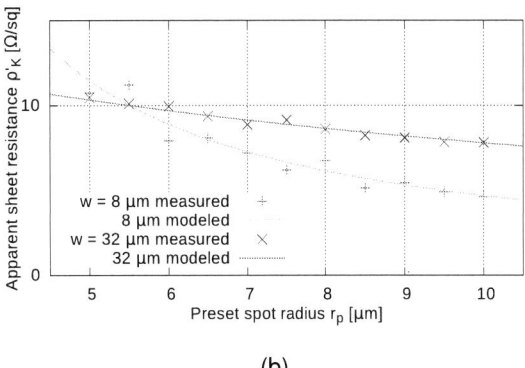

(b)

Fig. 7. Measured and model fit sheet resistance characteristics under various preset spot sizes r_p: (a) Sheet resistance vs. designed wire width. (b) Sheet resistance vs. spot size with fixed line widths, 32 μm and 8 μm. $L = 2$mm for both cases.

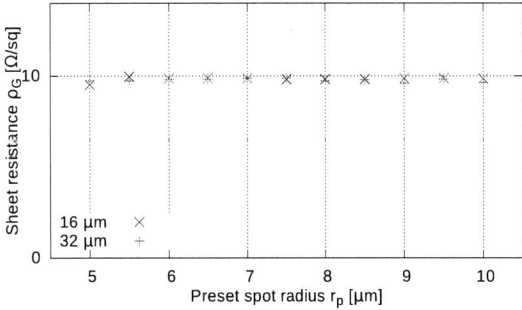

Fig. 8. Sheet resistance vs. processing-spot size characteristics from Greek cross structures with arm widths $w = 16$ and 32 μm.

spot size.

For further investigation about the effective spot size diminishing, we fabricated the test structures in changing preset laser spot radii (r_p=5, 5.5, 6, 6.5, 7, 7.5, 8, 8.5, 9, 9.5, and 10 μm). Figure 7(a) shows apparent sheet resistance vs. designed wire width characteristics derived from the line width test structures. Some narrow patterns ($w = 2$ and 4 μm) with small radii ($r_p = 5, 5.5, 6.5$ μm) could not be fabricated because of the random waving influence (Fig. 1), resulting a couple of missing points. The apparent sheet resistance ρ'_K could also be well represented with equation (2), by including initial offset: $\Delta w \rightarrow \Delta w - 2(10 - r_p)$ [μm] while keeping the same intrinsic sheet resistance $\rho_K = 10\,\Omega$ and offset value $\Delta w = 8.98$ μm. Flat dependency of apparent sheet resistance ρ'_K on structure widths w was found around $r_p = 5.5$ μm, which is coherent with the extracted offset value ($\Delta w = 8.98$ μm). Fig 7(b) shows

sheet resistance vs. spot size with fixed line widths, $w = 32$ μm and 8 μm. The dependence can also be explained by the same equation:

$$\rho'_K = \rho_K \frac{w}{w + \Delta w - 2(10 - r_p)}. \qquad (3)$$

Figure 8 show sheet resistances obtained from Greek cross. As clearly seen, the value was constant independently of process spot sizes and designed arm widths of Greek crosses. The sheet resistance value derived from Greek cross was $\rho_G = 10\Omega$. The

978-1-5386-5072-1/18 $31.00 © 2018 IEEE

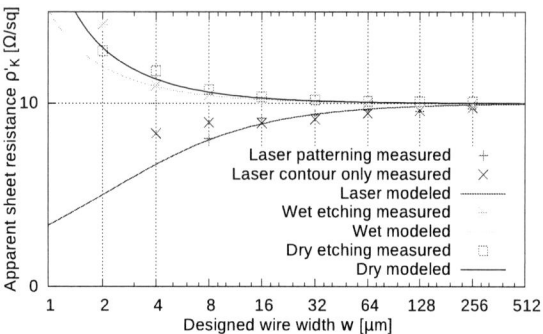

Fig. 9. Apparent sheet resistance ρ_K vs. wire width w relationships under various patterning methods.

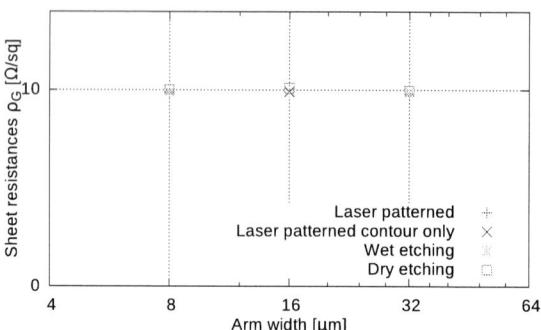

Fig. 10. Sheet resistances ρ_G characteristics from Greek cross structures under various processes.

value matched to the ρ_K obtained from the line width test structures, thereby validating the method and measurements.

Then dry and wet processed samples with standard photolithography were measured and compared with laser patterned one. In addition, another sample was fabricated in which only a contour of the structure was laser cut, without rubbing-out the rest of the chip. The last chip was made to asses the potential heat damage of the ITO film by the rubout step after contour insulating. For the laser processes, the preset radius was set to $r_p = 6.5\,\mu m$. Fig. 9 presents apparent sheet resistances ρ'_K vs. wire width characteristics derived from the line width test structure. The measurement results are shown in Fig. 9; There were no difference in apparent sheet resistances in between rubout and non-rubout samples. Again, the relationship could be represented by equation (2) with $\rho_K = 10\,\Omega$ and $w = 8.98\,\mu m$. The apparent sheet resistances of dry- and wet-etched samples increased as the line width decreased, showing line width reduction (thinning) during the process. The thinning offset was least-square fit from measurement data to be $\Delta w = -0.34\,\mu m$ for wet process and $\Delta w = -0.47\,\mu m$ for dry process. The discrepancies in between modeled line and measured points were larger for smaller w; it can be attributed to pattern fluctuation due to the laser scanning process. On the contrary, Fig. 10 presents sheet resistances ρ_G derived from the Greek cross structures with different technologies; All sheet resistance values were very close each other, which suggests that the sheet resistance was not changed by any of the processes. These series results indicate that heat damage was small.

IV. CONCLUSIONS

In this paper, we reported the electrical critical dimension assessment of laser direct patterned ITO film on a glass substrate. The measurements of the laser-patterned ITO wires and analyses on apparent sheet resistances ρ_K with simple width offset model extracted a real laser spot width that gives electrical critical dimension. The Greek cross sheet resistances were constant independently of their width and technologies, showing that the heat damage was not pronouncing for the tested machine conditions. Both sheet resistances from line

width test structure and Greek cross were coherent, thus validating the measurements and analyses. In conclusion, it is possible to precisely and quickly determine effective laser spot radii and its fluctuations. The method is applicable to various new materials associated with laser direct processing conditions, thereby providing semiconductor manufacturing with a short turn-around-time devices and materials development capability.

ACKNOWLEDGMENTS

A part of this work is supported by Japan Society for the Promotion of Science (JSPS) KAKENHI (grant nos. 16H04345, 17J04382), Japan Science and Technology Agency(JST) Collaboration Hubs for International Research Program, ISAS/JAXA Strategic Research Grant and French National Research Agency Grant (ANR-16-CE33-0022). Test Structure GDS design was supported by VLSI Design and Education Center (UTokyo VDEC), the University of Tokyo, in collaboration with Cadence Design Systems, Inc. The series processes were done by using the open facilities of VDEC accessible through MEXT's Nanotechnology Platform Program in UTokyo VDEC.

REFERENCES

[1] S. Naithani, R. Mandamparambil, H. Fledderus, D. Schaubroeck, and G. V. Steenberge, "Fabrication of a laser patterned flexible organic light-emitting diode on an optimized multilayered barrier," *Applied Optics*, vol. 53, no. 12, pp. 2638–2645, 2014.

[2] H. Peng, Y. Jiang, and S. Chen, "Efficient vacuum-free-processed quantum dot light-emitting diodes with printable liquid metal cathodes," *Nanoscale*, vol. 8, no. 41, pp. 17765–17773, 2016.

[3] M. F. Chen, Y. P. Chen, W. T. Hsiao, and Z. P. Gu, "Laser direct write patterning technique of indium tin oxide film," *Thin Solid Films*, vol. 515, no. 24, pp. 8515–8518, 2007.

[4] Z. H. Li, E. S. Cho, and S. J. Kwon, "Laser direct patterning of the T-shaped ITO electrode for high-efficiency alternative current plasma display panels," *Applied Surface Science*, vol. 257, no. 3, pp. 776–780, 2010.

[5] Y. Uenaga and S. Takayama, "Abnormal resistivity change in indium-tin oxide films," *Journal of Japan Institute of Metals*, vol. 71, no. 9, pp. 751–757, 2007.

[6] D. A. Willis, "Thermal mechanisms of laser micromachining of indium tin oxide," in *Lasers and Applications in Science and Engineering*, 2004, pp. 313–320.

[7] J. G. Lunney, R. R. O'Neill, and K. Schulmeister, "Excimer laser etching of transparent conducting oxides," *Applied Physics Letters*, vol. 59, no. 6, pp. 647–649, 1991.

[8] S. Enderling, C. L. Brown, S. Smith, M. H. Dicks, J. T. M. Stevenson, M. Mitkova, M. N. Kozicki, and A. J. Walton, "Sheet resistance measurement of non-standard cleanroom materials using suspended Greek cross test structures," *IEEE Transactions on Semiconductor Manufacturing*, vol. 19, no. 1, pp. 2–9, 2006.

978-1-5386-5072-1/18 $31.00 © 2018 IEEE

2018 INTERNATIONAL CONFERENCE ON MICROELECTRONIC TEST STRUCTURES, MARCH 19-22, 2018, AUSTIN, TEXAS, USA

Open Model for External Mechanical Stress of Semiconductors and MEMS

R. T. Buhler[1] and R. C. Giacomini[1]

[1]Department of Electrical Engineering
Centro Universitário FEI, São Bernardo do Campo, SP, Brazil
Email: rtbuhler@ieee.org

Abstract— **This paper defines the details of the bending equipment solution and the calibration required for characterization of external mechanical stress in semiconductors and MEMS. The equipment is suited for use in probe station for electrical characterization of devices under controlled external mechanical stress.**

Keywords—external mechanical stress, semiconductor, MEMS.

I. INTRODUCTION

The adoption of strained materials to enhance the carriers' mobility by the semiconductor industry has become standard for high performance applications, especially during the last decade and the recent rise of the IOT trend. By the year of 2020 it is predicted that 20 billion devices using the strained silicon technology will be connected to the internet [1]–[5]. The development of a bending setup that allows the experimental study of strained materials, without having access to a foundry is covered in this paper. It allows the study of elastic properties of crystalline substrate [6], the repetitive bending effects [7], bandgap change in optical devices [8] and MEMS [9], to cite a few examples. Some previous research approaches [2], [10], [11] were not developed to be a permanent equipment solution and, to the best of our knowledge, there is no commercial solution available specific in microelectronics. This paper intends to give the start into the direction of defining the details of the bending equipment and the calibration used for characterization of mechanical stress in semiconductor materials by releasing it as an open model under the Creative Commons license [14], with all the required files freely available to download, use, improve and shared again by the scientific community.

II. MECHANICAL STRESS PRINCIPLE

In semiconductors, the mechanical stress alters the crystallographic structure, changing the carriers' mobility, given by (1) and (2), bandgap, scattering rate, among other parameters. "m_l" and "m_t" are the longitudinal and transversal masses, respectively.

$$\mu = \frac{q\tau}{m^*} \tag{1}$$

where:

$$m^* = \left[\frac{1}{6}\frac{2}{m_l} + \frac{4}{m_t}\right]^{-1} \tag{2}$$

In n-MOSFETs, the biaxial tensile stress (Fig. 1a) reduces the effective mass of electrons in-place, while increasing the out-of-plane mass [3]. For p-MOSFETs, the biaxial compressive stress (Fig. 1b) reduces the hole's masses, enhancing its mobility.

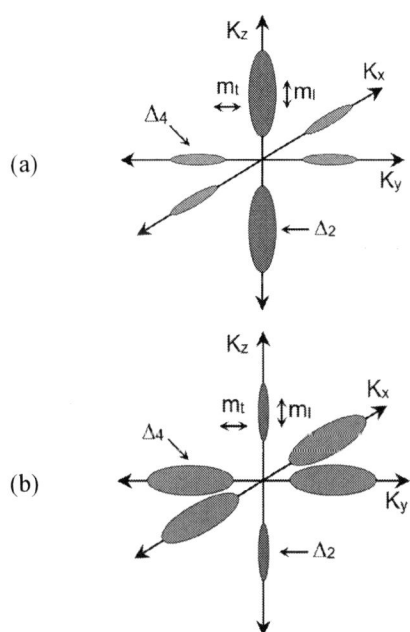

Fig. 1. Ellipsoids of constant energy in k-space for biaxial (a) tensile stress and (b) compressive stress in silicon.

The stress can be uniaxial or biaxial, and when applied in silicon there is a slip in the conduction band energy, with the in-plane (Δ_4) and out-of-plane (Δ_2) ellipsoids changing its size. The uniaxial stress configuration uses the four-point bending technique presented in fig. 2 to produce a large portion of maximum stress above the two central bearings (identified in red) as the moment, and therefore the radius of the curvature of the beam, is nearly constant between the load points [6], [12].

978-1-5386-5072-1/18 $31.00 © 2018 IEEE

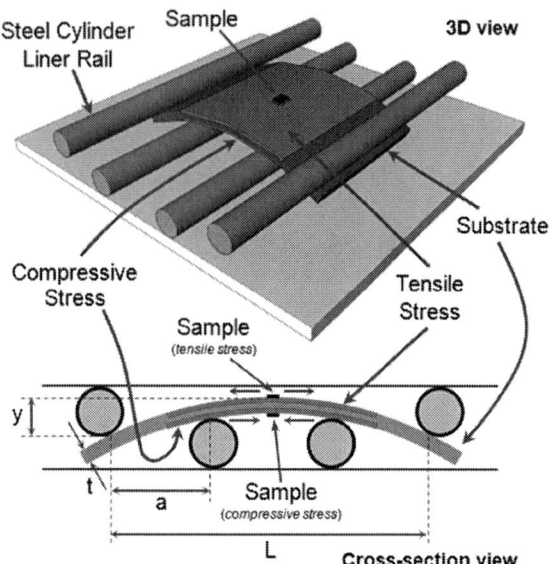

Fig. 2. 3D and cross-section view of the four-point bending principle.

The biaxial stress use two concentric rings with different diameters to bend the wafer in two directions. Here, the uniaxial stress will be demonstrated.

III. PROJECT DEVELOPMENT AND LICENSING

The project was developed using the SketchUp 3D modeling software, from Trimble Inc. [13]. The bending equipment and the probe support were designed for use in the Cascade Microtech REL 3600 probe station as it was the one available in our laboratory. However, the design files can be easily adapted to other probe stations.

All original design files are made available in the SketchUp ".skp" and the universal ".stl" file format under the Creative Commons – Attribution – Non Commercial – Share Alike 3.0 Unported (CC BY-NC-SA 3.0) license [14] in the MakerBot's Thingiverse platform [15]. This Creative Commons license means that:

You are free to:

- Share: copy and redistribute the material in any medium or format

- Adapt: remix, transform, and build upon the material

Under the following terms:

- Attribution: You must give appropriate credit, provide a link to the license, and indicate if changes were made. You may do so in any reasonable manner, but not in any way that suggests the licensor endorses you or your use.

- Non-Commercial: You may not use the material for commercial purposes.

- Share Alike: If you remix, transform, or build upon the material, you must distribute your contributions under the same license as the original.

- No additional restrictions: You may not apply legal terms or technological measures that legally restrict others from doing anything the license permits.

The CC BY-NC-SA 3.0 license was chosen to ensure that it can continue to be openly developed and distributed to the scientific community. This paper intends to give the direction in defining the details of the bending equipment and the calibration used for characterization of mechanical stress in semiconductor materials and MEMS. The current design is available in the following link:

https://www.thingiverse.com/thing:2765348

The Thingiverse platform allows the continuous update of the project, interaction with users and the creation of derivative files from the original project (a.k.a. remix). As the current design is updated or additional ones are developed, the collection page will be updated accordingly.

IV. THE BENDING EQUIPMENT

Having non-encapsulated IC samples in mind, the present bending setup can dose the amount of uniaxial stress induced in the sample. The mechanical bending parts and materials required are described in table 1.

Table 1. Parts and materials required for the mechanical bending setup.

Quantity	Description
4 pcs	LM8UU steel linear bearing bushing
4 pcs	90x12mm steel cylinder liner rail (vertical guides)
4 pcs	10x6mm steel cylinder liner rail
2 pcs	10mm long 3/8" threaded rod stainless steel
6 pcs	3/8" stainless steel finished hex nut
2 pcs	3/8" stainless steel nylon insert lock finished hex nuts
20 pcs	M3x6mm stainless steel zinc-plated phillips pan-head machine screw
4 pcs	M3x20mm stainless steel zinc-plated phillips pan-head machine screw
4 pcs	M3 stainless steel nylon insert lock finished hex nuts
4 pcs	Inside 3/8" outside 1" stainless steel flat washer
372g	PLA (Polylactic Acid) filament of 1.75mm
2 pcs	Step motor nema 17 – 3.5kgf.cm – 1.2A
2 pcs	DRV8825 Stepper Motor Driver Carrier, High Current
1 pc	Arduino Nano board based on the ATmega328

Table 2 describes the required parts and materials for the elevated probes support.

Table 2. Parts and materials required for the elevated probes support.

Quantity	Description
12 pcs	M4 stainless steel finished hex nut
8 pcs	M4x20mm stainless steel zinc-plated phillips pan-head machine screw
4 pcs	M4x10mm stainless steel zinc-plated phillips pan-head machine screw
132g	PLA (Polylactic Acid) filament of 1.75mm

In fig. 2, the sample attached to the upper part of the substrate will sense the longitudinal tensile stress transferred to it. When attached to the lower part it will sense the compressive stress.

The bending equipment and the elevated probe supports are projected to fit in the Cascade Microtech REL 3600 probe station and is presented in fig. 6 in the orange color, fully assembled in the probe station with the sample and probes in place, without losing the use of the microscope.

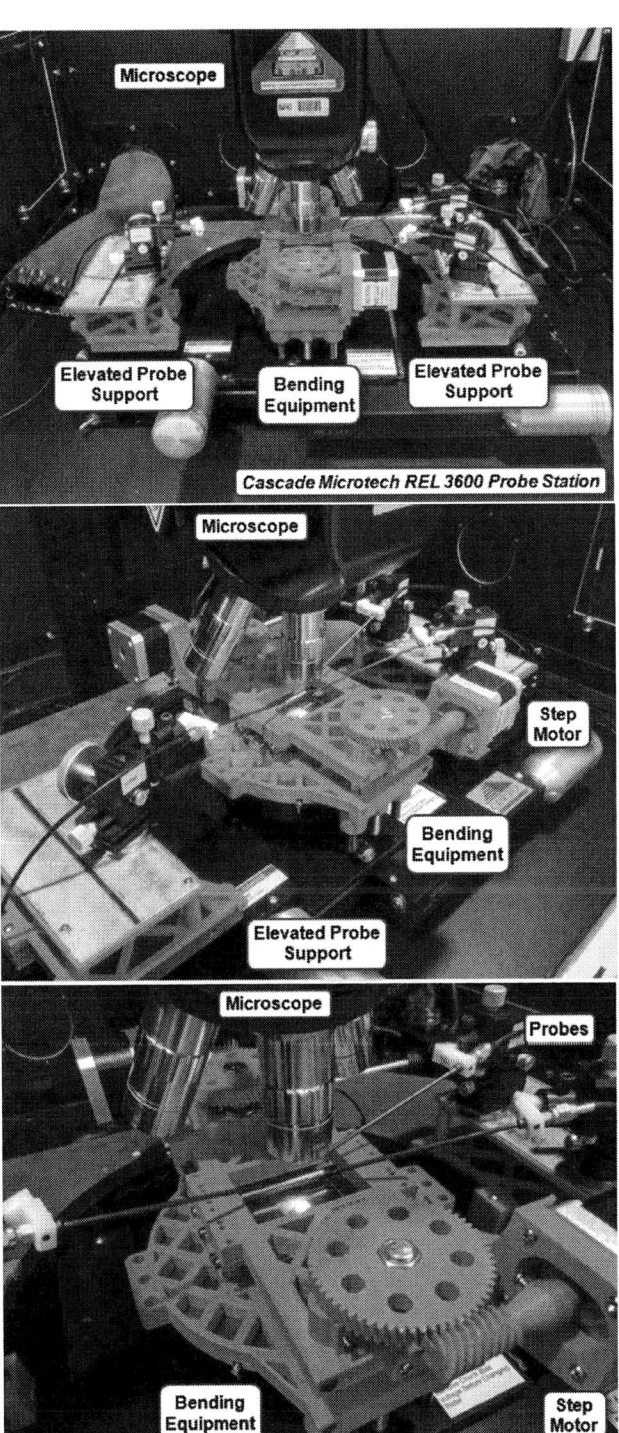

Fig. 3. Semiconductor bending equipment and the elevated probes support in the Cascade Microtech REL 3600 probe station.

All the main parts were designed and 3D printed with PLA (Polylactic Acid) filament of 1.75mm, reinforced with steel cylinder liner rails to ensure the stability of the equipment. Two attached step motors are controlled by an Arduino Nano board to control the movement of the bending equipment remotely from outside of the probe station.

V. STRAIN AND STRESS CALIBRATION

The deformation caused by the beam bending is measured using the strain gauge attached to the substrate by a suitable adhesive. As the substrate is bend, the tension or compression is transferred to the strain gauge, changing its resistance. The strain (ε) and the resistance (R) are related by the gauge factor (GF), defined by (3):

$$GF = \frac{\frac{\Delta R}{R} / \frac{\Delta L}{L}}{\frac{\Delta R}{R} / \varepsilon} \tag{3}$$

The strain is dimensionless and is defined as a fractional change in the length of the strain gauge. In semiconductor materials, such as silicon, a deformation of 1% displacement in Si lattice is comparable to the strain obtained from process fabrication flow using strain technology, enough to increase the electron and hole mobility [10], [16].

For our strain transfer test, the polycarbonate compact sheet substrate of 45x32mm, 2mm thick was used and the BF350-3AA strain gauge (fig. 4) with size of 7.1x4.5mm, resistance of 350Ω and tolerance lower than ±0.1%, gauge factor of 2.0~2.2 and strain limit of 2.0%. It is attached to the upper part of the substrate using the adhesive ethylcyanoacrylate, *Tekbond 793* (cyanoacrylate family) and submitted to a tensile strain.

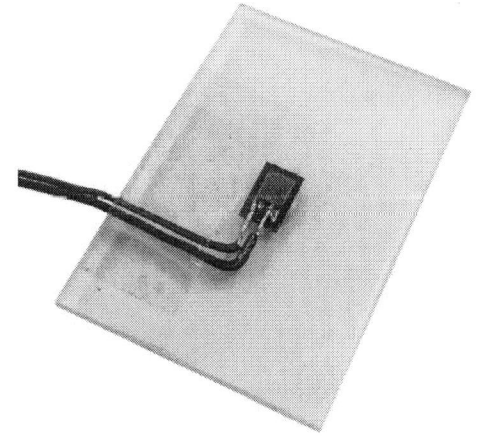

Fig. 4. Strain gauge BF350-3AA attached to the polycarbonate substrate.

The longitudinal stress (σ) generated in the bending substrate can be calculated using (4) from references [12], [17], [18] where "E" is the Young`s modulus and the other variables are shown in fig. 2. The distances "L" and "a" can be changed in the bending equipment and the variables "t" and "E" will depend on the chosen substrate.

$$\sigma = \frac{E \cdot y \cdot t}{2 \cdot a \left(\frac{L}{2} - 2\frac{a}{3}\right)} \quad (4)$$

The longitudinal strain obtained as a function of the bending's equipment scale is presented in fig. 5, as the average of three separate measurements, alongside the standard deviation, not exceeding the strain gauge's 2.0% strain limit.

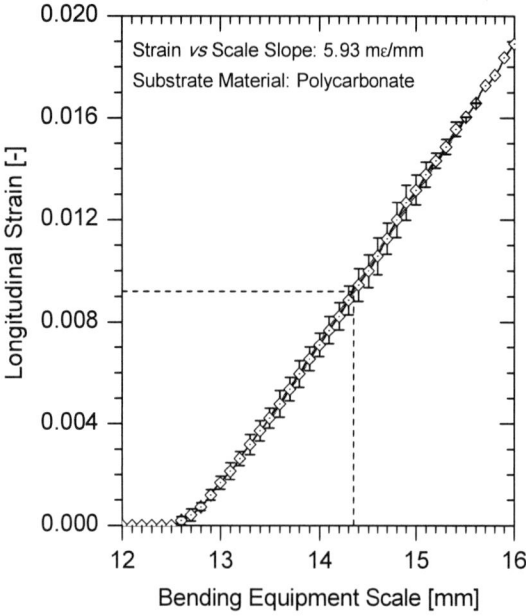

Fig. 5. Longitudinal strain and standard deviation as a function of micrometer scale.

Strains of this magnitude, around 1%~2%, are compatible to the ones observed experimentally in Ge MOSFETs using embedded source/drain SiGe stressors [16], [19]. The expected theoretical stress using the last eq. and the material's properties from table 3 with the bending equipment scale position set to 14.35mm is σ=28.3MPa.

Table 3. Variables and properties of equipment setup and material.

Variable:	Material: Polycarbonate
Pos. (mm)	14.35
E (GPa)	2.6
y (m)	1.7×10^3
t (m)	2×10^3
a (m)	12×10^3
L (m)	42×10^3
σ (Pa)	2.83×10^7

Comparing it with the strain gauge result from fig. 5, a 7.3% to 27.6% difference is observed within the standard deviation from measurements, as result of the strain transferred through the adhesive and eventual deviations in measured values. Different substrate materials and thickness will return different stress transferred to the sample. As example, a silicon wafer in <110> orientation has E=169GPa (65 times larger than

polycarbonate) and thickness of 600µm (almost a quarter) [20], [21] and could reach a few GPa, even cracking the wafer.

VI. SAMPLE PREPARATION AND POSITIONING

The integrated circuit (IC) sample used here as an example is glued to the substrate following the same steps used for the strain gauge. The access to the sample using the probes and its visualization using the microscope are not compromised by the bending setup, as seen in fig. 6.

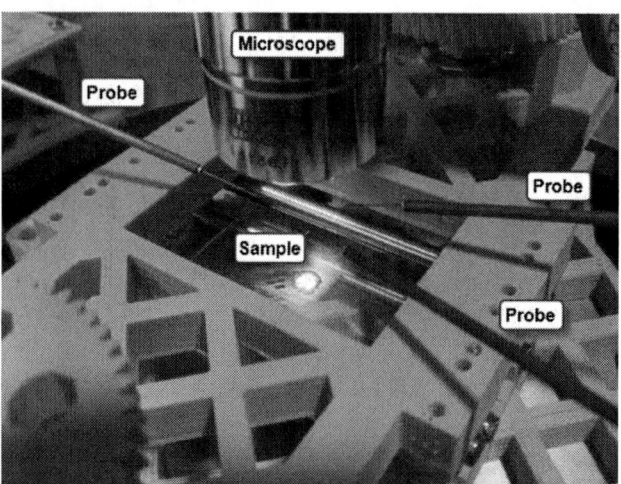

Fig. 6. IC sample positioned in the semiconductor bending equipment with probes and microscope in place.

Fig. 7 presents the sample under tensive strain, with the lower cylinder rails placed closer to the center and the two upper ones placed near the extremities of the substrate. The compressive strain would be achieved placing the cylinder rails in opposite positions, the lower ones near the extremities and upper ones closer to the center.

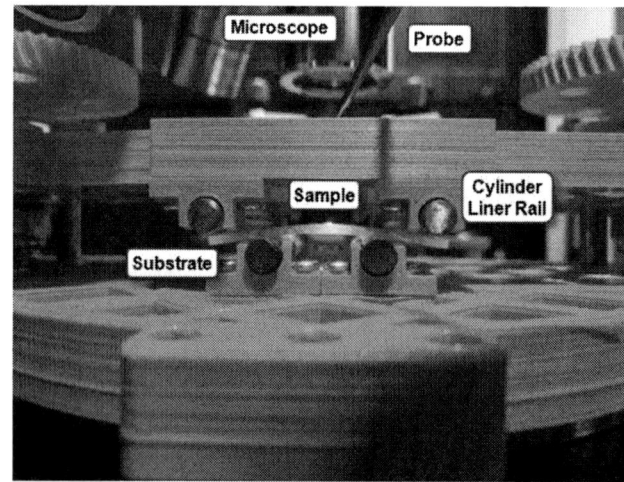

Fig. 7. Lateral view of the sample in the semiconductor bending equipment under tensile strain.

VII. Conclusions

A wide range of samples can be submitted to controlled mechanical strain for proper characterizations using the proposed bending setup. The goal is to define the details to develop a reliable open model that is simple to implement, freely available to download, use, improve and shared again by any research team, with easiness to adapt for different characterization equipment.

Acknowledgment

The authors would like to thank CAPES and CNPq for the financial support.

References

[1] L. Collins, "Silicon takes the strain," *IEE Rev.*, vol. 49, no. 11, pp. 46–49, Dec. 2003.

[2] S. E. Thompson, G. Sun, K. Wu, J. Lim, and T. Nishida, "Key differences for process-induced uniaxial vs. substrate-induced biaxial stressed Si and Ge channel MOSFETs," in *IEDM Technical Digest. IEEE International Electron Devices Meeting, 2004.*, 2004, pp. 221–224.

[3] T. Sun, Yongke, Thompson, Scott E., Nishida, *Strain Effect in Semiconductors.* 2010.

[4] N. Mohta and S. E. Thompson, "Mobility enhancement," *IEEE Circuits Devices Mag.*, vol. 21, no. 5, pp. 18–23, Sep. 2005.

[5] "Tutorial T9: Devices and Circuits to Address the Challenges in IOT," in *2017 30th International Conference on VLSI Design and 2017 16th International Conference on Embedded Systems (VLSID)*, 2017, pp. xliii–xliii.

[6] M. Radovic, E. Lara-Curzio, and L. Riester, "Comparison of different experimental techniques for determination of elastic properties of solids," *Mater. Sci. Eng. A*, vol. 368, no. 1–2, pp. 56–70, Mar. 2004.

[7] S. Shetty and T. Reinikainen, "Three- and four-point bend testing for electronic packages," *J. Electron. Packag.*, vol. 125, no. 4, pp. 556–561, 2003.

[8] J. Ohta, *Smart CMOS Image Sensors and Applications.* CRC Press, 2007.

[9] J.-Q. Huang, Q.-A. Huang, M. Qin, W. Dong, and X. Chen, "Strain Effect of the Dielectric Constant in Silicon Dioxide," *J.*

[10] *Microelectromechanical Syst.*, vol. 19, no. 6, pp. 1521–1523, 2010.

S. E. Thompson, S. Suthram, Y. Sun, G. Sun, M. Parthasarathy, M. Chu, and T. Nishida, "Future of Strained Si/Semiconductors in Nanoscale MOSFETs," in *2006 International Electron Devices Meeting*, 2006, pp. 1–4.

[11] Y. S. Choi, T. Numata, T. Nishida, R. Harris, and S. E. Thompson, "Impact of mechanical stress on gate tunneling currents of germanium and silicon p-type metal-oxide-semiconductor field-effect transistors and metal gate work function," *J. Appl. Phys.*, vol. 103, no. 6, p. 64510, 2008.

[12] G. W. HOLLENBERG, G. R. TERWILLIGER, and R. S. GORDON, "Calculation of Stresses and Strains in Four-Point Bending Creep Tests," *J. Am. Ceram. Soc.*, vol. 54, no. 4, pp. 196–199, Apr. 1971.

[13] T. Inc., "3D modeling for everyone | SketchUp," 2018. [Online]. Available: https://www.sketchup.com/. [Accessed: 22-Jan-2018].

[14] C. Commons, "Creative Commons — Attribution-NonCommercial-ShareAlike 3.0 Unported — CC BY-NC-SA 3.0," 2018. [Online]. Available: https://creativecommons.org/licenses/by-nc-sa/3.0/. [Accessed: 22-Jan-2018].

[15] M. Thingiverse, "Thingiverse - Digital Designs for Physical Objects," 2018. [Online]. Available: https://www.thingiverse.com/. [Accessed: 22-Jan-2018].

[16] R. T. Buhler, G. Eneman, P. Favia, L. J. Witters, B. Vincent, A. Hikavyy, R. Loo, H. Bender, N. Collaert, E. Simoen, J. A. Martino, and C. Claeys, "TCAD Strain Calibration Versus Nanobeam Diffraction of Source/Drain Stressors for Ge MOSFETs," *IEEE Trans. Electron Devices*, vol. 62, no. 4, pp. 1079–1084, Apr. 2015.

[17] S. Suthram, J. C. Ziegert, T. Nishida, and S. E. Thompson, "Piezoresistance Coefficients of (100) Silicon nMOSFETs Measured at Low and Highl (~ 1.5 GPa) Channel Stress," *IEEE Electron Device Lett.*, vol. 28, no. 1, pp. 58–61, Jan. 2007.

[18] S. Timoshenko, "Strength of Materials - Part 1." 1957.

[19] R. Bühler, G. Eneman, P. Favia, H. Bender, B. Vincent, A. Hikavyy, R. Loo, J. a. Martino, C. Claeys, E. Simoen, N. Collaert, and A. Thean, "Comparison between experimental and simulated strain profiles in Ge channels with embedded source/drain stressors," *Phys. status solidi*, vol. 11, no. 11–12, pp. 1578–1582, Nov. 2014.

[20] M. a. Hopcroft, W. D. Nix, and T. W. Kenny, "What is the Young's Modulus of Silicon?," *J. Microelectromechanical Syst.*, vol. 19, no. 2, pp. 229–238, 2010.

[21] J. J. Wortman and R. A. Evans, "Young's Modulus, Shear Modulus, and Poisson's Ratio in Silicon and Germanium," *J. Appl. Phys.*, vol. 36, no. 1, p. 153, 1965.

978-1-5386-5072-1/18 $31.00 © 2018 IEEE 194

Session 10

Noise and RF

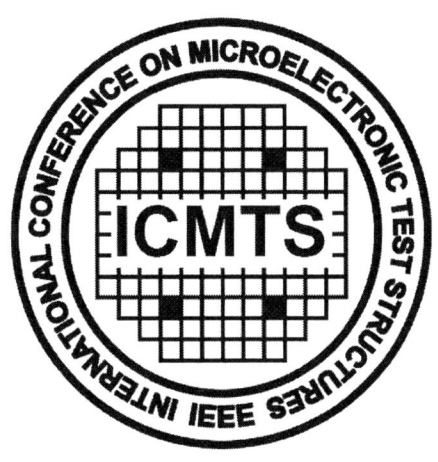

March 22, 2018

10:50-12:10

Co-Chairs:

Hans P. TUINHOUT, NXP Semiconductors, The Netherlands

Anthony J. WALTON, University of Edinburgh, Scotland

978-1-5386-5072-1/18 $31.00 © 2018 IEEE

Importance of complete characterization setup on on-wafer TRL calibration in sub-THz range

Chandan Yadav, Marina Deng, Magali De Matos, Sebastien Fregonese, and Thomas Zimmer

IMS Laboratory, University of Bordeaux
351 cours de la Libération – 33405 Talence cedex, France

email: marina.deng@ims-bordeaux.fr

Abstract— In this paper, we present the effect of different sub-mm and mm-wave probe geometry and topology on the measurement results of dedicated test-structures calibrated with on-wafer TRL. These results are compared against 3D EM simulation of the intrinsic test-structures. To analyze difference between the measured and intrinsic EM simulation results, on-wafer TRL calibration performed on EM simulation results of a dedicated test-structure is also presented.

Keywords—Test-Structures; TRL calibration; characterization; EM simulation; HFSS; Sub-THz; SPICE .

I. INTRODUCTION

With time, millimeter-wave and sub-millimeter wave are found suitable for diverse applications [1], [2] and require extensive research focus from high speed integrated devices to the system level design for full utilization of these frequency bands. To design a system working in the mm- and sub-mm range, development of an accurate SPICE model of the elementary devices is required. In development of a robust and precise SPICE model for the state-of-the-art high speed devices, its validation against device experimental data is indispensable. Accuracy of the used experimental data is essential to extract correct values of the compact model parameters and to use them further for circuit design. To obtain accurate experimental data, design of test-structures, de-embedding/calibration methods along with characterization techniques play a very important role at the sub-THz frequencies [3-5].

Calibration on Impedance Standard Substrate (ISS)-standard combined with Open-Short de-embedding method is an industry *de-facto* standard but it becomes less accurate above 30 to 40 GHz [6, 7]. More advanced de-embedding methods can be adopted to improve the accuracy [8-10] in cost of the complexity. To reduce reliance from less accurate off-wafer calibrations, to improve accuracy and repeatability of the measured data, on-wafer calibration technique enabling accurate characterization of high speed devices in sub-THz frequency range is highly desirable [11]. In this work, we have adopted thru-reflect-line (TRL) method for the on-wafer calibration of device-under-test (DUT), which is the reference

for on-wafer measurements at sub-mm wave and mm wave frequencies [12].

II. ON-WAFER CALIBRATION AND VERIFICATION STANDARDS

In order to perform on-wafer TRL calibrations, STMicroelectronics BiCMOS55 (B55) technology is used to design and fabricate all the test-structures [13]. The developed test-structures consist of nine metallic layers (copper in M1-M8 + Aluminum layer on top of M8 to realize contact pads) in back-end-of-line (BEOL). Our test-structures are compatible to perform characterization using 50 μm pitch probes and 100 μm pitch probes. In this work, used on-wafer test-structures (see Fig. 1 – Fig. 3) are as follows:

- Classical LINE
- THRU
- REFLECT
- LOAD (for impedance correction)
- Open-M1 (DUT)
- Short-M1 (DUT)
- Meander LINE (DUT).

To obtain high reflection, SHORT and OPEN are used as REFLECT standards. The fabricated classical straight LINE standard and THRU standard have line length equal to 115 μm and 35 μm, respectively.
To carry out 3D electromagnetic (EM) simulation of intrinsic test-structures and to perform on-wafer TRL calibration based on 3D EM simulations, test-structures layout design is imported into EM simulator.

III. EM SIMULATION SETUP

3D EM simulations have been carried out using Ansoft HFSS. In the simulation setup, test-structures placed on silicon substrate are encapsulated inside an air box and absorbing radiation boundary condition is assigned to the faces of the box. The radiation boundary condition emulates

978-1-5386-5072-1/18 $31.00 © 2018 IEEE

2018 INTERNATIONAL CONFERENCE ON MICROELECTRONIC TEST STRUCTURES, MARCH 19-22, 2018, AUSTIN, TEXAS, USA

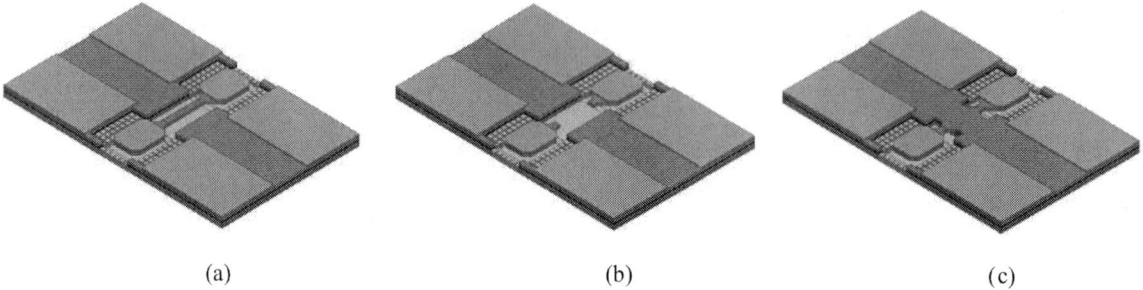

(a)   (b)   (c)

Fig 1: 3D view of THRU, OPEN and SHORT standards used in on-wafer TRL calibration are shown in panel (a), (b), and (c), respectively. Length of the THRU standard is 35 μm.

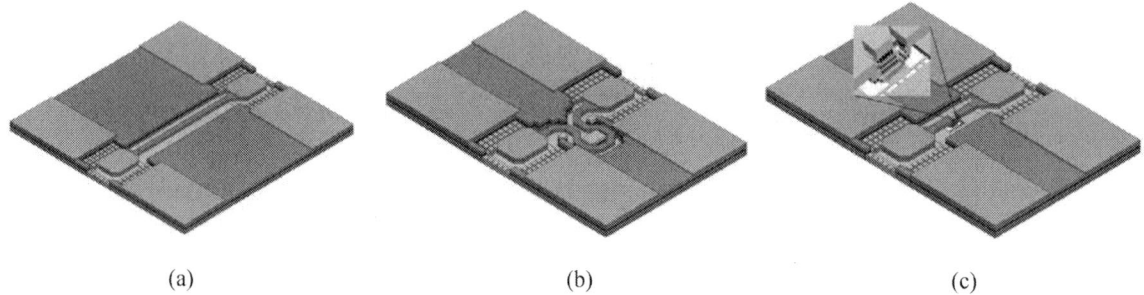

(a)   (b)   (c)

Fig 2: Test-structures used in on-wafer TRL calibration; (a) Classical LINE with 115 μm length, (b) Meander LINE, and (c) Short-M1 standard. In Short-M1, metal-1 (M1) is shorted to the ground as shown in the enlarged view. Meander LINE and Short-M1 are used as verification DUTs.

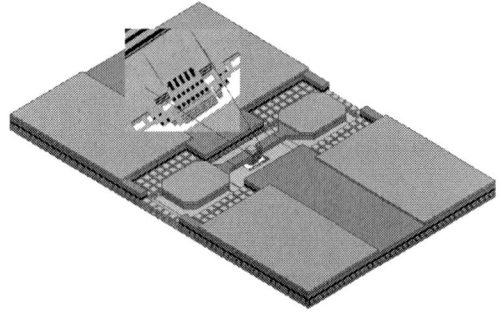

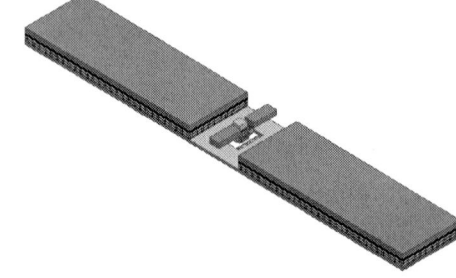

Fig 3: 3D view of the Open-M1 where metal-1 (M1) does not have connection with ground as shown in the enlarged view.

Fig 4: 3D view of intrinsic structure of Open-M1. Complete structure of Open-M1 including signal and ground pads is shown in Fig. 3.

infinite free space like environment during the simulation. In our test-structures design, intermediate insulator layers in back-end-of-line (BEOL) have different dielectric materials. However, in HFSS simulation, an average relative dielectric constant of 4.0 is set for all dielectric materials to reduce complexity. Other material parameter values, lateral and vertical dimensions of test-structures are chosen as per the structures' design. In the simulation, "wave port" is used to provide excitation at each port. To reduce simulation time in the simulation of symmetric test-structures, only half of the test-structure is used and "Perfect H" symmetry boundary

condition is assigned on the symmetry plane to obtain solution equivalent to complete structure. To perform EM simulation on intrinsic structures, only intrinsic parts of structures are taken into account (e.g. see intrinsic structure of Open-M1 shown in Fig. 4 in comparison to its complete structure shown in Fig. 3).

IV. RESULTS AND DISCUSSION

Using fabricated test-standards on the silicon wafer, first on-wafer TRL calibration is performed and the reference plane is

2018 INTERNATIONAL CONFERENCE ON MICROELECTRONIC TEST STRUCTURES, MARCH 19-22, 2018, AUSTIN, TEXAS, USA

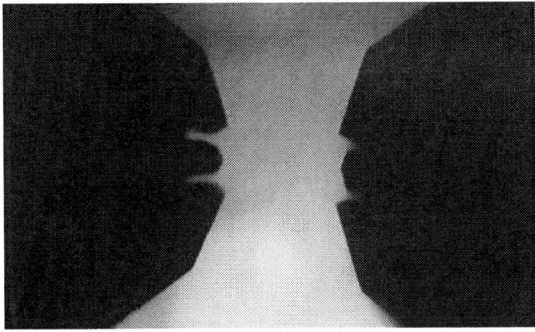

Fig 5: Image of the Picoprobe 50 μm pitch with waveguide used in G-band (140 GHz–220 GHz) measurement. Note that two other Picoprobes are also used in measurement but their images are not shown.

Fig 6: Image of the Cascade Infinity probe used in G-band (140 GHz – 220 GHz) measurement.

set after the pads (as can be recognized on the Open structure, see Fig. 1b). Next, test-structures such as the DUTs Open-M1, Short-M1 (dedicated to de-embedding) and Meander-LINE (see Fig. 2 and Fig. 3) are measured in two frequency bands i.e. 1 GHz – 110 GHz and 140 GHz – 220 GHz. To analyze the impact of the complete characterization setup combined with on-wafer TRL calibration, measurements are carried out using four different probes: 1) Picoprobe of 100 μm pitch (< 110 GHz), 2) Picoprobe of 50 μm pitch with coaxial cable (< 110 GHz), 3) Picoprobe of 50 μm pitch with waveguide (140 GHz - 220 GHz), and 4) Cascade Infinity probe of 100 μm pitch (140 GHz - 220 GHz) (see Fig. 5 and Fig. 6). In order to benchmark the on-wafer TRL calibration performed with different measurement configurations, 3D EM simulation is carried out first on the intrinsic structures. The results obtained after on-wafer TRL calibration and through intrinsic structures simulation in HFSS are shown in Fig. 7 – Fig. 11, where solid black line is used to represent the intrinsic structure EM simulation data and symbols are used to represent the DUT measured characteristics after on-wafer TRL calibration. In Fig. 7 and Fig. 8, magnitude and phase of reflection coefficient S_{11} of Open-M1 are shown up to 220 GHz. In the lower frequency range up to 60 GHz, all the results match reasonably, beyond 170 GHz a clear discrepancy can be obser-

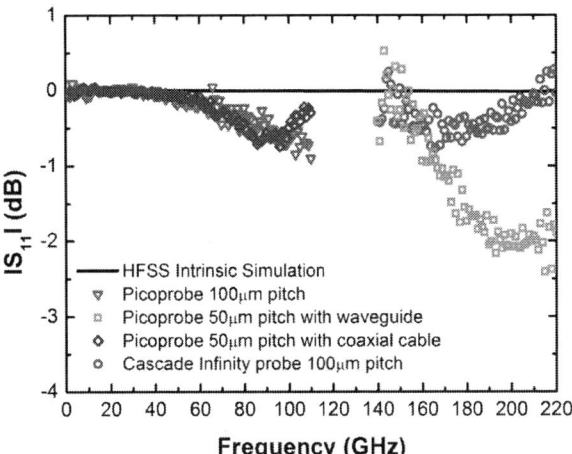

Fig 7: Magnitude of reflection coefficient S_{11} of Open-M1 w.r.t. frequency. Symbols are used to represent the measurement data and line is used for representation of intrinsic simulation performed in HFSS.

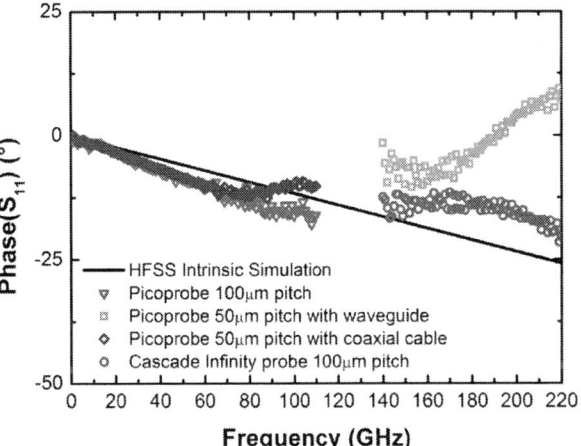

Fig 8: Phase of S_{11} of Open-M1 w.r.t. frequency. Symbols are used to represent the measurement data and line is used for representation of intrinsic simulation performed in HFSS.

-ved. In particular, the combination of Picoprobe with 50 μm pitch and TRL on-wafer calibration seems to be not sufficient for accurate representation of the DUT's intrinsic behavior. The magnitude and phase of transmission coefficient S_{21} of a meander-line is shown in Fig. 9 and Fig. 10 up 220 GHz. For this DUT, the different measurement configurations give similar results except in the low frequency range for the Picoprobe with 100 μm pitch. A coupling with neighboring structures may be the origin of this non-ideal behavior. Inductance of the Short-M1 w.r.t. frequency is shown in Fig. 11. From 50 to 110 GHz, the Picoprobe with 100 μm pitch, and from 140-220 GHz, the Cascade-probe with 100 μm pitch, are matching well the theoretically predicted intrinsic DUT characteristic.

978-1-5386-5072-1/18 $31.00 © 2018 IEEE

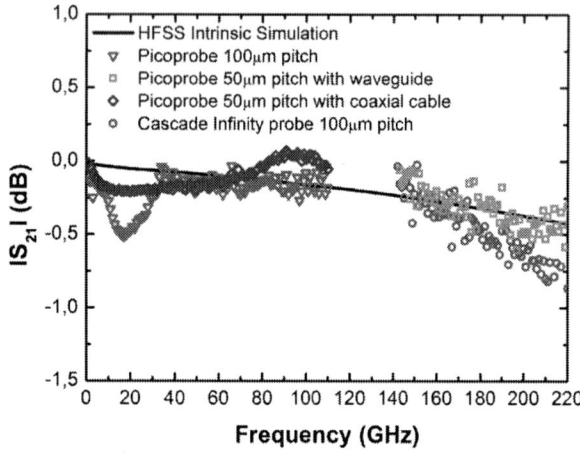

Fig 9: Magnitude of transmission coefficient S_{21} of meander-line w.r.t. frequency. Symbols are used to represent the measurement data and line is used for representation of intrinsic simulation.

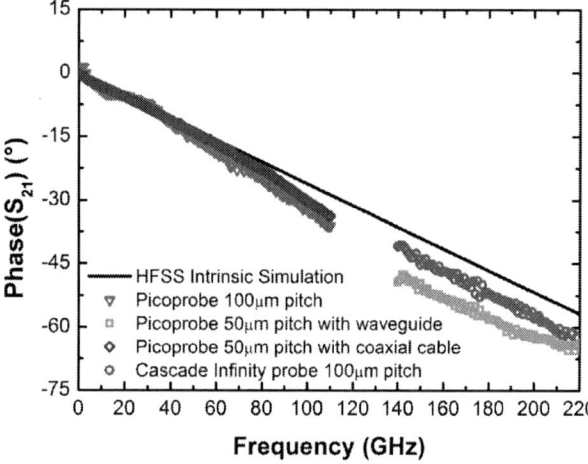

Fig 10: Phase of transmission coefficient S_{21} of Meander-LINE w.r.t. frequency. Symbols are used to represent the measurement data and line is used for representation of intrinsic simulation performed in HFSS.

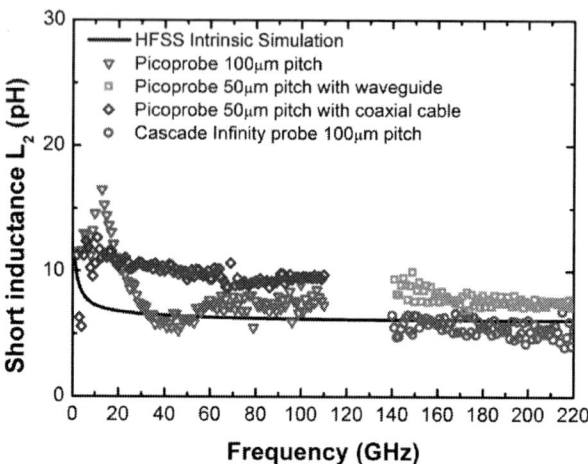

Fig 11: Inductance of the Short-M1 with respect to frequency. Symbols are used to represent the measurement data and line is used for representation of intrinsic simulation performed in HFSS.

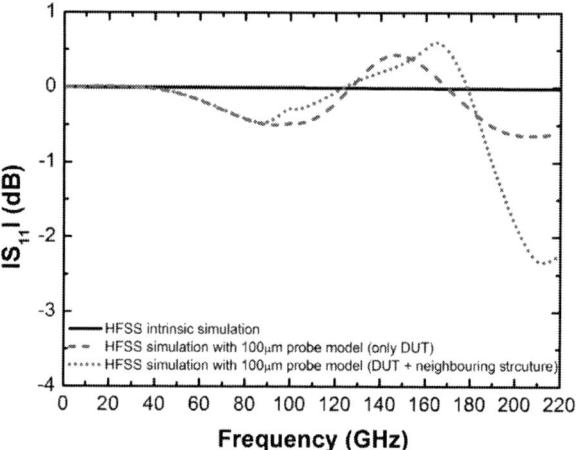

Fig 12: Comparison of reflection coefficient S_{11} of Open-M1 w.r.t. frequency. These results are obtained from EM simulation of intrinsic structure and on-wafer calibration performed on EM simulation of complete structure. Differences between intrinsic simulation and simulation using probe model has very similar pattern as observed in Fig. 7 between the measured data and intrinsic simulation. Simulation with and without neighboring structure around DUT indicates surrounding condition can impact the measurement.

In order to analyze differences between the intrinsic HFSS simulation results and measured data, we prepared 100 μm pitch probe model in HFSS based on the coplanar + coaxial topology (similar to Picoprobe with 100 μm pitch). Using the probe model, 3D EM simulations with and without neighboring structures are carried out on test-structures shown in Figs. 1 – 3. After performing on-wafer TRL calibration on EM simulation data, magnitude of reflection coefficient S_{11} of Open-M1 is shown in Fig.12. From Fig. 12, it can be observed that HFSS probe model gives different results than the intrinsic simulation and upto some extend mimic behavior of measurement data of Fig. 7. From Fig. 12, it can also be observed that measurement results are more prone to probe design and presence of neighboring structures around DUTs at higher frequencies. Therefore, effects predicted by calibrated

EM simulation results shown in Fig. 12 could be the possible reasons behind difference between the intrinsic simulation and measured data visible in Fig. 7 – Fig. 11. In addition, these results also show importance of the complete characterization setup in the on-wafer TRL calibration.

V. CONCLUSION

In conclusion, we want to point out that considering intrinsic simulation data as a reference, we observe that after on-wafer TRL calibration the measurements of dedicated test-structures using different probes topologies (i.e. based on coplanar and microstrip) give different results. It is interesting to see that after on-wafer TRL calibration the measurements done on the same DUT with same probe topology but with different pitch also shows variation in results as frequency increases. These results show the importance to take into account the complete characterization setup when performing on-wafer TRL calibration. In particular, depending on the design of the test-structures together with the probe geometry, on-wafer TRL calibration is not sufficient for some particular cases and coupling correction becomes mandatory.

ACKNOWLEDGMENT

This work is partly funded by the French Nouvelle-Aquitaine Authorities through the SUBTILE and FAST project. The authors also acknowledge financial support from the EU under Project Taranto (No. 737454). The authors would like to thank STMicroelectronics for supplying Silicon wafer.

REFERENCES

[1] T. W. Crowe, W. R. Deal, M. Schröter, C. K. Clive Tzuang and K. Wu, "Terahertz RF Electronics and System Integration [Scanning the Issue]," in *Proceedings of the IEEE*, vol. 105, no. 6, pp. 985-989, June 2017.

[2] M. Tonouchi, "Cutting-edge terahertz technology" *Nature photonics*, pp. 97-105, Feb. 2007.

[3] A. M. Mangan, S. P. Voinigescu, Ming-Ta Yang and M. Tazlauanu, "De-embedding transmission line measurements for accurate modeling of IC designs," in *IEEE Transactions on Electron Devices*, vol. 53, no. 2, pp. 235-241, Feb. 2006.

[4] A. Issaoun, Y. Z. Xiong, J. Shi, J. Brinkhoff and F. Lin, "On the Deembedding Issue of CMOS Multigigahertz Measurements," in *IEEE*

Transactions on Microwave Theory and Techniques, vol. 55, no. 9, pp. 1813-1823, Sept. 2007.

[5] P. Manuel, S. Fregonese, A. Curutchet, P. Baureis and T. Zimmer, "New 3D-TRL structures for on-wafer calibration For high frequency S-parameter measurement," *2015 European Microwave Conference (EuMC)*, Paris, 2015, pp. 167-170.

[6] M. Potereau, C. Raya, M. D. Matos, S. Fregonese, A. Curutchet, M. Zhang, B. Ardouin, et T. Zimmer, "Limitations of On-Wafer Calibration and De-embedding Methods in the Sub-THz Range", *J. Comput. Commun.*, vol. 01, no. 06, pp. 25-29, 2013.

[7] C. Raya and B. Ardouin, "Méthodes de caractérisation en hautes fréquences des technologies de circuits intégrés en silicium dédiées aux applications Téra hertz et sub-Téra hertz, " Proceedings of Journées Scientifiques 2013, Paris, France.

[8] F. Pourchon, C. Raya, N. Derrier, P. Chevalier, D. Gloria, S. Pruvost, et D. Céli "From measurement to intrinsic device characteristics: Test structures and parasitic determination," *2008 IEEE Bipolar/BiCMOS Circuits and Technology Meeting*, Monteray, CA, 2008, pp. 232-239.

[9] N. Derrier, A. Rumiantsev, and D. Celi, "State-of-the-art and future perspectives in calibration and de-embedding techniques for characterization of advanced SiGe HBTs featuring sub-THz fT/fMAX", in *2012 IEEE Bipolar/BiCMOS Circuits and Technology Meeting (BCTM)*, 2012, pp. 1-8.

[10] M. Potéreau, M. Deng, C. Raya, B. Ardouin, K. Aufinger, C. Ayela, M. De Matos, A. Curutchet, S. Frégonèse, T. Zimmer, "Meander type transmission line design for on-wafer TRL calibration," *2016 46th European Microwave Conference (EuMC)*, London, 2016, pp. 381-384.

[11] A. Rumiantsev, S. L. Sweeney and P. L. Corson, "Comparison of on-wafer multiline TRL and LRM+ calibrations for RF CMOS applications," *2008 72nd ARFTG Microwave Measurement Symposium*, Portland, OR, 2008, pp. 132-136.

[12] D. F. Williams, P. Corson, J. Sharma, H. Krishnaswamy, W. Tai, Z. George, D. Ricketts, P. Watson, E. Dacquay, and S. P. Voinigescu, "Calibration-Kit Design for Millimeter-Wave Silicon Integrated Circuits," in *IEEE Transactions on Microwave Theory and Techniques*, vol. 61, no. 7, pp. 2685-2694, July 2013.

[13] P. Chevalier *et al.*, "A 55 nm triple gate oxide 9 metal layers SiGe BiCMOS technology featuring 320 GHz fT / 370 GHz fMAX HBT and high-Q millimeter-wave passives", in *Electron Devices Meeting (IEDM), 2014 IEEE International*, 2014, p. 3.9.1-3.9.3.

2018 INTERNATIONAL CONFERENCE ON MICROELECTRONIC TEST STRUCTURES, MARCH 19-22, 2018, AUSTIN, TEXAS, USA

Measurement Time Reduction Technique
for Input Referred Noise of Dynamic Comparator

Yuki Ishijima[1], Shuya Nakagawa[2] and Hiroki Ishikuro[3]

[1,2,3]Department of Electronics and Electrical Engineering
Keio University, Yokohama, Japan

3-14-1, Hiyoshi, Kohoku-ku, Yokohama, 223-8522 Japan
Telephone: (045) 566–1443

[1]Email: ishijima@iskr.elec.keio.ac.jp
[2]Email: nakagawa@iskr.elec.keio.ac.jp
[3]Email: ishikuro@elec.keio.ac.jp

Abstract—Time reduction technique for the measurement of input referred noise of dynamic comparator is presented. By using binary search technique, the proposed method can reduce the measurement time of comparator input referred noise to $(\log_2 n)/n$, where n is a required resolution. Experimental results obtained by the developed measurement system shows good correspondence with the simulated input referred noise.

Keywords—dynamic comparator, input referred noise, binary search, DAC

I. INTRODUCTION

Dynamic comparators play important role in mixed-signal circuits, especially in analog-to-digital converters (ADCs). With the advancement of high precision and low voltage sensors for IoT era, the required resolution of ADC also increases. As a result, the voltage step of 1LSB of ADC becomes smaller than 1mV and sometimes reaches micro-volt range. The input referred noise of a comparator cannot be ignored (Fig.1) in such high precision ADCs and precise noise modeling is required in circuit design [1]. When the comparator clock is "L", the differential outputs of dynamic comparator are "H". When the comparator clock rises, both the outputs gradually fall and then split into "H" and "L". The state before split is called metastable and the point of the split is called operating point. Different from usual amplifiers, dynamic comparators use cross coupled regenerative latch and are not stable at comparison timing. Therefore, noise measurement becomes difficult and time consuming, which increases IC manufacturing test cost. To provide an efficient noise modeling method, this paper proposes fast measurement technique of input referred noise of dynamic comparator. This technique can also be applied to the measurement of setup time of flip flop or other positive feedback circuits with metastability.

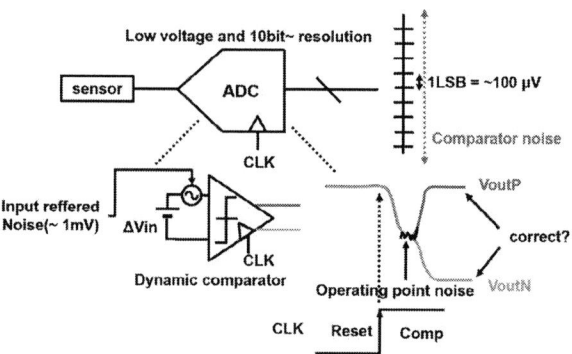

Fig.1 The effect of comparator noise

II. PROPOSED MEASUREMENT TECHNIQUE FOR INPUT REFERRE NOISE OF DYNAMIC COMPARATOR

As shown in Fig.2, comparator output ("H" or "L") becomes random pattern if the input signal level (ΔV_{in}) is comparable to the input referred noise (V_{noise}). The probability density function of input referred noise is Gaussian function without offset voltage (1), where ϕ is probability density function and σ is standard deviation. The probability density function of the comparator output data of "H" becomes cumulative distribution function of the input referred noise. Therefore, the input referred noise can be measured by sweeping the input signal level and counting the "H" of the comparator output. To precisely measure the probability function, several thousand times must be counted at each input level. Furthermore, in a real comparator, offset voltage (V_{offset}) also exists and probability function is shifted (Fig.3). And the probability distribution function is (2). As the technology scales, the offset voltage increases and the sweep range of input signal should be expanded. This brings a dramatic increase of measurement time.

978-1-5386-5072-1/18 $31.00 © 2018 IEEE

2018 INTERNATIONAL CONFERENCE ON MICROELECTRONIC TEST STRUCTURES, MARCH 19-22, 2018, AUSTIN, TEXAS, USA

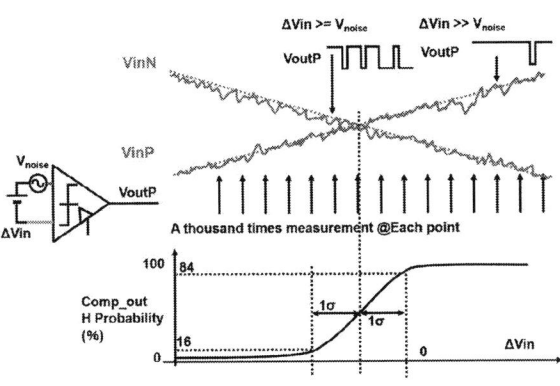

Fig.2 Input referred noise characterics

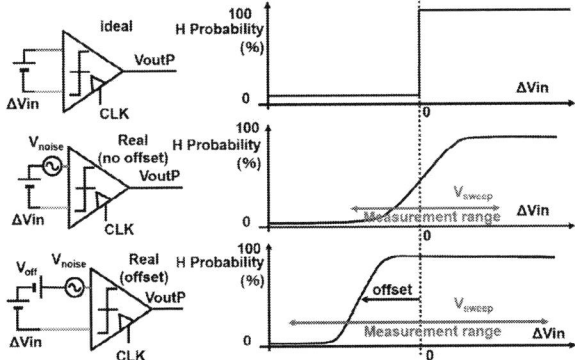

Fig.3 Ideal, real (without offset), and real (with offset) comparator

input-output characteristics

$$\emptyset(\Delta V_{in}) = \frac{1}{\sqrt{2\pi}\sigma} exp\left(-\frac{\Delta V_{in}^2}{2\sigma^2}\right) \qquad (1)$$

$$\emptyset(\Delta V_{in}) = \frac{1}{\sqrt{2\pi}\sigma} exp\left\{-\frac{(\Delta V_{in} - \Delta V_{offset})^2}{2\sigma^2}\right\} \qquad (2)$$

To reduce the measurement time, we propose to use a binary search technique (Fig.4) to detect the standard deviation of the input referred noise (σV_{noise}) and offset voltage of a comparator. If the probability density function of comparator output follows Gaussian distribution, the probability of occurrence of "H" becomes 16% and 84%, when the input signal level (ΔV_{in}) is equal to $-V_{offset} -\sigma V_{noise}$ and $-V_{offset} +\sigma V_{noise}$, respectively. By using binary search technique, the σV_{noise} can be detected with a precision of $V_{sersch}/2^n$ by n step, and V_{offset} can also be determined.

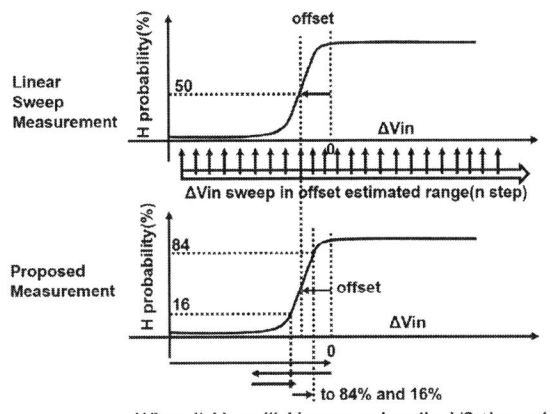

Fig.4 Comparison of linear sweep amd proposed binary search method

III. MEASUREMENT SETUP AND RESULTS

Fig.5 shows the block diagram of the measurement system of dynamic comparator input referred noise. For precise measurement the environmental noise must be enough lower than the input referred noise of comparator. For this purpose, high resolution (16bit) low noise DAC ICs are used to generate the input signal for comparator. To reduce the noise of the DACs, low-pass filters are placed at output ports of the DACs and noise bandwidth are limited at 10MHz. There is a tradeoff between the measurement time and accuracy. When the bandwidth of the LPF is narrow, the DAC output noise can be further reduced. However, the settling time of the DAC and measurement time increases. The lower the noise of the comparator, the environmental noise should be more suppressed by limiting the noise bandwidth. From this viewpoint, the proposed technique becomes more effective for measurement of low noise comparator.

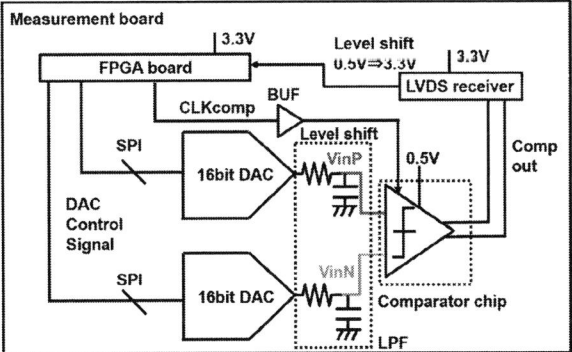

Fig.5 Block diagram of measurement system

978-1-5386-5072-1/18 $31.00 © 2018 IEEE   203

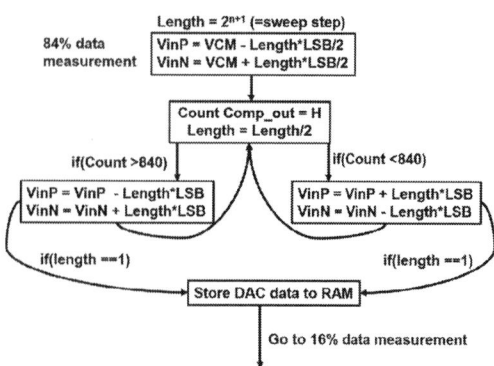

Fig.6 Flow chart of proposed method

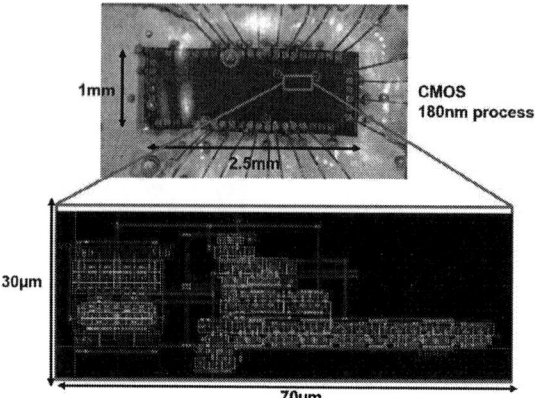

Fig.8 Comparator test chip photograph and layout

The functions of the comparator output data ("H") counting, binary search algorithm, and DAC control are implemented in FPGA. A clock for the comparator is also generated from the FPGA for synchronous operation between the comparator, DAC ICs, and FPGA. To make it possible to measure an ultra-low voltage (0.5V) comparator, the comparator's output level is shifted to FPGA logic input level (3.3V) by using a LVDS receiver IC. Then, FPGA board receives the data and stores to RAM. Fig.6 shows flow chart of proposed binary search method. In this case, ΔV_{in} is changed toward the voltage that makes the probability of "H" 84%, with its step decreasing by half of the previous step.

A photograph of this measurement system is shown in Fig.7. This system consists of two boards. One is measurement board whose circuits contains whole system except comparator. The other is device-under-test (DUT) board on which the CMOS comparator test chip is mounted. An ultra-low supply voltage (0.5V) comparator was designed and fabricated in 180-nm CMOS process (Fig.8). Fig.9 shows the schematic of comparator. The signal inputs and the clock inputs of this comparator are NMOS transistor. When comparator clock is low, the comparator operates in reset mode and the outputs are

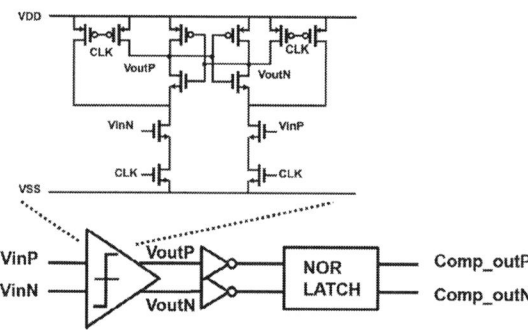

Fig.9 Comparator schematic on chip

pulled up at VDD. On the other hand, when comparator clock is high, the comparator operates in comparison mode and the outputs split into VDD and VSS. In order to hold the comparison result in reset mode, inverters and NOR latch gate are placed at the output of comparator circuits.

Fig.10 shows comparison between the simulated and measured probability density function. In the measurement, the power supply voltage for the comparator was 0.5V, and driven at 1.25 MHz clock frequency. The measured data is almost consistent with the simulated results. The environmental noise of the developed measurement system was measured separately by using low noise differential amplifier module. DAC output differential voltages connected to the module, and the noise level of the amplifier module output is measured by oscilloscope, and found to be 1mV with bandwidth of 10MHz. This module has 46dB gain and 3µV input referred noise with bandwidth of 10MHz, for that reason the environmental noise was found to be 50µV, which is small enough compared to the comparator input referred voltage. And the 1LSB voltage level of the DAC IC is 22µV, so it is also small enough. Therefore, it can be confirmed that the probability density function of comparator output follows Gaussian distribution.

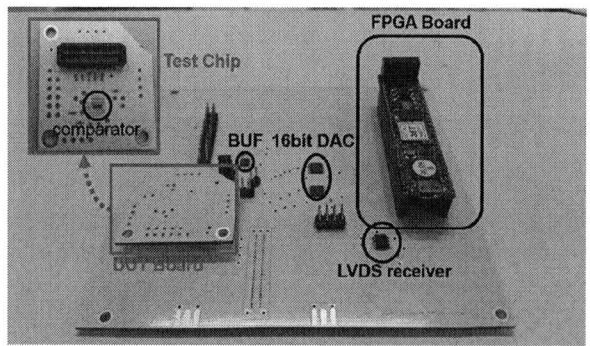

Fig.7 Photograph of measurement board and test board

2018 INTERNATIONAL CONFERENCE ON MICROELECTRONIC TEST STRUCTURES, MARCH 19-22, 2018, AUSTIN, TEXAS, USA

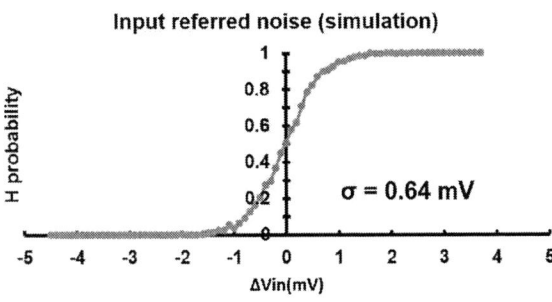

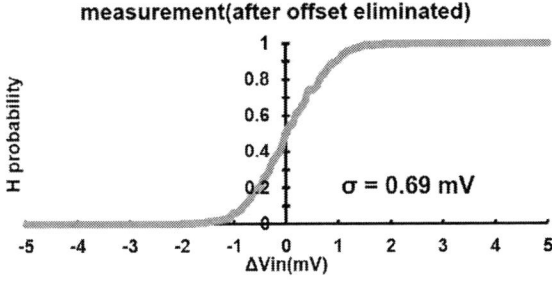

Fig.10 Simulation and measurement result

Fig. 11 compares the required measurement time by linear sweep method and proposed binary search technique. In this measurement we repeated measurement ten times and averaged their results in using binary search method. Therefore, the number of measurement steps of binary search method is $10 \times 2 \times \log_2(n)$ when that of linear sweep method is n. Even if the offset voltage becomes large, the required measured time slightly increases when the proposed technique is used. On the other hand, the required measurement time linearly increases when the linear sweep technique is used. Therefore, if the required sweep range becomes large, the advantage of proposed technique is more enhanced.

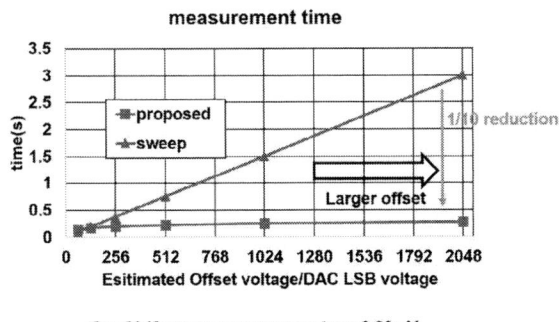

@n=2048 sweep measurement σ = 0.69mV
proposed measurement σ = 0.70mV

Fig.11 Measurement time of linear sweep and proposed binary search method

IV. CONCLUSION

Time reduction technique for the measurement of input referred noise of dynamic comparator was proposed and experimentally demonstrated by developed measurement system. A comparator test chip fabricated in 180nm-CMOS was used for noise measurement, and showed good correspondence between the simulation results and experimental results.

REFERENCES

[1] Long Chen, Xiyuan Tang, Arindam Sanyal, Yeonam Yoon, Jie Cong and Nan Sun, "A 0.7-V 0.6-µW 100-kS/s Low-Power SAR ADC With Statistical Estimation-Based Noise Reduction ," IEEE JSSC, vol.57 pp. 1388-1398, May 2017

[2] Shufeng Zheng and Juha Kostamovaara, "Statistical behavior of a comparator with weak repetitive signal and additive white Gaussian noise," IEEE I2MTC, May 2016.

[3] Bob Verbruggen, Jorgo Tsouhlarakis, Takaya Yamamoto, Masao Iriguchi, Ewout Martens and Jan Craninckx, " A 60 dB SNDR 35 MS/s SAR ADC With Comparator-Noise-Based Stochastic Residue Estimation," IEEE JSSC, vol.50 pp. 2002-2012, September 2015

System Aware DUT Design for Optimum On-Wafer Noise Measurement

Chih-Hung Chen[1], Benson Yang[1,2], Pei-Hsien Chu[1], Graham Brown[1], and Saswati Das[1]

[1]Department of Electrical and Computer Engineering
McMaster University, Hamilton, ON, Canada
Email: see http://www.ece.mcmaster.ca/~chihhung/index.html
[2]Physical Sciences Platform, Sunnybrook Research Institute, Toronto, ON, Canada
Email: benson.yang@sunnybrook.ca

Abstract—**This paper presents a system-aware design of device-under-tests (DUT) for optimum high-frequency (HF) on-wafer noise measurement. It overcomes the challenges in modeling the bias and geometry dependence of noise sources due to the voltage drop in the interconnections at the output port of a large DUT. It also prevents the measurement inaccuracy resulted from insufficient noise from a small DUT. Experimental data and suggested device sizes for different technologies are presented.**

Keywords—*device-under-test design, thermal noise, high-frequency noise characterization, on-wafer noise measurement*

I. INTRODUCTION

The high-level of integration and continuous downscaling of CMOS processes enable modern nanoscale MOSFETs with cut-off frequencies (f_T) and maximum oscillation frequencies (f_{max}) in several hundreds of GHz, and make MOSFETs a very good candidate for emerging low-power, mobile, Internet-of-Things (IoT), and radio-frequency (RF) applications [1]-[3]. In these low-power and high-frequency applications, the noise performance of the device used in the design of the front-end receiver, especially for the first-stage low-noise amplifier (LNA) becomes critically important because it determines the noise performance of the over-all receiver. Therefore, accurate high-frequency noise characterization becomes important to provide insights of the noise behavior, and enable accurate and scalable noise models to facilitate the design process. One of the challenges in the high-frequency noise characterization is the design of the device-under-test (DUT), especially for the size of the DUT. The HF DUT is usually laid out in a Ground-Signal-Ground (GSG) configuration as shown in [4] and [5]. When the channel length of the DUT scaled down but the total channel width kept the same, the total current might be too high and cause considerable voltage drop between the external power supply and the intrinsic device terminal. This creates challenges and inaccuracies when modeling the bias and geometry dependence for the noise sources of interest. On the

other hand, if we reduce the total channel width, it resolves the issue of the voltage drop at the output port of the DUT, but increases the inaccuracy in the HF noise measurement because of the reduced noise power from the DUT compared to those from the instrument. This paper overcomes this dilemma in choosing the DUT size by considering the noise contribution of the instrument in the DUT design.

II. NOISE MEASUREMENT SYSTEM

The high-frequency noise measurement system, as shown in Fig. 1, consists of (a) a probe station, an HP 346C noise source, source and load tuners, a low noise amplifier (LNA), an Agilent 4142B DC power supply, (b) a Maury tuner controller, an Agilent 11713A switch driver, an Agilent N5230A Vector Network Analyzer (VNA), and an Agilent N8975A Noise Figure Analyzer (NFA). Fig. 2 shows the equivalent circuit for the HF noise measurement system, and the noise reference plan is at the input of the LNA as shown in Fig. 1. Based on the representation of noise in linear two-ports described in [6] and [7], here u_{Rec} and i_{Rec} represent the input-referred, cross-correlated noise voltage and current sources, respectively for the noise receiver, where $i_{Rec} = i_{un,Rec} + Y_{cor} u_{Rec}$, and $Y_{cor} = G_{cor} + j \cdot B_{cor}$ is the correlation admittance. The noise current source i_1 represents the short-circuited noise current of the DUT at its input port, which includes, for example, the induced gate noise and the thermal noise from the gate resistance in the case of MOSFETs. On the other hand, the noise current source i_2 represents the short-circuited noise current of the DUT at its output port, which includes the channel noise i_d and the thermal noise from the substrate parasitics in MOSFETs. Here Z_S represents the source impedance experienced by the DUT at its input port, $[S^{DUT}]$ represents the two-port S-parameters of the DUT, $Z_{out,DUT}$ is the output impedance seen at the output of the DUT, and $Z_{in,Rec}$ is the input impedance of the noise receiver.

Based on the equivalent circuit model shown in Fig. 2, the four noise parameters of the HF noise measurement system,

978-1-5386-5072-1/18 $31.00 © 2018 IEEE

2018 INTERNATIONAL CONFERENCE ON MICROELECTRONIC TEST STRUCTURES, MARCH 19-22, 2018, AUSTIN, TEXAS, USA

(a)

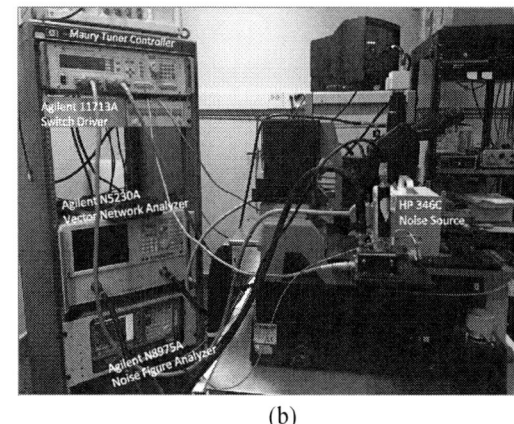
(b)

Fig. 1. The high-frequency (HF) noise measurement system in this study consists of (a) a probe station, an HP 346C noise source, source and load tuners, a low noise amplifier (LNA), an Agilent 4142B DC power supply, (b) a Maury tuner controller, an Agilent 11713A switch driver, an Agilent N5230A Vector Network Analyzer (VNA), and an Agilent N8975A Noise Figure Analyzer (NFA).

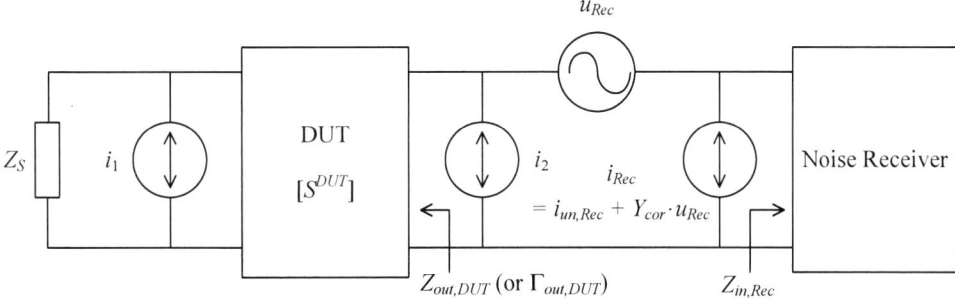

Fig. 2. The equivalent circuit model for the HF noise measurement system shown in Fig. 1.

namely the minimum noise factor F_{min} (or minimum noise figure NF_{min} in dB), the equivalent noise resistance R_n, the optimum source conductance G_{opt}, and the optimum source susceptance B_{opt} can be characterized in the calibration phase using the procedure described in [8] and [9]. The measured noise parameters vs. frequency characteristics of the HF noise measurement system used in this study are shown in Table I.

III. INPUT-REFERRED NOISE POWER FROM INSTRUMENT

For accurate HF noise measurement, the DUT needs to provide sufficiently large noise compared to the total noise resulted from the instrument. With the measured four noise parameters of the noise receiver obtained in Section II, based on (28) to (30) in [6], we can obtain G_{cor}, B_{cor}, and G_u by

$$G_{cor} = \frac{F_{min}-1}{2R_n} - G_{opt}, \tag{1}$$

$$B_{cor} = -B_{opt}, \tag{2}$$

and

$$G_u = R_n \cdot \left(G_{opt}^2 - G_{cor}^2 \right) \tag{3}$$

where G_u is the equivalent noise conductance that results in the uncorrelated noise current $i_{un,Rec}$. Based on (24) in [6], we can calculate the uncorrelated noise current of the receiver $i_{un,Rec}$ by

$$\overline{i_{un,Rec}^2} = 4kT_0 G_u \Delta f \tag{4}$$

where k is Boltzmann's constant, T_o is 290 K, and Δf is the noise bandwidth. Based on (23) and (25) in [6], we can calculate the input-referred noise voltage and current sources of the noise receiver, respectively by

$$\overline{u_{Rec}^2} = 4kT_o R_n \Delta f \tag{5}$$

and

978-1-5386-5072-1/18 $31.00 © 2018 IEEE

2018 INTERNATIONAL CONFERENCE ON MICROELECTRONIC TEST STRUCTURES, MARCH 19-22, 2018, AUSTIN, TEXAS, USA

Table I: Measured Four Noise Parameters and its Resulting Noise Sources of the Noise Receiver

Frequency (GHz)	4	6	8	10	12	14	16	18	20	22	24	26
NF_{min} (dB)	3.70	4.08	3.81	3.63	3.62	3.62	3.58	3.60	3.79	3.69	3.74	4.25
R_n (Ω)	22.6	35.7	20.9	31.7	27.6	23.3	23.5	22.1	22.3	37.9	20.6	43.8
G_{opt} (mS)	19.6	14.3	22.4	13.8	16.1	18.8	18.4	19.9	21.0	12.2	23.2	12.6
B_{opt} (mS)	-4.4	3.1	-11.9	11.6	-10.4	7.9	-3.1	-1.5	7.5	-8.4	19.7	-7.2
$\overline{i_{Rec}^2}$ (10^{-22} A^2/Hz)	1.46	1.22	2.15	1.66	1.62	1.54	1.31	1.41	1.77	1.33	3.05	1.48
$\overline{u_{Rec}^2}$ (10^{-19} V^2/Hz)	3.63	5.71	3.35	5.08	4.42	3.73	3.76	3.54	3.57	6.07	3.30	7.02

$$\overline{i_{Rec}^2} = \overline{\left| i_{Rec} - i_{un,Rec} \right|^2} + \overline{i_{un,Rec}^2} = 4kT_o \left[\left| Y_{cor} \right|^2 \cdot R_n + G_u \right] \Delta f . \quad (6)$$

The calculated noise current and voltage sources vs. freqeuncy characteristics are shown in the last two rows of Table I. We noticed that the noise voltage source $\overline{u_{Rec}^2}$ is in the order of 10^{-19} V^2/Hz, which is much larger than that of the noise current source $\overline{i_{Rec}^2}$, which is in the order of 10^{-22} A^2/Hz. However, the noise current from the noise voltage source u_{Rec} flowing in to the noise receiver depends on the output impedance Z_{out} seen at the output of the DUT. We can compute the equivalent input-referred noise current $i_{ueq,Rec}$ from u_{Rec} by equating

$$i_{ueq,Rec} \cdot \frac{Z_{out,DUT}}{Z_{out,DUT} + Z_{in,Rec}} = \frac{u_{Rec}}{Z_{out,DUT} + Z_{in,Rec}} \quad (7)$$

or

$$i_{ueq,Rec} = \frac{u_{Rec}}{Z_{out,DUT}} . \quad (8)$$

The total input-referred noise current from the noise receiver $\overline{i_{in,Rec}^2}$ for the DUT to overcome is then given by

$$\overline{i_{in,Rec}^2} = \overline{\left| i_{ueq,Rec} + i_{Rec} \right|^2} \\ = \overline{i_{ueq,Rec}^2} + \overline{i_{Rec}^2} + \mathrm{Re}\left(\overline{i_{ueq,Rec} i_{Rec}^*} + \overline{i_{Rec} i_{ueq,Rec}^*} \right) \quad (9)$$

where Re( ) represents the real part of the complex power. According to the definition of i_{Rec} in [6], as shown in Fig. 1, the input-referred noise current i_{Rec} is $i_{Rec} = i_{un,Rec} + Y_{cor} u_{Rec}$,

and we can then further simplify $\overline{i_{in,Rec}^2}$ by

$$\overline{i_{in,Rec}^2} = \frac{\overline{u_{Rec}^2}}{\left| Z_{out,DUT} \right|^2} + \overline{i_{Rec}^2} + \overline{u_{Rec}^2} \cdot 2 \cdot \mathrm{Re}\left(\frac{Y_{cor}^*}{Z_{out,DUT}} \right) \\ = \overline{u_{Rec}^2} \cdot \left[\frac{1}{\left| Z_{out,DUT} \right|^2} + 2 \cdot \mathrm{Re}\left(\frac{Y_{cor}^*}{Z_{out,DUT}} \right) \right] + \overline{i_{Rec}^2} . \quad (10)$$

Because $Z_{out,DUT}$, in general, is a function of device sizes, bias conditions, operation frequencies, and source impedance Z_s. In this paper, we assume $Z_{out,DUT} = 50\ \Omega$ to compute the largest $\overline{i_{in,Rec}^2}$ for the worst case scenario in practice. Then, we can compute the largest $\overline{i_{inmax,Rec}^2}$ at each operation frequency by

$$\overline{i_{inmax,Rec}^2} = \overline{u_{Rec}^2} \cdot \left[\frac{1}{50^2} + \frac{G_{cor}}{25} \right] + \overline{i_{Rec}^2} . \quad (11)$$

Fig. 3 shows the measured total input-referred noise current $\overline{i_{inmax,Rec}^2}$ (▲) and its individual contribution from $\overline{u_{Rec}^2} / 50^2$ (●), $\overline{u_{Rec}^2} \cdot (G_{cor} / 25)$ (◇), and $\overline{i_{Rec}^2}$ (■), respectively for a 50 Ω DUT output impedance. We can see that each term in (11) contributes about 1/3 of the total noise power in $\overline{i_{inmax,Rec}^2}$. In this study, to characterize the noise up to 26 GHz, the noise power from the DUT has to be higher than 6.2×10^{-22} A^2/Hz.

IV. DUT Design for On-Wafer Noise Measurement

To ensure an accurate noise measurement, we suggest that the noise contributed from i_1 and i_2 of the DUT to the input of the noise receiver is at least 10 times larger than $\overline{i_{inmax,Rec}^2}$, i.e., 6.2×10^{-21} A^2/Hz at 26 GHz in this study. To estimate the DUT

978-1-5386-5072-1/18 $31.00 © 2018 IEEE

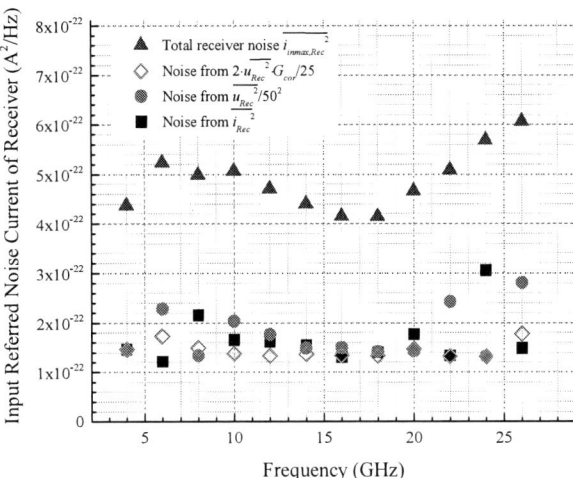

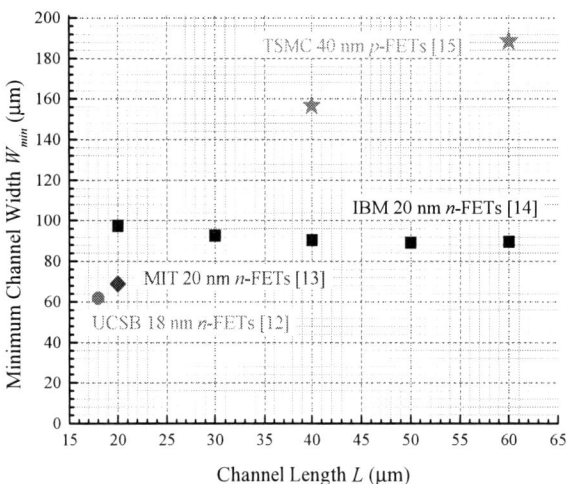

Fig. 3. Measured input-referred noise current (A^2/Hz) of the noise receiver with a 50 Ω DUT output impedance.

Fig. 4. Suggested minimum channel width W_{min} for different technologies in [12]-[15].

size, we assign half of the required noise current contributed from i_1, i.e., 3.1×10^{-21} A^2/Hz, and half of it from i_2. For i_1, in addition to the induced gate noise, the noise contribution from i_1 can be controlled by the noise from the gate resistance, which can be adjusted by the number of fingers [11]. For i_2, since it is mainly contributed from the channel noise, we can change the channel width and use the long channel theory, i.e., $\overline{i_d^2} = 4kTg_m$ to estimate i_2. Because $4kTg_m$ underestimates the channel noise in modern short-channel devices, this can ensure us to have sufficiently wide DUT to produce enough $\overline{i_d^2}$ for accurate HF noise measurement. For the technologies in [12]–[15], since only $g_{m,peak(nor)}$, the peak transconductance g_m normalized to $W = 1$ μm is reported, we use $g_{m,peak(nor)}$ in the $\overline{i_d^2}$ equation to demonstrate the calculation of the required minimum channel width W_{min}. Fig. 4 presents the calculated W_{min} using $W_{min} = 3.1 \times 10^{-21} / 4kTg_{m,peak(nor)}$. We notice that, as expected, the technologies having lower g_m requires wider channel width to achieve the same measurement accuracy. In addition, p-type FETs requires wider channel widths than that of n-type FETs to achieve the same measurement accuracy because of smaller g_m caused by the lower hole mobility.

V. CONCLUSION

In this paper, we derived the equations and presented a detailed procedure to obtain the total input-referred noise power of the HF noise measurement system, which can guide the design of a DUT to ensure the measurement accuracy. The suggested W_{min} is the minimum channel width required at the bias conditions around the peak g_m. For emerging low-power applications requiring devices working in subthreshold regions, much wider channel width than the suggested W_{min} is required.

REFERENCES

[1] R. Carter, *et al.*, "22nm FDSOI technology for emerging mobile, Internet-of-Things, and RF applications," *IEDM Tech. Dig.*, 2016, pp. 27–30.

[2] E.-Y. Jeong, *et al.*, "High performance 14nm FinFET technology for low power mobile RF application," *2017 Symp. VLSI Technol.*, vol. 1853, no. 2007, 2017, pp. T142–T143.

[3] J. Singh, *et al.*, "14-nm FinFET technology for analog and RF applications," *IEEE Trans. Electron Devices*, vol. 65, pp. 31–37, 2018.

[4] C. H. Chen and M. J. Deen, "High frequency noise of MOSFET's I— Modeling," *Solid-State Electronics*, vol. 42, pp. 2069–2081, Nov. 1998.

[5] C. H. Chen and M. J. Deen, "A general noise and s-parameter de-embedding procedure for on-wafer high-frequency noise measurements of MOSFETs," *IEEE Trans. Microwave Theory and Techniques*, vol. 49, no. 5, pp. 1004–1005, May 2001.

[6] "Representation of noise in linear two ports," *Proc. IRE*, vol. 48, no. 1, pp. 69–74, Jan. 1960.

[7] "Representation of noise in linear two ports," *Proc. IEEE*, vol. 57, no. 6, p. 1211, June 1969.

[8] C.-H. Chen, Y. L. Wang, M. Bakr, and Z. Zeng, "Novel noise parameter determination for on-wafer microwave noise measurements," *IEEE Trans. Instrumentation & Measurement*, vol. 57, no. 11, pp. 2462–2471, Nov. 2008.

[9] X. Chen, N. Misljenovic, B. Hosein, C.-H. Chen, and C. Tsironis, "Calibration of a noise receiver taking care of its gain variations," *23rd International Conference on Noise and Fluctuations (ICNF 2015)*, Xi'an China, 2015, pp. 1–4.

[10] H. Hillbrand and P. H. Russer, "An efficient method for computer-aided noise analysis of linear amplifier networks," *IEEE Trans. Circuit Syst.*, vol. CAS-23, pp. 235–238, Apr. 1976.

[11] M. J. Deen, C. H. Chen, and Y. Cheng, "MOSFET modeling for low noise, RF circuit design," *IEEE Custom Integrated Circuits Conference (CICC 2002)*, Orlando, Florida, 2002, pp. 201–208.

[12] S. Lee *et al.* "Highly scalable raised source/drain InAs quantum well MOSFETs exhibiting I_{ON} = 482 μA/μm at I_{OFF} =100 nA/μm and V_{DD} =0.5 V," *IEEE Electron Device Lett.*, vol. 35, pp. 621–623, 2014.

[13] J. Lin *et al.*, "A new self-aligned quantum-well MOSFET architecture fabricated by a scalable tight-pitch process," *IEDM Tech. Dig.*, 2013, pp. 421–424.

[14] Y. Sun, *et al.*, "High-performance CMOS-compatible self-aligned In$_{0.53}$Ga$_{0.47}$As MOSFETs with G_{MSAT} over 2200 μS/μm at V_{DD} = 0.5 V," *IEDM Tech. Dig.*, 2014, pp. 582–585.

[15] M. J. H. van Dal *et al.*, "Demonstration of scaled Ge p-channel FinFETs integrated on Si," *IEDM Tech. Dig.*, 2012, pp. 521–524.

Measurement of Temperature Effect on Random Telegraph Noise Induced Delay Fluctuation

A.K.M. Mahfuzul Islam*, Masashi Oka[†], and Hidetoshi Onodera[†]

* Institute of Industrial Science, The University of Tokyo, 4-6-1 Komaba, Meguro-ku, Tokyo 153-8505, JAPAN.
e-mail: mahfuzul@iis.u-tokyo.ac.jp
Tel: +81-3-5452-6253, Fax: +81-3-552-6632

[†] Graduate School of Informatics, Kyoto University, Yoshida-honmachi, Sakyo-ku, Kyoto 606-8501, JAPAN.
e-mail: {oka,onodera}@vlsi.kuee.kyoto-u.ac.jp
Tel: +81-75-753-5353, Fax: +81-75-753-5343

Abstract—**We present detailed measurement results of temperature effect on Random Telegraph Noise (RTN) induced delay fluctuation using a test chip fabricated in a 65-nm Silicon-On-Thin-Buried-Oxide process. Skewed ring oscillators (ROs) are used to characterize pMOFSET and nMOSFET specific RTN effects. Distributions of overall threshold fluctuation of a device have been extracted such that the simulated delay distribution matches with the measured delay distribution. For worst-case delay prediction, circuit analysis with ΔV_T distribution model for low temperature is necessary. Estimation results reveal that RTN amplitude decreases slightly with the increase of temperature. However, low correlation of 0.3 to 0.4 has been observed across temperatures ranging from $0\,°\mathrm{C}$ to $80\,°\mathrm{C}$ for delay paths. We find appearing and disappearing of traps across the temperature range causing the low correlation. Low correlation poses challenges in realizing robust runtime performance compensation techniques such as replica critical path based delay compensation.**

I. INTRODUCTION

With technology scaling, the increase of low frequency noise $(1/f)$ and random telegraph noise (RTN) have become a concern. RTN and low frequency noise are the results of trapping and de-trapping of channel carriers into the Si-SiO$_2$ interface [1]. RTN is reported to cause large delay fluctuations especially when the transistors operate at weak inversion region [2]. 50 % of delay fluctuation has been estimated for a 14 nm high-κ metal gate process [3]. One of the key reasons of large RTN amplitudes is the surface potential fluctuation in the channel [2, 4]. In order to assess the impact of RTN amplitude to circuit reliability, the following three phenomena need to be understood and modeled accordingly. Firstly, an appropriate distribution function is required. Empirically, the Exponential distribution is widely used to model single trap induced threshold voltage fluctuation, ΔV_T [5]. However, there are reports mentioning that the Lognormal distribution better represents the ΔV_T distributions [3]. Secondly, we need to know how the distributions change according to gate bias and temperature. Thirdly, the predictability of RTN for a single delay path across gate bias and temperature conditions. as RTN is a dynamic phenomenon compared with the static process variations, Low correlation of delay fluctuation across the operating conditions means that post-silicon timing correction cannot be performed by methods such as delay tuning.

In the case of static random dopant fluctuation (RDF) based V_T variability, detailed measurements have been performed showing that V_T distribution follows Gaussian distribution up to 5σ [6]. Furthermore, it has been investigated that temperature has very small effect on static random ΔV_T variability [7]. In a conventional DC I–V based RTN characterization, the drain bias is kept small so that the channel carrier density remains nearly flat throughout the channel. However, during a switching operation of transistors in a digital circuit, transistor gate voltage switches from low to high and vice versa. Furthermore, strong drain bias applies making delay prediction difficult. It is reported that RTN amplitude varies depending on the previous state of gate bias [3]. For the case of RTN, there are few results on the temperature effect on RTN amplitudes. We find that depending on the temperature, not only the time constants of a trap change but also appearing and disappearing of traps occur. As a result, low correlations between delay fluctuations across temperatures have been observed.

In this paper, we present detailed measurement results of temperature effect on Random Telegraph Noise (RTN) induced delay fluctuation. Skewed ring oscillators (ROs) are used to evaluate pMOFSET and nMOSFET specific RTN effects. Furthermore, threshold voltage distributions have been extracted such that the simulated delay distribution matches with the measured delay distributions. In Sec. I, we give a general discussion and some simple models for RTN time constants and amplitude to explain the temperature effects. We describe our test structure and design methodology of ROs in Sec. III. Section IV gives detailed measurement results and some discussions on the possible reasons behind the observations. We provide some conclusions in Sec. V.

II. TEMPERATURE EFFECT ON RTN

A. Low Frequency Noise and RTN

Low frequency noise or the $1/f$ noise is considered to be the superposition of multiple RTNs of different time constants. Temperature effects on low frequency noise in scaled silicon-on-insulator MOSFETs are reported [8]. According to the report, the noise variability decreases with the increase of temperature. The noise variability also depends on the operating frequency. Especially for a logic gate, the RTN amplitude may differ depending on the wait time of the operation [3]. Thus, hysteric effect may occur. As a delay path consists of multiple transistors, in-situ measurements with ROs are helpful in understanding the RTN effects.

B. Time Constant

Historically, tunneling mechanism is thought to be the determinant factor in determining the time to capture, τ_c. However, recent observations show that tunneling mechanism based model is inadequate in explaining the time constants. Oxide relaxation is now considered to be the determinant factor for a carrier to be trapped [9, 10]. The average time to capture, $\bar{\tau}_c$ is modeled as

$$\bar{\tau}_c = \frac{1}{n_s \, \sigma \, \bar{v}_{th}} = \frac{1}{n_s \, \sigma_0 \, \bar{v}_{th}} e^{E_B/kT}. \tag{1}$$

Here, n_s is the channel carrier density, v_{th} is mean thermal voltage of carriers, and σ is capture cross-section. σ_0 is a scaling parameter. k is Boltzmann constant and T is absolute temperature. E_B is the energy barrier for a carrier to be trapped. With the increase of temperature, an exponential decrease of τ_c is expected according to the model. Time to emission, τ_e is modeled as

$$\tau_e = \tau_c \, e^{-(E_T - E_F)/kT}. \tag{2}$$

Here, E_T is trap energy level, E_F is Fermi level. However, it is also reported that the above model fails to predict the measured τ_e in some cases. A more accurate expression as in Eq. (3) is proposed where two emission mechanisms are considered simultaneously [11].

$$\frac{1}{\tau_e} = \frac{1}{\tau_{e1}} + \frac{1}{\tau_{e2}}, \tag{3}$$

$$\tau_{e1} = \tau_c \cdot e^{-\frac{E_T - E_F}{kT}}, \tag{4}$$

$$\tau_{e2} = \frac{1}{\sigma \, \bar{v}_{th} \, N} e^{\Delta E/kT}. \tag{5}$$

Here, N is the density-of-states in the band to which the carrier is emitted. ΔE is related to the activation energy for emission.

C. Trap Density

The number of traps, N_T is reported to follow Poisson distribution [5].

$$a_{N_T} = \frac{e^{-\lambda} \lambda^{N_T}}{N_T!}. \tag{6}$$

λ is the mean number of traps which scales with channel area. However, with the increase of channel area, the effect of traps on the amplitude becomes smaller. As a result, traps with small amplitude may not be characterized.

D. Amplitude

Trapping of a charge results in fluctuation of the surface potential, which in turn modulates the channel carrier density. In an ideal transistor where the silicon surface potential is flat, the change of threshold voltage, ΔV_T due to the trapping of a carrier into the oxide can be expressed by Eq. (7).

$$\Delta V_T = \frac{q}{C_{ox} W L}. \tag{7}$$

Here, q is the elementary charge, C_{ox} is the oxide capacitance per area, W is channel width, and L is channel length. However, the simple expression of Eq. (7) cannot explain the large amplitudes reported in the literature. RTN induced ΔV_T distribution has a long tail. [2] has shown by simulation that the reason for large amplitude originates from the surface potential, ϕ_s unevenness. [4] has given a simple model that models the relationship between RTN amplitude variability and surface potential, ϕ_s variability by Eq. (8).

$$\Delta V_T = \frac{q}{C_{ox} W L} e^{\frac{q(V_T - V_{Tj})}{\eta kT}}. \tag{8}$$

Here, V_{Tj} is analogous to local threshold voltage when the channel is virtually divided into many smaller channels. Because of the surface potential fluctuation, V_{Tj} varies randomly in the channel. V_{Tj} variation is related to ϕ_s by Eq. (9) [12].

$$\sigma_{V_{Tj}} = \eta \sigma_{\phi_s}. \tag{9}$$

σ_{ϕ_s} is related to random dopant fluctuation induced threshold voltage variability, σ_{V_T}. If ϕ_s distribution can be considered to be Gaussian, Eq. (8) implies that single trap induced ΔV_T distribution would follow a Lognormal distribution. However, in real devices, other factors such as drain bias effect, trap locations may play a significant role. Thus, it is essential to measure in-situ RTN effects for accurate modeling.

E. Effect on Delay Variation

A delay path consists of multiple logic gates with transistors of various gate widths. Some delay paths have higher activity rate whereas some may have very low activity rate. As multiple transistors are included in a delay path, the probability of occurring RTNs with large amplitudes gets higher. Furthermore, the effective trap density increases under wide temperature range as some traps may be active at low power and some traps may be active at high temperature. This phenomenon is confirmed in our measurement results as will be discussed in Sec. IV. The total delay variation is the result of complex RTN occurring either in a transistor or single RTNs over multiple transistors. In order to estimate the delay variation, the scaling effect on trap density and amplitude need to be modeled. After knowing the trap density distribution and single trap induced RTN amplitude distribution, probability distribution function (PDF) of overall amplitude with varying number of traps can

2018 INTERNATIONAL CONFERENCE ON MICROELECTRONIC TEST STRUCTURES, MARCH 19-22, 2018, AUSTIN, TEXAS, USA

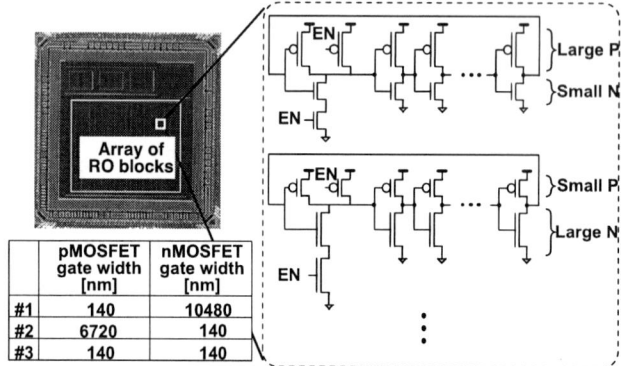

Fig. 1: Test structure containing an array of identical RO blocks. Each block contains ROs with different gate widths.

	pMOSFET gate width [nm]	nMOSFET gate width [nm]
#1	140	10480
#2	6720	140
#3	140	140

be statistically obtained from Eq. (6) and Eqs. (10) to (13).

$$P_1(x) = \frac{1}{x\sigma\sqrt{2\pi}}e^{-\frac{(\log x - \mu)^2}{2\sigma^2}}, \tag{10}$$

$$P_N(x) = \int_{-\infty}^{\infty} P_{N-1}(x-t)P_1(t)dt, \tag{11}$$

$$P(x) = a_0\delta(x) + \sum_{i=1}^{N} a_i P_i(x), \tag{12}$$

$$\mu = \log\left(\frac{q}{C_{ox}WL}\right), \sigma = \frac{q}{kT} \times \sigma_{\phi_s}. \tag{13}$$

Here, $P(x)$ is the PDF of overall RTN amplitude distribution. μ and σ are the parameters for single trap induced RTN amplitude Lognormal distribution. It would be useful for the designers to approximate $P(x)$ with an equivalent closed form function that does not need statistical simulation.

III. DESIGN OF TEST STRUCTURE

A. Test Chip

Figure 1 shows the test structure implemented in a 65 nm Silicon-on-Thin-Buried-Oxide (SOTB) Process. The channel is lightly doped to suppress RDF induced V_T variation. Threshold voltages of both the pMOSFET and nMOSFET, when the delay is modeled by α-power law delay model [13], are around 0.3 V. Transistor with more than 0.4 V gate bias operates at strong inversion region. We use ROs to evaluate RTN induced delay variation. The test chip consists of 16×13 array of RO blocks. Several ROs with different topologies and gate widths are implemented in each block. Therefore, from a single chip, we obtain 208 samples of frequency measurements for a particular RO topology.

B. RO Design

To characterize pMOSFET and nMOSFET specific characteristics, largely skewed inverters are used. ROs consisting of inverters with balanced gate width sizing are also implemented to evaluate RTN effects on actual delay paths. All ROs are 7 staged where the first stage is a NAND gate. The gate width sizing of the NAND gate transistors is the same as those used in the other inverters. In this study, three types of ROs as shown in Fig. 1 are measured. RO with index

#1 is a pMOSFET dominant RO meaning the observed RTN induced delay fluctuation can be attributed to pMOSFETs only. Similarly, RO with index #2 is an nMOSFET dominant RO. Delay fluctuations of RO #3 are contributed by both the pMOSFETs and nMOSFETs. All the inverter stages share the same sizing resulting in the structure as a homogeneous one.

Previously, we have proposed inhomogeneous structures to identify RTN effects on single transistors [14]. Here, we use a homogeneous one which has the following advantages. In this paper, our primary goal is to characterize the overall RTN amplitudes and the correlation across the temperatures. In an actual delay paths, multiple transistors contribute to the delay and RTN with large amplitude may occur in any of those transistors. Thus, a homogeneous RO structure can be treated as a representative delay path. However, because of the involvement of multiple transistors, extraction of transistor ΔV_T distribution becomes difficult. We use the extraction methods used in [15] to overcome this difficulty. In RO #1, nMOSFET gate width is 75 times larger than the pMOSFETs. In RO #2, pMOSFET gate width is 48 times larger than the nMOSFETs. Fro the RO #3, minimum gate widths for both of the pMOSFETs and nMOSFETs are used. The RO structure and test chip used here is the same as in [16]. Frequency of each RO sample is measured over a time period of 10 s with an integration time of 1 ms. Supply voltage, V_{DD} is set to 0.6 V which ensures strong inversion operation. The rise delay of a pMOSFET dominant inverter is in the order of ns.

IV. MEASUREMENT RESULTS

A. Delay Fluctuations at Different Temperatures

Figure 2 shows measured $\Delta d/d$ against time for a pMOSFET dominant RO sample for 5 different temperatures of $0\,°C$, $20\,°C$, $40\,°C$, $60\,°C$, and $80\,°C$. Discrete delay fluctuations are observed at $0\,°C$. Overall delay fluctuation is 0.6 % in this case. With the increase of temperature, the average time to capture, τ_c decreases as is suggested by Eq. (1). At $80\,°C$, the discrete delay levels become indistinguishable, and the delay fluctuation is reduced to 0.2 %. One possible reason can be related to the integration time required to obtain a sample. When the trapping and de-trapping time constants become smaller than the integration time, the delay fluctuation may get averaged out because of the integration. Further investigation is required to clarify this phenomenon.

Figure 3 shows RTN profiles at 5 different temperatures for another sample of a pMOSFET dominant RO. In the figure, new traps have appeared at temperatures of $20\,°C$, $40\,°C$, and $60\,°C$. the number of traps involved increases compared with that at $0\,°C$. In this case, large delay fluctuations are observed at a higher temperature compared with that at $0\,°C$. Thus, because of the difference in trap profiles and the number of active traps at different temperatures, the delay fluctuations of a particular delay path is expected to show less correlation across the temperatures.

Figures 4 and 5 show RTN profiles for two samples of nMOSFET dominant ROs. Similar to the case of pMOSFET, τ_c get smaller with the increase of temperature. In Fig. 5,

978-1-5386-5072-1/18 $31.00 © 2018 IEEE

2018 INTERNATIONAL CONFERENCE ON MICROELECTRONIC TEST STRUCTURES, MARCH 19-22, 2018, AUSTIN, TEXAS, USA

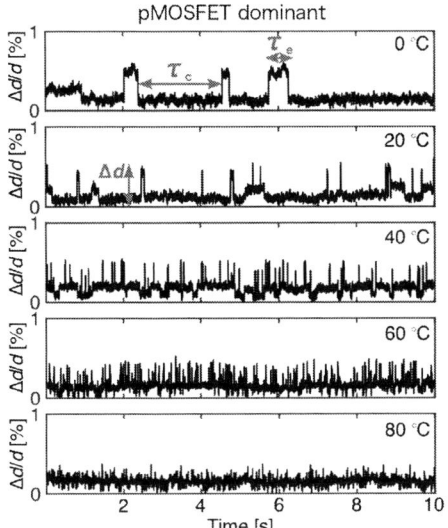

Fig. 2: RTN induced delay fluctuation for a pMOSFET dominant RO sample at different temperatures.

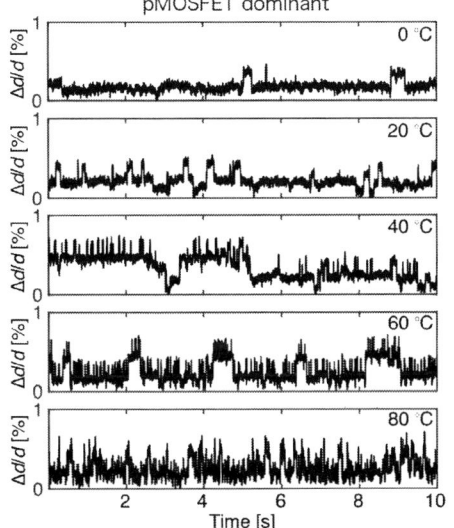

Fig. 3: RTN induced delay fluctuation for a pMOSFET dominant RO sample at different temperatures.

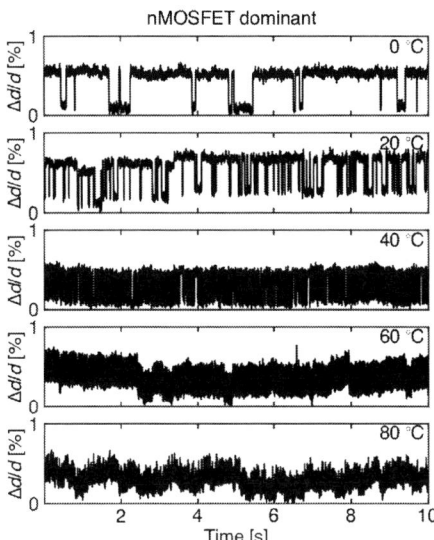

Fig. 4: RTN induced delay fluctuation for an nMOSFET dominant RO sample at different temperatures.

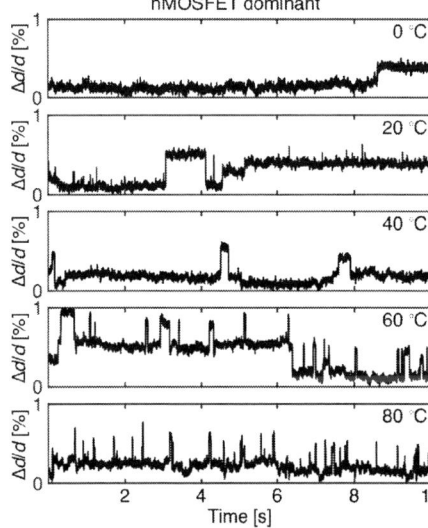

Fig. 5: RTN induced delay fluctuation for an nMOSFET dominant RO sample at different temperatures.

new traps are being observed at higher temperatures of $60\,^{\circ}$C and $80\,^{\circ}$C. Discrete fluctuations get indistinguishable making it more like flicker noise.

In Fig. 6, we show delay fluctuation over time and its power spectrum density (PSD) for a sample of pMOSFET dominant RO where large amplitude is observed at $80\,^{\circ}$C because of the appearance of new traps. Figures 6(a) and 6(b) represents delay fluctuations and its PSD for $0\,^{\circ}$C. Figures 6(c) and 6(d) represents delay fluctuations and its PSD for $80\,^{\circ}$C. From the PSD results we observe that both the delay waveforms at $0\,^{\circ}$C and $80\,^{\circ}$C follows $1/f^2$ at low frequencies.

B. Delay Correlation Across Temperatures

From the above discussions, it is clear that not only the trap characteristics change over temperature, but also appearing or disappearing of traps occur. As a result, low correlation be-

tween delay fluctuation amplitude is expected for a particular delay across temperatures. Figure 7 shows the correlation between delay fluctuations at two different temperatures of $0\,^{\circ}$C and $80\,^{\circ}$C for three different ROs. No significant correlation is found for the pMOSFET dominant RO. The delay fluctuation is much larger at $0\,^{\circ}$C compared with that at $80\,^{\circ}$C. nMOSFET dominant ROs, and ROs where both the pMOSFETs and nMOSFETs have small gate widths show a similar trend. No significant correlation has been observed.

C. Amplitude Distribution

As explained in Sec. II, one key concern is the distribution of overall RTN amplitudes. Figure 8 shows the QQ plot of delay variations for temperature range from $0\,^{\circ}$C to $80\,^{\circ}$C. The delay distribution gets slightly smaller with the increase of temperature. There are two reasons for the decrease of

978-1-5386-5072-1/18 $31.00 © 2018 IEEE   213

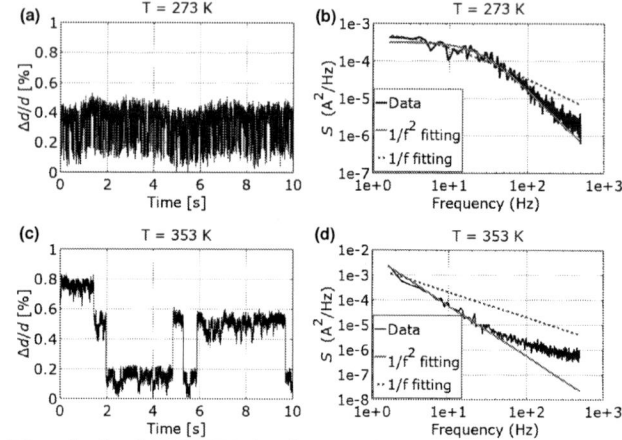

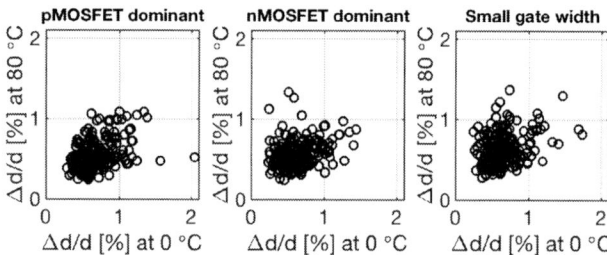

Fig. 6: A pMOSFET dominant RO sample with new trap appearing at high temperature.

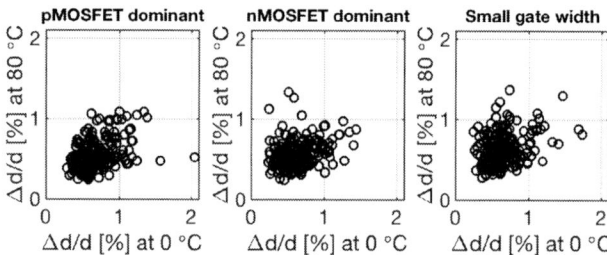

Fig. 7: Correlation between RTN induced delay fluctuations between 0 °C and 80 °C for three different ROs. Correlation coefficient varies from 0.3 to 0.4.

delay fluctuation. One is the lowering of threshold voltage with the increase of temperature causing less delay fluctuation for the same amount of ΔV_T. The other is the lowering of the RTN amplitude ΔV_T itself. The effect of temperature on ΔV_T distribution will be discussed next. One key observation here is that delay distributions follow Lognormal distributions.

D. Estimation of Total ΔV_T Distribution

Characterization of transistor overall ΔV_T distribution due to RTN is a challenge for RO based measurements as multiple transistors are involved simultaneously. However, overall ΔV_T distribution of transistors can be estimated under the following two assumptions.

1) Single trap ΔV_T distribution follows Lognormal distribution as suggested by Eq. (8), and
2) Overall ΔV_T distribution considering trap numbers can be approximated by a Lognormal distribution.

The use of pMOSFET and nMOSFET dominant RO based delay fluctuation measurement allows us to estimate overall ΔV_T distributions for pMOSFET and nMOSFET separately [16]. Figure 9 shows the estimated ΔV_T distributions for pMOSFET and nMOSFET at 3 different temperatures. With the increase of temperature, ΔV_T variability gets reduced. This can be attributed to Eq. (13) where thermal voltage fluctuation causes less threshold voltage fluctuation due to ϕ_s.

Next, we verify the validity of the estimated ΔV_{TP} and ΔV_{TN} distributions. We estimate the overall $\Delta d/d$ distribu-

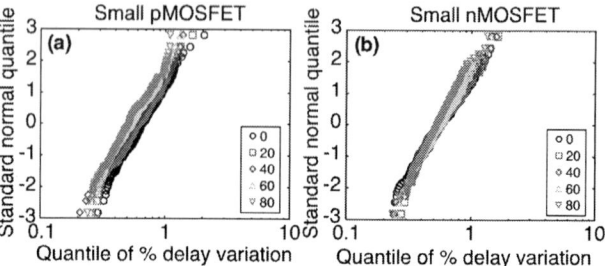

Fig. 8: Measured distributions of $\Delta d/d$ at different temperatures.

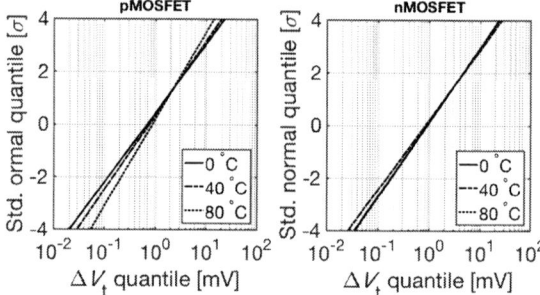

Fig. 9: Estimated ΔV_{TN} distributions at different temperatures.

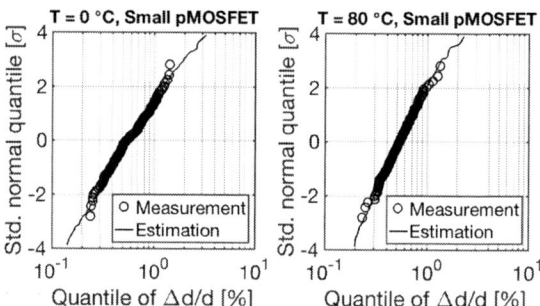

Fig. 10: Comparison between measured and simulated distributions of $\Delta d/d$ for pMOSFET dominant RO.

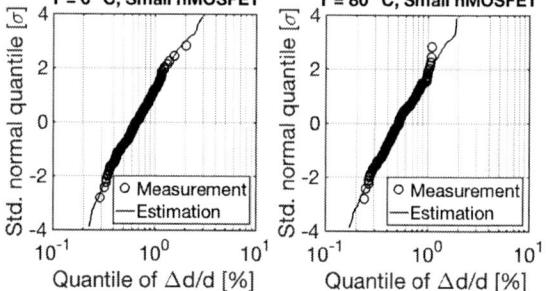

Fig. 11: Comparison between measured and simulated distributions of $\Delta d/d$ for nMOSFET dominant RO.

tions for each RO by Monte Carlo simulation using the estimated ΔV_{TP} and ΔV_{TN} distributions. We then compare the simulated $\Delta d/d$ distributions with the measured distributions. Figures 10 and 11 show the comparison results for pMOSFET and nMOSFET dominant ROs respectively. The estimated and measured delay distributions match well showing that the assumptions made above are reasonable.

E. Impact on Circuit Reliability

The measurement results show that increase in temperature decreases RTN induced delay fluctuation to a small extent. The worst-case RTN effect occurs at the low temperature region. The dynamic nature of RTN results in very low correlation of delay fluctuation of a delay path at different temperatures. The low correlation has severe effect on the use of spare cells or replica critical paths for performance compensation.

V. Conclusion

Measurement results from a 65 nm SOTB process reveal that RTNs with significant discrete delay levels occur quite frequently over a wide temperate range. Maximum Delay fluctuation of 2% has been observed at 0 °C. In general, τ_c and τ_e decrease rapidly with the increase of temperature. However, for some samples, new slow traps appear at high temperatures. Also, the number of active traps differ from temperature to temperature, but in general more traps are active at a higher temperature. All these mean that delay fluctuation of a particular delay path has small correlation across temperature. The overall delay fluctuation distribution tends to get smaller with the increase of temperature. Worst case delay fluctuation may occur at the low temperature. The measured observation poses the following impacts on circuit design. Firstly, statistical analysis considering low temperature distribution is required and may be enough. Secondly, as delay correlation is low across the temperatures, post-silicon performance compensation techniques are not reliable.

Acknowledgment

The VLSI chip in this study was fabricated in the chip fabrication program of the VLSI Design and Education Center (VDEC), the University of Tokyo, in collaboration with Renesas Electronics. This work has been supported in part by JSPS KAKENHI 25280014 and 16H01713.

References

[1] M. Kirton and M. Uren, "Noise in solid-state microstructures: a new perspective on individual defects, interface states, and low-frequency noise," *Adv. Phys.*, vol. 38, no. 4, pp. 367–468, 1989.

[2] A. Asenov, R. Balasubramaniam, A. R. Brown, and J. H. Davies, "RTS amplitudes in decananometer MOSFETs: 3-D simulation study," *IEEE Trans. Electron Devices*, vol. 50, no. 3, pp. 839–845, 2003.

[3] H. Miki, M. Yamaoka, D. J. Frank, K. Cheng, D. G. Park, E. Leobandung, and K. Torii, "Voltage and temperature dependence of random telegraph noise in highly scaled HKMG ETSOI nFETs and its impact on logic delay uncertainty," *Dig. Tech. Pap. - Symp. VLSI Technol.*, vol. 12, pp. 137–138, 2012.

[4] K. Sonoda, K. Ishikawa, T. Eimori, and O. Tsuchiya, "Discrete dopant effects on statistical variation of random telegraph signal magnitude," *IEEE Trans. Electron Devices*, vol. 54, no. 8, pp. 1918–1925, 2007.

[5] K. Takeuchi, T. Nagumo, S. Yokogawa, K. Imai, and Y. Hayashi, "Single-Charge-Based Modeling of Transistor Characteristics Fluctuations Based on Statistical Measurement of RTN Amplitude," in *Symp. VLSI Technol.*, 2009, pp. 54–55.

[6] T. Tsunomura, A. Nishida, F. Yano, A. T. Putra, K. Takeuchi, S. Inaba, S. Kamohara, K. Terada, T. Hiramoto, and T. Mogami, "Analyses of 5σ Vth Fluctuation in 65nm-MOSFETs using Takeuchi Plot," in *Symp. VLSI Technol.*, 2008, pp. 156–157.

[7] T. Tsunomura, A. Nishida, and T. Hiramoto, "Investigation of Threshold Voltage Variability at High Temperature Using Takeuchi Plot," *Jpn. J. Appl. Phys.*, vol. 49, no. 5, p. 054101, may 2010.

[8] C. G. Theodorou, E. G. Ioannidis, S. Haendler, E. Josse, C. A. Dimitriadis, and G. Ghibaudo, "Low frequency noise variability in ultra scaled FD-SOI n-MOSFETs: Dependence on gate bias, frequency and temperature," *Solid. State. Electron.*, vol. 117, pp. 88–93, 2016.

[9] D. Veksler, G. Bersuker, H. Park, C. Young, K. Y. Lim, S. Lee, and H. Shin, "The critical role of defect structural relaxation in interpreting noise measurements in MOSFETs Introduction & Motivation," in *IEEE Int. Integr. Reliab. Work. Final Rep.*, 2009, pp. 102–105.

[10] T. Nagumo, K. Takeuchi, T. Hase, and Y. Hayashi, "Statistical Characterization of Trap Position, Energy, Amplitude and Time Constants by RTN Measurement of Multiple Individual Traps," in *Int. Electron Devices Meet.*, dec 2010, pp. 28.3.1–28.3.4.

[11] Y. Son, T. Kang, S. Park, and H. Shin, "A simple model for capture and emission time constants of random telegraph signal noise," *IEEE Trans. Nanotechnol.*, vol. 10, no. 6, pp. 1352–1356, 2011.

[12] G. Slavcheva, J. H. Davies, A. R. Brown, and A. Asenov, "Potential fluctuations in metal-oxide-semiconductor field-effect transistors generated by random impurities in the depletion layer," *J. Appl. Phys.*, vol. 91, no. 7, pp. 4326–4334, 2002.

[13] T. Sakurai and A. R. Newton, "Alpha-Power Law MOSFET Model and its Applications to CMOS Inverter Delay and Other Formulas," *IEEE J. Solid-State Circuits*, vol. 25, no. 2, pp. 584–594, apr 1990.

[14] A. K. M. Mahfuzul Islam, T. Nakai, and H. Onodera, "Statistical analysis and modeling of Random Telegraph Noise based on gate delay variation measurement," *IEEE Trans. Semicond. Manuf.*, vol. PP, no. 99, pp. 1–12, 2017.

[15] A. K. M. M. Islam and H. Onodera, "Effect of Supply Voltage on Random Telegraph Noise of Transistors under Switching Condition," in *Int. Symp. Power Timing Model. Optim. Simul.*, 2017.

[16] A. M. Islam, T. Nakai, and H. Onodera, "A Statistical Modeling Methodology of RTN Gate Size Dependency Based on Skewed Ring Oscillators," in *IEEE Int. Conf. Microelectron. Test Struct.*, 2017, pp. 159–164.

Author Index

Agnihotri, P. 153	Grund, E. 48
Akiyama, M. 79	Hariharan, S. 27
Ang, B. 122	Hess, C. 7
Arai, Y. 149	Higo, A. 185
Araki, T. 79	Hiramoto, T. 157
Ashton, R. 48	Hiromoto, M. 35
Baker, C. 16	Hokimoto, S. 128
Balakrishnan, S. 122	Hong, W. 153
Bergen, A. 48	Hurley, P.K. 161
Bertello, B. 27	Hwan Kim, D. 13
Blair, E.O. 179	Ida, J. 149
Bolshakov, P. 161	Ikeda, K. 23
Bonmann, M. 75	Imi, H. 93
Borghello, G. 166	Ishihara, T. 111, 128
Brown, G. 206	Ishijima, Y. 202
Brunets, I. 87	Ishikuro, H. 202
Bucher, M. 166	Ito, K. 97
Buchoux, A. 179	Itou, K. 157
Buehler, R.T. 189	Jeppson, K. 75
Chang, J. 3, 117	Katariay, S. 57
Chen, C.-H. 206	Katragadda, V. 142
Chen, I.-R. 117	Khosravi, A. 161
Cheng, D. 117	Kim, S.-Y. 13
Chevas, L. 166	Kishimoto, T. 111
Cho, J. 153	Knezevic, T. 69
Choi, C.-H. 13	Koch, H.D. 166
Choi, S. 13	Kondo, M. 79
Chong, N. 117	Konno, T. 97
Chua, P.-H. 206	Kwon, S.-K. 13
Ciavatti, J. 153	Lee, H.-C. 43
Compton, C. 102	Lee, H.-D. 13
Das, S. 206	Lee, J.G. 153
De Matos, M. 197	Lee, T.J. 153
DeBruler, L. 137	Lemke, S. 27
Demircan, E. 43	Li, X. 137
Deng, M. 197	Liao, J.-H. 142
Do, N. 27	Lim, D.-H. 13
Driussi, F. 57	Lin, Q. 3
Dunare, C. 179	Liu, B.W. 153
Faccio, F. 166	Liu, X. 69
Fairbanks, S. 48	Mackenzie Dover, C.M. 63
Fregonese, S. 197	Mahajan, V. 153
Fukui, M. 157	Mahfuzul Islam, A.K.M. 210
Gahoiy, A. 57	Majumdar, A. 117
Gasasira, A. 142	Makris, N. 166
Giacomini, R.C. 189	Marland, J.R.K. 179

Author List (cont'd)

Marr, K.	137	Stake, J.	75
McQuirk, D.	16	Stokes, A.A.	179
Mita, Y.	173, 185	Sugiyama, M.	79
Mizushima, A.	185	Suzuki, S.	157
Mori, T.	149	Tadayoni, M.	27
Mount, A.R.	63	Takakura, T.	157
Mujumdar, S.	137	Takeda, K.	149
Muthee, M.	142	Takeuchi, K.	157
Myong Kim, D.	13	Tanaka, C.	23
Nakagawa, S.	202	Terry, J.G.	63
Nakamura, K.	93	Tiwari, V.	27
Namba, N.	79	Tsiamis, A.	179
Nanver, L.K.	69	Tuinhout, H.	87
Nikolaou, A.	166	Uemura, T.	79
Nishikata, D.	93	Usami, N.	185
Nishizawa, S.	97	Venica, S.	57
Noda, Y.	79	Violette, M.	137
Normand, M.	179	Vorobiev, A.	75
Oh, D.-J.	13	Wallace, R.M.	161
Oka, M.	210	Walton, A.J.	63, 179
Okamoto, Y.	173, 185	Wells, J.	31
Onodera, H.	210, 128	Wong, C.Y.	153
Otrokov, M.	31	Yadav, C.	197
Pan, H.	3	Yamaguchi, S.	93
Papadopoulou, A.	166	Yang, B.	206
Parameswaran, S.	122	Yang, X.	75
Parga Basanta, L.	179	Yeh, P.-C.	117
Pate-Cazal, T.	27	Yeol Mun, S.Y.	153
Peddenpohl, B.	31	Yoshimoto, S.	79
Peterson, D.	137	Young, C.D.	161
Pretti, D.	137	Yu, S.	7
Ranga Reddy, R.R.	173	Zegers-van Duijnhoven, A.	87
Ross, A.W.S.	63	Zhao, P.	161
Samavedam, S.B.	153	Zhu, B.	153
Saraya, T.	157	Zimmer, T.	197
Sato, T.	35		
Schmueser, I.	179		
Seelmann, F.	142		
Sekitani, T.	79		
Shi, Y.J.	153		
Shintani, M.	35		
Shiomi, J.	128		
Shroff, M.D.	43		
Smith, B.	16		
Smith, S.	63, 179		
Sohn, D.K.	153		
Song, H.-S.	13		